Fiber Reinforced Polymers - The Technology Applied for Concrete Repair

Fiber Reinforced Polymers - The Technology Applied for Concrete Repair

Editor

Alano Riley

scitus
academics

Fiber Reinforced Polymers - The Technology Applied for Concrete Repair

Edited by **Alano Riley**

ISBN: 978-1-68117-272-9
Library of Congress Control Number: 2016937442

www.scitusacademics.com
Box No. 4766, 616 Corporate Way,
Suite 2, Valley Cottage,
NY 10989

Printed in United States of America on Acid Free Paper

Preface

Fiber reinforced polymers for concrete and building materials have been introduced as a substitute to predictable steel reinforcement because of their high toughness and their high strength-to-weight percentage under severe environmental circumstances. Composite materials are made of two or more component materials with substantially different physical or chemical properties that remain largely indistinguishable within the completed structure. Fiber reinforced polymer composites (FRPs) are progressively considered to improve or replace infrastructure materials or systems made of traditional civil engineering components, namely concrete and steel. FRP composites are versatile, non-corrosive, have high relevant strength and rigidity, are easy to build and can be customized to selection criteria. The use of natural fibers as a reinforcement for different materials has already been recorded in ancient Egypt, but its rediscovery can be dated to the early 20th century. Although considerable research successes have been achieved and several codes and design protocols have been developed globally, there is still an increasingly urgent need for innovative and rational modelling under realistic environmental conditions and for precise long - term performance modelling. This book provides an accurate and complete information of the manufacture, mechanical properties, resistance and tolerance power, and use of 3D fiber reinforced composites in concrete reinforcement and repair. The world's aging concrete infrastructure forced research to focus on providing cost-effective solutions to increase service life and improve the seismic and structural performance of concrete and building structures. This book is dedicated to the pivotal role of natural fibers in reinforcing different degradable and non - biodegradable polymer matrices. This book identifies a number of major reasons for the use in polymer composites of natural fillers. The use of natural fillers can be seen as an approach to adjusting the material performance of polymer composites, assuming that filler / matrix interactions are optimized and the hygroscopicity of natural fillers is inhibited.

Table of Contents

Chapter 1

Introduction of Fibre-Reinforced Polymers – Polymers and Composites: Concepts, Properties and Processes

Martin Alberto Masuelli[1]

[1] *Instituto de Física Aplicada, CONICET. Cátedra de Química Física II, Área de Química Física. Facultad de Química, Bioquímica y Farmacia. Universidad Nacional de San Luis. Chacabuco 917 (CP: 5700), San Luis, Argentina*

1. INTRODUCTION

Fibre-reinforced polymer (FRP), also *Fibre-reinforced plastic*, is a composite material made of a polymer matrix reinforced with fibres. The fibres are usually glass, carbon, or aramid, although other fibres such as paper or wood or asbestos have been sometimes used. The polymer is usually an epoxy, vinylester or polyester thermosetting plastic, and phenol formaldehyde resins are still in use. FRPs are commonly used in the aerospace, automotive, marine, and construction industries.

Composite materials are engineered or naturally occurring materials made from two or more constituent materials with significantly different physical or chemical properties which remain separate and distinct within the finished structure. Most composites have strong, stiff fibres in a matrix which is weaker and less stiff. The objective is usually to make a component which is strong and stiff, often with a low density. Commercial material commonly has glass or carbon fibres in matrices based on thermosetting polymers, such as epoxy or polyester resins. Sometimes, thermoplastic polymers may be preferred, since they are moldable after initial production. There are further classes of composite in which the matrix is a metal or a ceramic. For the most part, these are still in a developmental stage, with problems of high manufacturing costs yet to be overcome [1]. Furthermore, in these composites the reasons for adding the fibres (or, in some cases, particles) are often rather complex; for example, improvements may be sought in creep, wear, fracture toughness, thermal stability, etc [2].

Fibre reinforced polymer (FRP) are composites used in almost every type of advanced engineering structure, with their usage ranging from aircraft, helicopters and spacecraft through to boats, ships and offshore platforms and to automobiles, sports goods, chemical processing equipment and civil infrastructure such as bridges and buildings. The usage of FRP composites continues to grow at an impressive rate as these materials are used more in their existing markets and become established in relatively new markets such as biomedical devices and civil structures. A key factor driving the increased applications of composites over the recent years is the development of new advanced forms of FRP materials. This includes developments in high performance resin systems and new styles of reinforcement, such as carbon nanotubes and nanoparticles. This book provides an up-to-date account of the fabrication, mechanical properties, delamination resistance, impact tolerance and applications of 3D FRP composites [3].

The fibre reinforced polymer composites (FRPs) are increasingly being considered as an enhancement to and/or substitute for infrastructure components or systems that are constructed of traditional civil engineering materials, namely concrete and steel. FRP composites are lightweight, no-corrosive, exhibit high specific strength and specific stiffness, are easily constructed, and can be tailored to satisfy performance requirements. Due to these advantageous characteristics, FRP composites have been included in new construction and rehabilitation of structures through its use as reinforcement in concrete, bridge decks, modular structures, formwork, and external reinforcement for strengthening and seismic upgrade [4].

The applicability of Fiber Reinforced Polymer (FRP) reinforcements to concrete structures as a substitute for steel bars or prestressing tendons has been actively studied in numerous research laboratories and professional organizations around the world. FRP reinforcements offer a number of advantages such as corrosion resistance, non-magnetic properties, high tensile strength, lightweight and ease of handling. However, they generally have a linear elastic response in tension up to failure (described as a brittle failure) and a relatively poor transverse or shear resistance. They also have poor resistance to fire and when exposed to high temperatures. They loose significant strength upon bending, and they are sensitive to stress-rupture effects. Moreover, their cost, whether considered per unit weight or on the basis of force carrying capacity, is high in comparison to conventional steel reinforcing bars or prestressing tendons. From a structural engineering viewpoint, the most serious problems with FRP reinforcements are the lack of plastic behavior and the very low shear strength in the transverse direction. Such characteristics may lead to premature tendon rupture, particularly when combined effects are present, such as at shear-cracking planes in reinforced concrete beams where dowel action exists. The dowel action reduces residual tensile and shear resistance in the tendon. Solutions and limitations of use have been offered and continuous improvements are expected in the future. The unit cost of FRP reinforcements is expected to decrease significantly with increased market share and demand. However, even today, there are applications where FRP reinforcements are cost effective and justifiable. Such cases include the use of bonded FRP sheets or plates in repair and strengthening of concrete structures, and the use of FRP meshes or textiles or fabrics in thin cement products. The cost of repair and rehabilitation of a structure is always, in relative terms, substantially higher than the cost of the initial structure. Repair generally requires a relatively small volume of repair materials but a relatively high commitment in labor. Moreover the cost of labor in developed countries is so high that the cost of material becomes secondary. Thus the highest the performance and durability of the repair material is, the more cost-effective is the repair. This implies that material cost is not really an issue in repair and that the fact that FRP repair materials are costly is not a constraining drawback [5].

When considering only energy and material resources it appears, on the surface, the argument for FRP composites in a sustainable built environment is questionable. However, such a conclusion needs to be evaluated in terms of potential advantages present in use of FRP composites related to considerations such as:

- Higher strength
- Lighter weight
- Higher performance
- Longer lasting
- Rehabilitating existing structures and extending their life
- Seismic upgrades
- Defense systems
- Space systems
- Ocean environments

In the case of FRP composites, environmental concerns appear to be a barrier to its feasibility as a sustainable material especially when considering fossil fuel depletion, air pollution, smog, and acidification associated with its production. In addition, the ability to recycle FRP composites is limited and, unlike steel and timber, structural components cannot be reused to perform a similar function in another structure. However, evaluating the environmental impact of FRP composites in infrastructure applications, specifically through life cycle analysis, may reveal direct and indirect benefits that are more competitive than conventional materials.

Composite materials have developed greatly since they were first introduced. However, before composite materials can be used as an alternative to conventional materials as part of a sustainable environment a number of needs remain.

- Availability of standardized durability characterization data for FRP composite materials.
- Integration of durability data and methods for service life prediction of structural members utilizing FRP composites.
- Development of methods and techniques for materials selection based on life cycle assessments of structural components and systems.

Ultimately, in order for composites to truly be considered a viable alternative, they must be structurally and economically feasible. Numerous studies regarding the structural feasibility of composite materials are widely available in literature [6]. However, limited studies are available on the economic and environmental feasibility of these materials from the perspective of a life cycle approach, since short term data is available or only economic costs are considered in the comparison. Additionally, the long term affects of using composite materials needs to be determined. The byproducts of the production, the sustainability of the constituent materials, and the potential to recycle composite materials needs to be assessed in order to determine of composite materials can be part of a sustainable environment. Therefore in this chapter describe the physicochemical properties of polymers and composites more used in Civil Engineering. The theme will be addressed in a simple and basic for better understanding.

2. MANUFACTURED PROCESS AND BASIC CONCEPTS

The synthetic polymers are generally manufactured by polycondensation, polymerization or polyaddition. The polymers combined with various agents to enhance or in any way alter the material properties of polymers the result is referred to as a plastic. The Composite plastics can be of homogeneous or heterogeneous mix. Composite plastics refer to those types of plastics that result from bonding two or more homogeneous materials with different material properties to derive a final product with certain desired material and mechanical properties. The Fibre reinforced plastics (or fiber reinforced polymers) are a category of composite plastics that specifically use fibre materials (not mix with polymer) to mechanically enhance the strength and elasticity of plastics. The original plastic material without fibre reinforcement is known as the matrix. The matrix is a tough but relatively weak plastic that is reinforced by stronger stiffer reinforcing filaments or fibres. The extent that strength and elasticity are enhanced in a fibre reinforced plastic depends on the mechanical properties of the fibre and matrix, their volume relative to one another, and the fibre length and orientation within the matrix. Reinforcement of the matrix occurs by definition when the FRP material exhibits increased strength or elasticity relative to the strength and elasticity of the matrix alone.

Polymers are different from other construction materials like ceramics and metals, because of their macromolecular nature. The covalently bonded, long chain structure makes them macromolecules and determines, via the weight averaged molecular weight, Mw, their processability, like spin-, blow-, deep draw-, generally melt-formability. The number averaged molecular weight, Mn, determines the mechanical strength, and high molecular weights are beneficial for properties like strain-to-break, impact resistance, wear, etc. Thus, natural limits are met, since too high molecular weights yield too high shear and elongational viscosities that make polymers inprocessable. Prime examples are the very useful poly-tetra-fluor-ethylenes, PTFE's, and ultrahigh-molecular-weight-poly-ethylenes, UHMWPE's, and not only garbage bags are made of polyethylene, PE, but also high-performance fibers that are even used for bullet proof vests (alternatively made from, also inprocessable in the melt, rigid aromatic polyamides). The resulting mechanical properties of these high performance fibers, with moduli of 150 GPa and strengths of up to 4 GPa, represent the optimal use of what the potential of the molecular structure of polymers yields, combined with their low density. Thinking about polymers, it becomes clear why living nature used the polymeric concept to build its structures, and not only in high strength applications like wood, silk or spider-webs [7].

2.1. Polymers

The linking of small molecules (monomers) to make larger molecules is a polymer. Polymerization requires that each small molecule have at least two reaction points or functional groups. There are two distinct major types of polymerization processes, condensation polymerization, in which the chain growth is accompanied by elimination of small molecules such as H_2O or CH_3OH, and addition polymerization, in which the polymer is formed without the loss of other materials. There are many variants and subclasses of polymerization reactions.

The polymer chains can be classified in linear polymer chain, branched polymer chain, and cross-linked polymer chain. The structure of the repeating unit is the difunctional monomeric unit, or "mer." In the presence of catalysts or initiators, the monomer yields a polymer by the joining together of n-mers. If n is a small number, 2–10, the products are dimers, trimers, tetramers, or oligomers, and the materials are usually gases, liquids, oils, or brittle solids. In most solid polymers, n has values ranging from a few score to several hundred thousand, and the corresponding molecular weights range from a few thousand to several million. The end groups of this example of addition polymers are shown to be fragments of the initiator. If only one monomer is polymerized, the product is called a homopolymer. The polymerization of a mixture of two monomers of suitable reactivity leads to the formation of a copolymer, a polymer in which the two types of mer units have entered the chain in a more or less random fashion. If chains of one homopolymer are chemically joined to chains of another, the product is called a block or graft copolymer.

Isotactic and syndiotactic (stereoregular) polymers are formed in the presence of complex catalysts, or by changing polymerization conditions, for example, by lowering the temperature. The groups attached to the chain in a stereoregular polymer are in a spatially ordered arrangement. The regular structures of the isotactic and syndiotactic forms make them often capable of crystallization. The crystalline melting points of isotactic polymers are often substantially higher than the softening points of the atactic product.

The spatially oriented polymers can be classified in atactic (random; dlldl or lddld, and so on), syndiotactic (alternating; dldl, and so on), and isotactic (right- or left-handed; dddd, or llll, and so on). For illustration, the heavily marked bonds are assumed to project up from the paper, and the dotted bonds down. Thus in a fully syndiotactic polymer, asymmetric carbons alternate in their left- or right-handedness (alternating d, l configurations), while in an isotactic polymer, successive carbons have the same steric configuration (d or l). Among the several kinds of polymerization catalysis, free-radical initiation has been most thoroughly studied and is most widely employed. Atactic polymers are readily formed by free-radical polymerization, at moderate temperatures, of vinyl and diene monomers and some of their derivatives. Some polymerizations can be initiated by materials, often called ionic catalysts, which contain highly polar reactive sites or complexes. The term heterogeneous catalyst is often applicable to these materials because many of the catalyst systems are insoluble in monomers and other solvents. These polymerizations are usually carried out in solution from which the polymer can be obtained by evaporation of the solvent or by precipitation on the addition of a nonsolvent. A distinguishing feature of complex catalysts is the ability of some representatives of each type to initiate stereoregular polymerization at ordinary temperatures or to cause the formation of polymers which can be crystallized [1, 6].

2.1.1. POLYMERIZATION

Polymerization, emulsion polymerization any process in which relatively small molecules, called monomers, combine chemically to produce a very large chainlike or network molecule, called a polymer. The monomer molecules may be all alike, or they may represent two, three, or more different compounds. Usually at least 100 monomer molecules must be combined to make a product that has certain unique physical properties-such as elasticity, high tensile strength, or the ability to form fibres-that differentiate polymers from substances composed of smaller and simpler molecules; often, many thousands of monomer units are incorporated in a single molecule of a polymer. The formation of stable covalent chemical bonds between the monomers sets polymerization apart from other processes, such as crystallization, in which large numbers of molecules aggregate under the influence of weak intermolecular forces.

Two classes of polymerization usually are distinguished. In condensation polymerization, each step of the process is accompanied by formation of a molecule of some simple compound, often water. In addition polymerization, monomers react to form a polymer without the formation of by-products. Addition polymerizations usually are carried out in the presence of catalysts, which in certain cases exert control over structural details that have important effects on the properties of the polymer [8].

Linear polymers, which are composed of chainlike molecules, may be viscous liquids or solids with varying degrees of crystallinity; a number of them can be dissolved in certain liquids, and they soften or melt upon heating. Cross-linked polymers, in which the molecular structure is a network, are thermosetting resins (i.e., they form under the influence of heat but, once formed, do not melt or soften upon reheating) that do not dissolve in solvents. Both linear and cross-linked polymers can be made by either addition or condensation polymerization.

2.1.2. POLYCONDENSATION

The polycondensation a process for the production of polymers from bifunctional and polyfunctional compounds (monomers), accompanied by the elimination of low-molecular weight by-products (for example, water, alcohols, and hydrogen halides). A typical example of polycondensation is the synthesis of complex polyester.

The process is called homopolycondensation if the minimum possible number of monomer types for a given case participates, and this number is usually two. If at least one monomer more than the number required for the given reaction participates in polycondensation, the process is called copolycondensation. Polycondensation in which only bifunctional compounds participate leads to the formation of linear macromolecules and is called linear polycondensation.

If molecules with three or more functional groups participate in polycondensation, three-dimensional structures are formed and the process is called three-dimensional polycondensation. In cases where the degree of completion of polycondensation and the mean length of the macromolecules are limited by the equilibrium concentration of the reagents and reaction products, the process is called equilibrium (reversible) polycondensation. If the limiting factors are kinetic rather than thermodynamic, the process is called nonequilibrium (irreversible) polycondensation.

Polycondensation is often complicated by side reactions, in which both the original monomers and the polycondensation products (oligomers and polymers) may participate. Such reactions include the reaction of monomer or oligomer with a mono-functional compound (which may be present as an impurity), intramolecular cyclization (ring closure), and degradation of the macromolecules of the resultant polymer. The rate competition of polycondensation and the side reactions determines the molecular weight, yield, and molecular weight distribution of the polycondensation polymer.

Polycondensation is characterized by disappearance of the monomer in the early stages of the process and a sharp increase in molecular weight, in spite of a slight change in the extent of conversion in the region of greater than 95-percent conversion.

A necessary condition for the formation of macro-molecular polymers in linear polycondensation is the equivalence of the initial functional groups that react with one another.

Polycondensation is accomplished by one of three methods:

1. in a melt, when a mixture of the initial compounds is heated for a long period to 10°-20°C above the melting (softening) point of the resultant polymer;
2. in solution, when the monomers are present in the same phase in the solute state;
3. on the phase boundary between two immiscible liquids, in which one of the initial compounds is found in each of the liquid phases (interphase polycondensation).

Polycondensation processes play an important role in nature and technology. Polycondensation or similar reactions are the basis for the biosynthesis of the most important biopolymers-proteins, nucleic acids, and cellulose. Polycondensation is widely used in industry for the production of polyesters (polyethylene terephthalate, polycarbonates, and alkyd resins), polyamides, phenol-formaldehyde resins, urea-formaldehyde resins, and certain silicones [9]. In the period 1965-70, polycondensation acquired great importance in connection with the development of industrial production of a series of new polymers, including heat-resistant polymers (polyarylates, aromatic polyimides, polyphe-nylene oxides, and polysulfones).

2.1.3. POLYADDITION

The polyaddition reactions are similar to polycondensation reactions because they are also step reactions, however without splitting off low molecular weight by-products. The reaction is exothermic rather than endothermic and therefore cannot be stopped at will. Typical for polyaddition reaction is that individual atoms, usually H-atoms, wander from one monomer to another as the two monomers combine through a covalent bond. The monomers, as in polycondensation reactions, have to be added in stoichiometric amounts. These reactions do not start spontaneously and they are slow.

Polyaddition does not play a significant role in the production of thermoplastics. It is commonly encountered with cross-linked polymers. Polyurethane, which can be either a thermoplastic or thermosets, is synthesized by the reaction of multi-functional isocyanates with multifunctional amines or alcohol. Thermosetting epoxy resins are formed by polyaddition of epoxides with curing agents, such as amines and acid anhydrides.

In comparing chain reaction polymerization with the other two types of polymerization the following principal differences should be noted: Chain reaction polymerization, or simply called polymerization, is a chain reaction as the name implies. Only individual monomer molecules add to a reactive growing chain end, except for recombination of two radical chain ends or reactions of a reactive chain end with an added modifier molecule. The activation energy for chain initiation is much grater than for the subsequent growth reaction and growth, therefore, occurs very rapidly.

2.2. Composites

Composite is any material made of more than one component. There are a lot of composites around you. Concrete is a composite. It's made of cement, gravel, and sand, and often has steel rods inside to reinforce it. Those shiny balloons you get in the hospital when you're sick are made of a composite, which consists of a polyester sheet and an aluminum foil sheet, made into a sandwich. The polymer composites made from polymers, or from polymers along with other kinds of materials [7]. But specifically the fiber-reinforced composites are materials in which a fiber made of one material is embedded in another material.

2.2.1. POLYMER COMPOSITES

The polymer composites are any of the combinations or compositions that comprise two or more materials as separate phases, at least one of which is a polymer. By combining a polymer with another material, such as glass, carbon, or another polymer, it is often possible to obtain unique combinations or levels of properties. Typical examples of synthetic polymeric composites include glass-, carbon-, or polymer-fiber-reinforced, thermoplastic or thermosetting resins, carbon-reinforced rubber, polymer blends, silica- or mica-reinforced resins, and polymer-bonded or -impregnated concrete or wood. It is also often useful to consider as composites such materials as coatings (pigment-binder combinations) and crystalline polymers (crystallites in a polymer matrix). Typical naturally occurring composites include wood (cellulosic fibers bonded with lignin) and bone (minerals bonded with collagen). On the other hand, polymeric compositions compounded with a plasticizer or very low proportions of pigments or processing aids are not ordinarily considered as composites.

Typically, the goal is to improve strength, stiffness, or toughness, or dimensional stability by embedding particles or fibers in a matrix or binding phase. A second goal is to use inexpensive, readily available fillers to extend a more expensive or scarce resin; this goal is increasingly important as petroleum supplies become costlier and less reliable. Still other applications include the use of some filler such as glass spheres to improve processability, the incorporation of dry-lubricant particles such as molybdenum sulfide to make a self-lubricating bearing, and the use of fillers to reduce permeability.

The most common fiber-reinforced polymer composites are based on glass fibers, cloth, mat, or roving embedded in a matrix of an epoxy or polyester resin. Reinforced thermosetting resins containing boron, polyaramids, and especially carbon fibers confer especially high levels of strength and stiffness. Carbon-fiber composites have a relative stiffness five times that of steel. Because of these excellent properties, many applications are uniquely suited for epoxy and polyester composites, such as components in new jet aircraft, parts for automobiles, boat hulls, rocket motor cases, and chemical reaction vessels.

Although the most dramatic properties are found with reinforced thermosetting resins such as epoxy and polyester resins, significant improvements can be obtained with many reinforced thermoplastic resins as well. Polycarbonates, polyethylene, and polyesters are among the resins available as glass-reinforced composition. The combination of inexpensive, one-step fabrication by injection molding, with improved properties has made it possible for reinforced thermoplastics to replace metals in many applications in appliances, instruments, automobiles, and tools.

In the development of other composite systems, various matrices are possible; for example, polyimide resins are excellent matrices for glass fibers, and give a high- performance composite. Different fibers are of potential interest, including polymers [such as poly(vinyl alcohol)], single-crystal ceramic whiskers (such as sapphire), and various metallic fibers.

Long ago, people living in South and Central America had used natural rubber latex, polyisoprene, to make things like gloves and boots, as well as rubber balls which they used to play games that were a lot like modern basketball. He took two layers of cotton fabric and embedded them in natural rubber, also known as polyisoprene, making a three-layered sandwich like the one you see on your right (Remember, cotton is made up of a natural polymer called cellulose). This made for good raincoats because, while the rubber made it waterproof, the cotton layers made it comfortable to wear, to make a material that has the properties of both its components. In this case, we combine the water-resistance of polyisoprene and the comfort of cotton.

Modern composites are usually made of two components, a fiber and matrix. The fiber is most often glass, but sometimes Kevlar, carbon fiber, or polyethylene. The matrix is usually a thermoset like an epoxy resin, polydicyclopentadiene, or a polyimide. The fiber is embedded in the matrix in order to make the matrix stronger. Fiber-reinforced composites have two things going for them. They are strong and light. They are often stronger than steel, but weigh much less. This means that composites can be used to make automobiles lighter, and thus much more fuel efficient.

A common fiber-reinforced composite is Fiberglas™. Its matrix is made by reacting polyester with carbon-carbon double bonds in its backbone, and styrene. We pour a mix of the styrene and polyester over a mass of glass fibers.

The styrene and the double bonds in the polyester react by free radical vinyl polymerization to form a crosslinked resin. The glass fibers are trapped inside, where they act as a reinforcement. In Fiberglas™ the fibers are not lined up in any particular direction. They are just a tangled mass, like you see on the right. But we can make the composite stronger by lining up all the fibers in the same direction. Oriented fibers do some weird things to the composite. When you pull on the composite in the direction of the fibers, the composite is very strong. But if you pull on it at right angles to the fiber direction, it is not very strong at all [8-9]. This is not always bad, because sometimes we only need the composite to be strong in one direction. Sometimes the item you are making will only be under stress in one direction. But sometimes we need strength in more than one direction. So we simply point the fibers in more than one direction. We often do this by using a woven fabric of the fibers to reinforce the composite. The woven fibers give a composite good strength in many directions.

The polymeric matrix holds the fibers together. A loose bundle of fibers would not be of much use. Also, though fibers are strong, they can be brittle. The matrix can absorb energy by deforming under stress. This is to say, the matrix adds toughness to the composite. And finally, while fibers have good tensile strength (that is, they are strong when you pull on them), they usually have awful compressional strength. That is, they buckle when you squash them. The matrix gives compressional strength to the composite.

Not all fibers are the same. Now it may seem strange that glass is used as reinforcement, as glass is really easy to break. But for some reason, when glass is spun into really tiny fibers, it acts very different. Glass fibers are strong, and flexible.

Still, there are stronger fibers out there. This is a good thing, because sometimes glass just isn't strong and tough enough. For some things, like airplane parts, that undergo a lot of stress, you need to break out the fancy fibers. When cost is no object, you can use stronger, but more expensive fibers, like Kevlar™, carbon fiber. Carbon fiber (Spectra™) is usually stronger than Kevlar™, that is, it can withstand more force without breaking. But Kevlar™ tends to be tougher. This means it can absorb more energy without breaking. It can stretch a little to keep from breaking, more so than carbon fiber can. But Spectra™, which is a kind of polyethylene, is stronger and tougher than both carbon fiber and Kevlar™.

Different jobs call for different matrices. The unsaturated polyester/styrene systems at are one example. They are fine for everyday applications. Chevrolet Corvette bodies are made from composites using unsaturated polyester matrices and glass fibers. But they have some drawbacks. They shrink a good deal when they're cured, they can absorb water very easily, and their impact strength is low.

2.2.2. BIOCOMPOSITES

For many decades, the residential construction field has used timber as its main source of building material for the frames of modern American homes. The American timber industry produced a record 49.5 billion board feet of lumber in 1999, and another 48.0 billion board feet in 2002. At the same time that lumber production is peaking, the home ownership rate reached a record high of 69.2%, with over 977,000 homes being sold in 2002. Because residential construction accounts for one-third of the total softwood lumber use in the United States, there is an increasing demand for alternate materials. Use of sawdust not only provides an alternative but also increases the use of the by product efficiently. Wood plastic composites (WPC) is a relatively new category of materials that covers a broad range of composite materials utilizing an organic resin binder (matrix) and fillers composed of cellulose materials. The new and rapidly developing biocomposite materials are high technology products, which have one unique advantage – the wood filler can include sawdust and scrap wood products. Consequently, no additional wood resources are needed to manufacture biocomposites. Waste products that would traditraditionally cost money for proper disposal, now become a beneficial resource, allowing recycling to be both profitable and environmentally conscious. The use of biocomposites and WPC has increased rapidly all over the world, with the end users for these composites in the construction, motor vehicle, and furniture industries. One of the primary problems related to the use of biocomposites is the flammability of the two main components (binder and filler). If a flame retardant were added, this would require the adhesion of the fiber and the matrix not to be disturbed by the retardant. The challenge is to develop a composite that will not burn and will maintain its level of mechanical performance. In lieu of organic matrix compounds, inorganic matrices can be utilized to improve the fire resistance. Inorganic-based wood composites are those that consist of a mineral mix as the binder system. Such inorganic binder systems include gypsum and Portland cement, both of which are highly resistant to fire and insects. The main disadvantage with these systems is the maximum amount of sawdust or fibers than can be incorporated is low. One relatively new type of inorganic matrix is potassium aluminosilicate, an environmentally friendly compound made from naturally occurring materials. The Federal Aviation Administration has investigated the feasibility of using this matrix in commercial aircraft due to its ability to resist temperatures of up to 1000 ºC without generating smoke, and its ability to enable carbon composites to withstand temperatures of 800 ºC and maintain 63% of its original flexural strength. Potassium aluminosilicate matrices are compatible with many common building material including clay brick, masonry, concrete, steel, titanium, balsa, oak, pine, and particleboard [10].

2.3. Fiberglass

Fiberglass refers to a group of products made from individual glass fibers combined into a variety of forms. Glass fibers can be divided into two major groups according to their geometry: continuous fibers used in yarns and textiles, and the discontinuous (short) fibers used as batts, blankets, or boards for insulation and filtration. Fiberglass can be formed into yarn much like wool or cotton, and woven into fabric which is sometimes used for draperies. Fiberglass textiles are commonly used as a reinforcement material for molded and laminated plastics. Fiberglass wool, a thick, fluffy material made from discontinuous fibers, is used for thermal insulation and sound absorption. It is commonly found in ship and submarine bulkheads and hulls; automobile engine compartments and body panel liners; in furnaces and air conditioning units; acoustical wall and ceiling panels; and architectural partitions. Fiberglass can be tailored for specific applications such as Type E (electrical), used as electrical insulation tape, textiles and reinforcement; Type C (chemical), which has superior acid resistance, and Type T, for thermal insulation [11].

Though commercial use of glass fiber is relatively recent, artisans created glass strands for decorating goblets and vases during the Renaissance. A French physicist, Rene-Antoine Ferchault de Reaumur, produced textiles decorated with fine glass strands in 1713. Glass wool, a fluffy mass of discontinuous fiber in random lengths, was first produced in Europe in 1900, using a process that involved drawing fibers from rods horizontally to a revolving drum [12].

The basic raw materials for fiberglass products are a variety of natural minerals and manufactured chemicals. The major ingredients are silica sand, limestone, and soda ash. Other ingredients may include calcined alumina, borax, feldspar, nepheline syenite, magnesite, and kaolin clay, among others. Silica sand is used as the glass former, and soda ash and limestone help primarily to lower the melting temperature. Other ingredients are used to improve certain properties, such as borax for chemical resistance. Waste glass, also called cullet, is also used as a raw material. The raw materials must be carefully weighed in exact quantities and thoroughly mixed together (called batching) before being melted into glass.

2.3.1. THE MANUFACTURING PROCESS

2.3.1.1. MELTING

Once the batch is prepared, it is fed into a furnace for melting. The furnace may be heated by electricity, fossil fuel, or a combination of the two. Temperature must be precisely controlled to maintain a smooth, steady flow of glass. The molten glass must be kept at a higher temperature (about 1371 °C) than other types of glass in order to be formed into fiber. Once the glass becomes molten, it is transferred to the forming equipment via a channel (forehearth) located at the end of the furnace [13].

2.3.1.2. FORMING INTO FIBERS

Several different processes are used to form fibers, depending on the type of fiber. Textile fibers may be formed from molten glass directly from the furnace, or the molten glass may be fed first to a machine that forms glass marbles of about 0.62 inch (1.6 cm) in diameter. These marbles allow the glass to be inspected visually for impurities. In both the direct melt and marble melt process, the glass or glass marbles are fed through electrically heated bushings (also called spinnerets). The bushing is made of platinum or metal alloy, with anywhere from 200 to 3,000 very fine orifices. The molten glass passes through the orifices and comes out as fine filaments [13].

2.3.1.3. CONTINUOUS-FILAMENT PROCESS

A long, continuous fiber can be produced through the continuous-filament process. After the glass flows through the holes in the bushing, multiple strands are caught up on a high-speed winder. The winder revolves at about 3 km a minute, much faster than the rate of flow from the bushings. The tension pulls out the filaments while still molten, forming strands a fraction of the diameter of the openings in the bushing. A chemical binder is applied, which helps keep the fiber from breaking during later processing. The filament is then wound onto tubes. It can now be twisted and plied into yarn [14].

2.3.1.4. STAPLE-FIBER PROCESS

An alternative method is the staplefiber process. As the molten glass flows through the bushings, jets of air rapidly cool the filaments. The turbulent bursts of air also break the filaments into lengths of 20-38 cm. These filaments fall through a spray of lubricant onto a revolving drum, where they form a thin web. The web is drawn from the drum and pulled into a continuous strand of loosely assembled fibers [15]. This strand can be processed into yarn by the same processes used for wool and cotton.

2.3.1.5. CHOPPED FIBER

Instead of being formed into yarn, the continuous or long-staple strand may be chopped into short lengths. The strand is mounted on a set of bobbins, called a creel, and pulled through a machine which chops it into short pieces. The chopped fiber is formed into mats to which a binder is added. After curing in an oven, the mat is rolled up. Various weights and thicknesses give products for shingles, built-up roofing, or decorative mats [16].

2.3.1.6. GLASS WOOL

The rotary or spinner process is used to make glass wool. In this process, molten glass from the furnace flows into a cylindrical container having small holes. As the container spins rapidly, horizontal streams of glass flow out of the holes. The molten glass streams are converted into fibers by a downward blast of air, hot gas, or both. The fibers fall onto a conveyor belt, where they interlace with each other in a fleecy mass. This can be used for insulation, or the wool can be sprayed with a binder, compressed into the desired thickness, and cured in an oven. The heat sets the binder, and the resulting product may be a rigid or semi-rigid board, or a flexible bat [15-16].

2.3.1.7. PROTECTIVE COATINGS

In addition to binders, other coatings are required for fiberglass products. Lubricants are used to reduce fiber abrasion and are either directly sprayed on the fiber or added into the binder. An anti-static composition is also sometimes sprayed onto the surface of fiberglass insulation mats during the cooling step. Cooling air drawn through the mat causes the anti-static agent to penetrate the entire thickness of the mat. The anti-static agent consists of two ingredients a material that minimizes the generation of static electricity, and a material that serves as a corrosion inhibitor and stabilizer.

Sizing is any coating applied to textile fibers in the forming operation, and may contain one or more components (lubricants, binders, or coupling agents). Coupling agents are used on strands that will be used for reinforcing plastics, to strengthen the bond to the reinforced material. Sometimes a finishing operation is required to remove these coatings, or to add another coating. For plastic reinforcements, sizings may be removed with heat or chemicals and a coupling agent applied. For decorative applications, fabrics must be heat treated to remove sizings and to set the weave. Dye base coatings are then applied before dying or printing [15-16].

2.3.1.8. FORMING INTO SHAPES

Fiberglass products come in a wide variety of shapes, made using several processes. For example, fiberglass pipe insulation is wound onto rod-like forms called mandrels directly from the forming units, prior to curing. The mold forms, in lengths of 91 cm or less, are then cured in an oven. The cured lengths are then de-molded lengthwise, and sawn into specified dimensions. Facings are applied if required, and the product is packaged for shipment [17].

2.4. CARBON FIBRE

Carbon-fiber-reinforced polymer or carbon-fiber-reinforced plastic (CFRP or CRP or often simply carbon fiber), is a very strong and light fiber-reinforced polymer which contains carbon fibers. Carbon fibres are created when polyacrylonitrile fibres (PAN), Pitch resins, or Rayon are carbonized (through oxidation and thermal pyrolysis) at high temperatures. Through further processes of graphitizing or stretching the fibres strength or elasticity can be enhanced respectively. Carbon fibres are manufactured in diameters analogous to glass fibres with diameters ranging from 9 to 17 μm. These fibres wound into larger threads for transportation and further production processes. Further production processes include weaving or braiding into carbon fabrics, cloths and mats analogous to those described for glass that can then be used in actual reinforcement processes. Carbon fibers are a new breed of high-strength materials. Carbon fiber has been described as a fiber containing at least 90% carbon obtained by the controlled pyrolysis of appropriate fibers. The existence of carbon fiber came into being in 1879 when Edison took out a patent for the manufacture of carbon filaments suitable for use in electric lamps [18].

2.4.1. CLASSIFICATION AND TYPES

Based on modulus, strength, and final heat treatment temperature, carbon fibers can be classified into the following categories:

1. Based on carbon fiber properties, carbon fibers can be grouped into:
    - Ultra-high-modulus, type UHM (modulus >450Gpa)
    - High-modulus, type HM (modulus between 350-450Gpa)
    - Intermediate-modulus, type IM (modulus between 200-350Gpa)

- Low modulus and high-tensile, type HT (modulus < 100Gpa, tensile strength > 3.0Gpa)
- Super high-tensile, type SHT (tensile strength > 4.5Gpa)

2. Based on precursor fiber materials, carbon fibers are classified into;
   - PAN-based carbon fibers
   - Pitch-based carbon fibers
   - Mesophase pitch-based carbon fibers
   - Isotropic pitch-based carbon fibers
   - Rayon-based carbon fibers
   - Gas-phase-grown carbon fibers
3. Based on final heat treatment temperature, carbon fibers are classified into:
   - Type-I, high-heat-treatment carbon fibers (HTT), where final heat treatment temperature should be above 2000°C and can be associated with high-modulus type fiber.
   - Type-II, intermediate-heat-treatment carbon fibers (IHT), where final heat treatment temperature should be around or above 1500 °C and can be associated with high-strength type fiber.
   - Type-III, low-heat-treatment carbon fibers, where final heat treatment temperatures not greater than 1000 °C. These are low modulus and low strength materials [19].

2.4.2. MANUFACTURE

In Textile Terms and Definitions, carbon fiber has been described as a fiber containing at least 90% carbon obtained by the controlled pyrolysis of appropriate fibers. The term "graphite fiber" is used to describe fibers that have carbon in excess of 99%. Large varieties of fibers called precursors are used to produce carbon fibers of different morphologies and different specific characteristics. The most prevalent precursors are polyacrylonitrile (PAN), cellulosic fibers (viscose rayon, cotton), petroleum or coal tar pitch and certain phenolic fibers.

Carbon fibers are manufactured by the controlled pyrolysis of organic precursors in fibrous form. It is a heat treatment of the precursor that removes the oxygen, nitrogen and hydrogen to form carbon fibers. It is well established in carbon fiber literature that the mechanical properties of the carbon fibers are improved by increasing the crystallinity and orientation, and by reducing defects in the fiber. The best way to achieve this is to start with a highly oriented precursor and then maintain the initial high orientation during the process of stabilization and carbonization through tension [18-19].

2.4.2.1. CARBON FIBERS FROM POLYACRYLONITRILE (PAN)

There are three successive stages in the conversion of PAN precursor into high-performance carbon fibers. Oxidative stabilization: The polyacrylonitrile precursor is first stretched and simultaneously oxidized in a temperature range of 200-300 °C. This treatment converts thermoplastic PAN to a non-plastic cyclic or ladder compound. Carbonization: After oxidation, the fibers are carbonized at about 1000 °C without tension in an inert atmosphere (normally nitrogen) for a few hours. During this process the non-carbon elements are removed as volatiles to give carbon fibers with a yield of about 50% of the mass of the original PAN. Graphitization: Depending on the type of fiber required, the fibers are treated at temperatures between 1500-3000 °C, which improves the ordering, and orientation of the crystallites in the direction of the fiber axis.

2.4.2.2. CARBON FIBERS FROM RAYON

a- The conversion of rayon fibers into carbon fibers is three phase process

Stabilization: Stabilization is an oxidative process that occurs through steps. In the first step, between 25-150 °C, there is physical desorption of water. The next step is a dehydration of the cellulosic unit between 150-240 °C. Finally, thermal cleavage of the cyclosidic linkage and scission of ether bonds and some C-C bonds via free radical reaction (240-400 °C) and, thereafter, aromatization takes place.

Carbonization: Between 400 and 700 °C, the carbonaceous residue is converted into a graphite-like layer.

Graphitization: Graphitization is carried out under strain at 700-2700 °C to obtain high modulus fiber through longitudinal orientation of the planes.

b- The carbon fiber fabrication from pitch generally consists of the following four steps:

Pitch preparation: It is an adjustment in the molecular weight, viscosity, and crystal orientation for spinning and further heating.

Spinning and drawing: In this stage, pitch is converted into filaments, with some alignment in the crystallites to achieve the directional characteristics.

Stabilization: In this step, some kind of thermosetting to maintain the filament shape during pyrolysis. The stabilization temperature is between 250 and 400 °C.

Carbonization: The carbonization temperature is between 1000-1500 °C.

2.4.2.3. CARBON FIBERS IN MELTBLOWN NONWOVENS

Carbon fibers made from the spinning of molten pitches are of interest because of the high carbon yield from the precursors and the relatively low cost of the starting materials. Stabilization in air and carbonization in nitrogen can follow the formation of melt-blown pitch webs. Processes have been developed with isotropic pitches and with anisotropic mesophase pitches. The mesophase pitch based and melt blown discontinuous carbon fibers have a peculiar structure. These fibers are characterized in that a large number of small domains, each domain having an average equivalent diameter from 0.03 mm to 1mm and a nearly unidirectional orientation of folded carbon layers, assemble to form a mosaic structure on the cross-section of the carbon fibers. The folded carbon layers of each domain are oriented at an angle to the direction of the folded carbon layers of the neighboring domains on the boundary [20].

2.4.2.4. CARBON FIBERS FROM ISOTROPIC PITCH

The isotropic pitch or pitch-like material, i.e., molten polyvinyl chloride, is melt spun at high strain rates to align the molecules parallel to the fiber axis. The thermoplastic fiber is then rapidly cooled and carefully oxidized at a low temperature (<100 °C). The oxidation process is rather slow, to ensure stabilization of the fiber by cross-linking and rendering it infusible. However, upon carbonization, relaxation of the molecules takes place, producing fibers with no significant preferred orientation. This process is not industrially attractive due to the lengthy oxidation step, and only low-quality carbon fibers with no graphitization are produced. These are used as fillers with various plastics as thermal insulation materials [20].

2.4.2.5. CARBON FIBERS FROM ANISOTROPIC MESOPHASE PITCH

High molecular weight aromatic pitches, mainly anisotropic in nature, are referred to as mesophase pitches. The pitch precursor is thermally treated above 350°C to convert it to mesophase pitch, which contains both isotropic and anisotropic phases. Due to the shear stress occurring during spinning, the mesophase molecules orient parallel to the fiber axis. After spinning, the isotropic part of the pitch is made infusible by thermosetting in air at a temperature below it's softening point. The fiber is then carbonized at temperatures up to 1000 °C. The main advantage of this process is that no tension is required during the stabilization or the graphitization, unlike the case of rayon or PANs precursors [21].

2.4.2.6. STRUCTURE

The characterization of carbon fiber microstructure has been mainly been performed by x-ray scattering and electron microscopy techniques. In contrast to graphite, the structure of carbon fiber lacks any three dimensional order. In PAN-based fibers, the linear chain structure is transformed to a planar structure during oxidative stabilization and subsequent carbonization. Basal planes oriented along the fiber axis are formed during the carbonization stage. Wide-angle x-ray data suggests an increase in stack height and orientation of basal planes with an increase in heat treatment temperature. A difference in structure between the sheath and the core was noticed in a fully stabilized fiber. The skin has a high axial preferred orientation and thick crystallite stacking. However, the core shows a lower preferred orientation and a lower crystallite height [22].

2.4.2.7. PROPERTIES

In general, it is seen that the higher the tensile strength of the precursor the higher is the tenacity of the carbon fiber. Tensile strength and modulus are significantly improved by carbonization under strain when moderate stabilization is used. X-ray and electron diffraction studies have shown that in high modulus type fibers, the crystallites are arranged around the longitudinal axis of the fiber with layer planes highly oriented parallel to the axis. Overall, the strength of a carbon fiber depends on the type of precursor, the processing conditions, heat treatment temperature and the presence of flaws and defects. With PAN based carbon fibers, the strength increases up to a maximum of 1300 ºC and then gradually decreases. The modulus has been shown to increase with increasing temperature. PAN based fibers typically buckle on compression and form kink bands at the innermost surface of the fiber. However, similar high modulus type pitch-based fibers deform by a shear mechanism with kink bands formed at 45° to the fiber axis. Carbon fibers are very brittle. The layers in the fibers are formed by strong covalent bonds. The sheet-like aggregations allow easy crack propagation. On bending, the fiber fails at very low strain [23].

2.4.2.8. APPLICATIONS

The two main applications of carbon fibers are in specialized technology, which includes aerospace and nuclear engineering, and in general engineering and transportation, which includes engineering components such as bearings, gears, cams, fan blades and automobile bodies. Recently, some new applications of carbon fibers have been found. Others include: decoration in automotive, marine, general aviation interiors, general entertainment and musical instruments and after-market transportation products. Conductivity in electronics technology provides additional new application [24].

The production of highly effective fibrous carbon adsorbents with low diameter, excluding or minimizing external and intra-diffusion resistance to mass transfer, and therefore, exhibiting high sorption rates is a challenging task. These carbon adsorbents can be converted into a wide variety of textile forms and nonwoven materials. Cheaper and newer versions of carbon fibers are being produced from new raw materials. Newer applications are also being developed for protective clothing (used in various chemical industries for work in extremely hostile environments), electromagnetic shielding and various other novel applications. The use of carbon fibers in Nonwovens is in a new possible application for high temperature fire-retardant insulation (eg: furnace material) [25].

2.5. Aramid-Definition

Aliphatic polyamides are macromolecules whose structural units are characteristically interlinked by the amide linkage -NH-CO-. The nature of the structural unit constitutes a basis for classification. Aliphatic polyamides with structural units derived predominantly from aliphatic monomers are members of the generic class of nylons, whereas aromatic polyamides in which at least 85% of the amide linkages are directly adjacent to aromatic structures have been designated aramids. The U.S. Federal Trade Commission defines nylon fibers as "a manufactured fiber in which the fiber forming substance is a long chain synthetic polyamide in which less than 85% of the amide linkages (-CO-NH-) are attached directly to two aliphatic groups." Polyamides that contain recurring amide groups as integral parts of the polymer backbone have been classified as condensation polymers regardless of the principal mechanisms entailed in the polymerization process. Though many reactions suitable for polyamide formation are known, commercially important nylons are obtained by processes related to either of two basic approaches: one entails the polycondensation of difunctional monomers utilizing either amino acids or stoichiometric pairs of dicarboxylic acids and diamines, and the other entails the ring-opening polymerization of lactams. The polyamides formed from diacids and diamines are generally described to be of the AABB format, whereas those derived from either amino acids or lactams are of the AB format.

The structure of polyamide fibers is defined by both chemical and physical parameters. The chemical parameters are related mainly to the constitution of the polyamide molecule and are concerned primarily with its monomeric units, end-groups, and molecular weight. The physical parameters are related essentially to chain conformation, orientation of both polymer molecule segments and aggregates, and to crystallinity [26]. This characteristic for single-chain aliphatic polyamides is determined by the structure of the monomeric units and the nature of end groups of the polymer molecules. The most important structural parameter of the noncrystalline (amorphous) phase is the glass transition temperature (T_g) since it has a considerable effect on both processing and properties of the polyamide fibers. It relates to a type of a glass–rubber transition and is defined as the temperature, or temperature range, at which mobility of chain segments or structural units commences. Thus it is a function of the chemical structure; in case of the linear aliphatic polyamides, it is a function of the number of CH_2 units (mean spacing) between the amide groups. As the number of CH_2 unit's increases, T_g decreases. Although T_g is further affected by the nature of the crystalline phase, orientation, and molecular weight, it is associated only with what may be considered the amorphous phase.

Any process affecting this phase exerts a corresponding effect on the glass transition temperature. This is particularly evident in its response to the concentration of water absorbed in polyamides. An increase in water content results in a steady decrease of T_g toward a limiting value. This phenomenon may be explained by a mechanism that entails successive replacement of intercatenary hydrogen bonds in the amorphous phase with water. It may involve a sorption mechanism, according to which 3 mol of water interact with two neighboring amide groups [27].

The properties of aromatic polyamides differ significantly from those of their aliphatic counterparts. This led the U.S. Federal Trade Commission to adopt the term ''aramid'' to designate fibers of the aromatic polyamide type in which at least 85% of the amide linkages are attached directly to two aromatic rings.

The search for materials with very good thermal properties was the original reason for research into aromatic polyamides. Bond dissociation energies of C-C and C-N bonds in aromatic polyamides are ~20% higher than those in aliphatic polyamides. This is the reason why the decomposition temperature of poly(m-phenylene isophthalamide) MPDI exceeds 450 ºC. Conjugation between the amide group and the aromatic ring in poly(p-phenylene terephthalamide) "PPTA" increases chain rigidity as well as the decomposition temperature, which exceeds 550 ºC.

Obviously, hydrogen bonding and chain rigidity of these polymers translates to very high glass transition temperatures. Using low-molecular-weight polymers, Aharoni [19] measured glass transition temperatures of 272 ºC for MPDI and over 295 ºC for PPTA (which in this case had low crystallinity). Others have reported values as high as 4928 ºC. In most cases the measurement of T_g is difficult because PPTA is essentially 100% crystalline. As one would expect, these values are not strongly dependent on the molecular weight of the polymer above a DP of ~10 [22].

The same structural characteristics that are responsible for the excellent thermal properties of these materials are responsible for their limited solubility as well as good chemical resistance. PPTA is soluble only in strong acids like H_2SO_4, HF, and methanesulfonic acid. Preparation of this polymer via solution polymerization in amide solvents is accompanied by polymer precipitation. As expected, based on its structure, MPDI is easier to solubilize then PPTA. It is soluble in neat amide solvents like N-methyl-2-pyrrolidone (NMP) and dimethylacetamide (DMAc), but adding salts like $CaCl_2$ or LiCl significantly enhances its solubility. The significant rigidity of the PPTA chain (as discussed above) leads to the formation of anisotropic solutions when the solvent is good enough to reach critical minimum solids concentration. The implications of this are discussed in greater detail later in this chapter. It is well known that chemical properties differ significantly between crystalline and noncrystalline materials of the same composition. In general, aramids have very good chemical resistance. Obviously, the amide bond is subject to a hydrolytic attack by acids and bases. Exposure to very strong oxidizing agents results in a significant strength loss of these fibers. In addition to crystallinity, structure consolidation affects the rate of degradation of these materials. The hydrophilicity of the amide group leads to a significant absorption of water by all aramids. While the chemistry is the driving factor, fiber structure also plays a very important role; for example, Kevlar 29 absorbs ~7% water, Kevlar 49~4%, and Kevlar 149 only 1%. Fukuda explored the relationship between fiber crystallinity and equilibrium moisture in great detail. Because of their aromatic character, aramids absorb UV light, which results in an oxidative color change. Substantial exposure can lead to the loss of yarn tensile properties. UV absorption by p-aramids is more pronounced than with m-aramids. In this case a self-screening phenomenon is observed, which makes thin structures more susceptible to degradation than thick ones. Very frequently p-aramids are covered with another material in the final application to protect them. The high degree of aromaticity of these materials also provides significant flame resistance. All commercial aramids have a limited oxygen index in the range of 28-32%, which compares with ~20% for aliphatic polyamides.

Typical properties of commercial aramid fibers are while yarns of m-aramids have tensile properties that are no greater than those of aliphatic polyamides, they do retain useful mechanical properties at significantly higher temperatures. The high glass transition temperature leads to low (less than 1%) shrinkage at temperatures below 250 ºC. In general, mechanical properties of m-aramid fibers are developed on drawing. This process produces fibers with a high degree of morphological homogeneity, which leads to very good fatigue properties. The mechanical properties of p-aramid fibers have been the subject of much study. This is because these fibers were the first examples of organic materials with a very high level of both strength and stiffness. These materials are practical confirmation that nearly perfect orientation and full chain extension are required to achieve mechanical properties approaching those predicted for chemical bonds. In general, the mechanical properties reflect a significant anisotropy of these fibers-covalent bonds in the direction of the fiber axis with hydrogen bonding and van der Waals forces in the lateral direction [26].

Aramid-based reinforcement has been viewed as a more specialty product for applications requiring high modulus and where the potential for electrical conductivity would preclude the use of carbon; for example, aramid sheet is used for all tunnel repairs. Product forms include dry fabrics or unidirectional sheets as well as pre-cured strips or bars. Fabrics or sheets are applied to a concrete surface that has been smoothed (by grinding or blasting) and wetted with a resin (usually epoxy). The composite materials used for concrete infrastructure repair that was initiated in the mid 1980s. After air pockets are removed using rollers or flat, flexible squeegees, a second resin coat might be applied. Reinforcement of concrete structures is important in earthquake prone areas, although steel plate is the primary material used to reinforce and repair concrete structures, higher priced fiber-based sheet structures offer advantages for small sites where ease of handling and corrosion resistance are important. The high strength, modulus, and damage tolerance of aramid-reinforced sheets makes the fiber especially suitable for protecting structures prone to seismic activity. The use of aramid sheet also simplifies the application process. Sheets are light in weight and can be easily handled without heavy machinery and can be applied in confined working spaces. Sheets are also flexible, so surface smoothing and corner rounding of columns are less critical than for carbon fiber sheets [28].

3. ALL PROCESS DESCRIPTION

FRP involves two distinct processes, the first is the process whereby the fibrous material is manufactured and formed, and the second is the process whereby fibrous materials are bonded with the matrix during the molding process.

3.1. Fibre Process

3.1.1. THE MANUFACTURE OF FIBRE FABRIC

Reinforcing Fibre is manufactured in both two dimensional and three dimensional orientations

1. Two Dimensional Fibre Reinforced Polymer are characterized by a laminated structure in which the fibres are only aligned along the plane in x-direction and y-direction of the material. This means that no fibres are aligned in the through thickness or the z-direction, this lack of alignment in the through thickness can create a disadvantage in cost and processing. Costs and labour increase because conventional processing techniques used to fabricate composites, such as wet hand lay-up, autoclave and resin transfer molding, require a high amount of skilled labour to cut, stack and consolidate into a preformed component.

2. Three-dimensional Fibre Reinforced Polymer composites are materials with three dimensional fibre structures that incorporate fibres in the x-direction, y-direction and z-direction. The development of three-dimensional orientations arose from industry's need to reduce fabrication costs, to increase through-thickness mechanical properties,

and to improve impact damage tolerance; all were problems associated with two dimensional fibre reinforced polymers [28].

3.1.2. THE MANUFACTURE OF FIBRE PREFORMS

Fibre preforms are how the fibres are manufactured before being bonded to the matrix. Fibre preforms are often manufactured in sheets, continuous mats, or as continuous filaments for spray applications. The four major ways to manufacture the fibre preform is though the textile processing techniques of Weaving, knitting, braiding and stitching.

1. Weaving can be done in a conventional manner to produce two-dimensional fibres as well in a multilayer weaving that can create three-dimensional fibres. However, multilayer weaving is required to have multiple layers of warp yarns to create fibres in the z- direction creating a few disadvantages in manufacturing, namely the time to set up all the warp yarns on the loom. Therefore most multilayer weaving is currently used to produce relatively narrow width products or high value products where the cost of the preform production is acceptable. Another Fibre-reinforced plastic 3D one of the main problems facing the use of multilayer woven fabrics is the difficulty in producing a fabric that contains fibres oriented with angles other than 0º and 90º to each other respectively.

2. The second major way of manufacturing fibre preforms is braiding. Braiding is suited to the manufacture of narrow width flat or tubular fabric and is not as capable as weaving in the production of large volumes of wide fabrics. Braiding is done over top of mandrels that vary in cross-sectional shape or dimension along their length. Braiding is limited to objects about a brick in size. Unlike the standard weaving process, braiding can produce fabric that contains fibres at 45 degrees angles to one another. Braiding three-dimensional fibres can be done using four steps, two-step or Multilayer Interlock Braiding. Four step or row and column braiding utilizes a flat bed containing rows and columns of yarn carriers that form the shape of the desired preform. Additional carriers are added to the outside of the array, the precise location and quantity of which depends upon the exact preform shape and structure required. There are four separate sequences of row and column motion, which act to interlock the yarns and produce the braided preform. The yarns are mechanically forced into the structure between each step to consolidate the structure in a similar process to the use of a reed in weaving.Two-step braiding is unlike the four step process because the two-step includes a large number of yarns fixed in the axial direction and a fewer number of braiding yarns. The process

consists of two steps in which the braiding carriers move completely through the structure between the axial carriers. This relatively simple sequence of motions is capable of forming performs of essentially any shape, including circular and hollow shapes. Unlike the four steps process the two steps process does not require mechanical compaction the motions involved in the process allows the braid to be pulled tight by yarn tension alone. The last type of braiding is multi-layer interlocking braiding that consists of a number of standard circular braiders being joined together to form a cylindrical braiding frame. This frame has a number of parallel braiding tracks around the circumference of the cylinder but the mechanism allows the transfer of yarn carriers between adjacent tracks forming a multilayer braided fabric with yarns interlocking to adjacent layers.

The multilayer interlock braid differs from both the four step and two-step braids in that the interlocking yarns are primarily in the plane of the structure and thus do not significantly reduce the in-plane properties of the perform. The four step and two step processes produce a greater degree of interlinking as the braiding yarns travel through the thickness of the preform, but therefore contribute less to the in-plane performance of the preform. A disadvantage of the multilayer interlock equipment is that due to the conventional sinusoidal movement of the yarn carriers to form the preform, the equipment is not able to have the density of yarn carriers that is possible with the two step and four step machines.

1. Knitting fibre preforms can be done with the traditional methods of Warp and [Weft] Knitting, and the fabric produced is often regarded by many as two-dimensional fabric, but machines with two or more needle beds are capable of producing multilayer fabrics with yams that traverse between the layers. Developments in electronic controls for needle selection and knit loop transfer and in the sophisticated mechanisms that allow specific areas of the fabric to be held and their movement controlled. This has allowed the fabric to form itself into the required three-dimensional perform shape with a minimum of material wastage.

2. Stitching is arguably the simplest of the four main textile manufacturing techniques and one that can be performed with the smallest investment in specialized machinery. Basically the stitching process consists of inserting a needle, carrying the stitch thread, through a stack of fabric layers to form a 3D structure. The advantages of stitching are that it is possible to stitch both dry and prepreg fabric, although the tackiness of prepare makes the process difficult and generally creates more damage within the prepreg material than in

the dry fabric. Stitching also utilizes the standard two-dimensional fabrics that are commonly in use within the composite industry therefore there is a sense of familiarity concerning the material systems. The use of standard fabric also allows a greater degree of flexibility in the fabric lay-up of the component than is possible with the other textile processes, which have restrictions on the fibre orientations that can be produced.

3.1.3. MOLDING PROCESSES

There are two distinct categories of molding processes using FRP plastics; this includes composite molding and wet molding. Composite molding uses Prepreg FRP, meaning the plastics are fibre reinforced before being put through further molding processes. Sheets of Prepreg FRP are heated or compressed in different ways to create geometric shapes. Wet molding combines fibre reinforcement and the matrix or resist during the molding process. The different forms of composite and wet molding, are listed below.

3.2. Composite Molding

3.2.1. BLADDER MOLDING

Individual sheets of prepreg material are laid -up and placed in a female-style mould along with a balloon-like bladder. The mould is closed and placed in a heated press. Finally, the bladder is pressurized forcing the layers of material against the mould walls. The part is cured and removed from the hot mould. Bladder molding is a closed molding process with a relatively short cure cycle between 15 and 60 minutes making it ideal for making complex hollow geometric shapes at competitive costs.

3.2.2. COMPRESSION MOLDING

A "preform" or "charge", of SMC, BMC or sometimes prepreg fabric, is placed into mould cavity. The mould is closed and the material is compacted & cured inside by pressure and heat. Compression molding offers excellent detailing for geometric shapes ranging from pattern and relief detailing to complex curves and creative forms, to precision engineering all within a maximum curing time of 20 minutes.

3.2.3. AUTOCLAVE – VACUUM BAG

Individual sheets of prepreg material are laid-up and placed in an open mold. The material is covered with release film, bleeder/breather material and a vacuum bag. A vacuum is pulled on part and the entire mould is placed into an autoclave (heated pressure vessel). The part is cured with a continuous vacuum to extract entrapped gasses from laminate. This is a very common process in the aerospace industry because it affords precise control over the molding process due to a long slow cure cycle that is anywhere from one to two hours. This precise control creates the exact laminate geometric forms needed to ensure strength and safety in the aerospace industry, but it is also slow and lab our intensive, meaning costs often confine it to the aerospace industry.

3.2.4. Mandrel wrapping

Sheets of prepreg material are wrapped around a steel or aluminum mandrel. The prepreg material is compacted by nylon or polypropylene cello tape. Parts are typically batch cured by hanging in an oven. After cure the cello and mandrel are removed leaving a hollow carbon tube. This process creates strong and robust hollow carbon tubes.

3.2.5. Wet layup

Fibre reinforcing fabric is placed in an open mould and then saturated with a wet (resin) by pouring it over the fabric and working it into the fabric and mould. The mould is then left so that the resin will cure, usually at room temperature, though heat is sometimes used to ensure a proper curing process. Glass fibres are most commonly used for this process, the results are widely known as fibreglass, and are used to make common products like skis, canoes, kayaks and surf boards.

3.2.6. Chopper gun

Continuous strand of fibreglass are pushed through a hand-held gun that both chops the strands and combines them with a catalyzed resin such as polyester. The impregnated chopped glass is shot onto the mould surface in whatever thickness the design and human operator think is appropriate. This process is good for large production runs at economical cost, but produces geometric shapes with less strength than other molding processes and has poor dimensional tolerance.

3.2.7. Filament winding

Machines pull fibre bundles through a wet bath of resin and wound over a rotating steel mandrel in specific orientations Parts are cured either room temperature or elevated temperatures. Mandrel is extracted, leaving a final geometric shape but can be left in some cases.

3.2.8. Pultrusion

Fibre bundles and slit fabrics are pulled through a wet bath of resin and formed into the rough part shape. Saturated material is extruded from a heated closed die curing while being continuously pulled through die. Some of the end products of the pultrusion process are structural shapes, i.e. beam, angle, channel and flat sheet. These materials can be used to create all sorts of fibreglass structures such as ladders, platforms, handrail systems tank, pipe, and pump supports.

3.3. Resin Infusion

Fabrics are placed into a mould which wet resin is then injected into. Resin is typically pressurized and forced into a cavity which is under vacuum in the RTM (Resin Transfer Molding) process. Resin is entirely pulled into cavity under vacuum in the VARTM (Vacuum Assisted Resin Transfer Molding) process. This molding process allows precise tolerances and detailed shaping but can sometimes fail to fully saturate the fabric leading to weak spots in the final shape.

3.3.1. Advantages and limitations

FRP allows the alignment of the glass fibres of thermoplastics to suit specific design programs. Specifying the orientation of reinforcing fibres can increase the strength and resistance to deformation of the polymer. Glass reinforced polymers are strongest and most resistive to deforming forces when the polymers fibres are parallel to the force being exerted, and are weakest when the fibres are perpendicular. Thus this ability is at once both an advantage and a limitation depending on the context of use. Weak spots of perpendicular fibres can be used for natural hinges and connections, but can also lead to material failure when production processes fail to properly orient the fibres parallel to expected forces. When forces are exerted perpendicular to the orientation of fibres the strength and elasticity of the polymer is less than the matrix alone. In cast resin components made of glass reinforced polymers such as UP and EP, the orientation of fibres can be oriented in two-dimensional and three-dimensional weaves. This means that when forces are possibly perpendicular to one orientation, they are parallel to another orientation; this eliminates the potential for weak spots in the polymer.

3.3.2. Failure modes

Structural failure can occur in FRP materials when:

- Tensile forces stretch the matrix more than the fibres, causing the material to shear at the interface between matrix and fibres.
- Tensile forces near the end of the fibres exceed the tolerances of the matrix, separating the fibres from the matrix.
- Tensile forces can also exceed the tolerances of the fibres causing the fibres themselves to fracture leading to material failure [29].

3.3.3. Material Requirements

The matrix must also meet certain requirements in order to first be suitable for the FRP process and ensure a successful reinforcement of it. The matrix must be able to properly saturate, and bond with the fibres within a suitable curing period. The matrix should preferably bond chemically with the fibre reinforcement for maximum adhesion. The matrix must also completely envelope the fibres to protect them from cuts and notches that would reduce their strength, and to transfer forces to the fibres. The fibres must also be kept separate from each other so that if failure occurs it is localized as much as possible, and if failure occurs the matrix must also debond from the fibre for similar reasons. Finally the matrix should be of a plastic that remains chemically and physically stable during and after reinforcement and molding processes. To be suitable for reinforcement material fibre additives must increase the tensile strength and modulus of elasticity of the matrix and meet the following conditions; fibres must exceed critical fibre content; the strength and rigidity of fibres itself must exceed the strength and rigidity of the matrix alone; and there must be optimum bonding between fibres and matrix.

3.4. GLASS FIBRE MATERIAL

FRPs use textile glass fibres; textile fibres are different from other forms of glass fibres used for insulating applications. Textile glass fibres begin as varying combinations of SiO_2, Al_2O_3, B_2O_3, CaO, or MgO in powder form. These mixtures are then heated through a direct melt process to temperatures around 1300 degrees Celsius, after which dies are used to extrude filaments of glass fibre in diameter ranging from 9 to 17 µm. These filaments are then wound into larger threads and spun onto bobbins for transportation and further processing. Glass fibre is by far the most popular means to reinforce plastic and thus enjoys a wealth of production processes, some of which are applicable to aramid and carbon fibres as well owing to their shared fibrous qualities. Roving is a process where filaments are spun into larger diameter threads. These threads are then commonly used for woven reinforcing glass fabrics and mats, and in spray applications. Fibre fabrics are web-form fabric reinforcing material that has both warped and weft directions. Fibre mats are web-form non-woven mats of glass fibres. Mats are manufactured in cut dimensions with chopped fibres, or in continuous mats using continuous fibres. Chopped fibre glass is used in processes where lengths of glass threads are cut between 3 and 26 mm, threads are then used in plastics most commonly intended for moulding processes. Glass fibre short strands are short 0.2–0.3 mm strands of glass fibres that are used to reinforce thermoplastics most commonly for injection moulding.

3.5. ARAMID FIBRE MATERIAL PROCESS

Aramid fibres are most commonly known Kevlar, Nomex and Technora. Aramids are generally prepared by the reaction between an amine group and a carboxylic acid halide group (aramid); commonly this occurs when an aromatic polyamide is spun from a liquid concentration of sulfuric acid into a crystallized fibre. Fibres are then spun into larger threads in order to weave into large ropes or woven fabrics (Aramid) [29]. Aramid fibres are manufactured with varying grades to base on varying qualities for strength and rigidity, so that the material can be somewhat tailored to specific design needs concerns, such as cutting the tough material during manufacture.

3.6. FRP, Applications

Fibre-reinforced plastics are best suited for any design program that demands weight savings, precision engineering, finite tolerances, and the simplification of parts in both production and operation. A molded polymer artifact is cheaper, faster, and easier to manufacture than cast aluminum or steel artifact, and maintains similar and sometimes better tolerances and material strengths. The Mitsubishi Lancer Evolution IV also used FRP for its spoiler material [30-32].

3.6.1. CARBON FIBRE REINFORCED POLYMERS

Rudder of commercial airplane

- Advantages over a traditional rudder made from sheet aluminum are:
- 25% reduction in weight
- 95% reduction in components by combining parts and forms into simpler molded parts.
- Overall reduction in production and operational costs, economy of parts results in lower production costs and the weight savings create fuel savings that lower the operational costs of flying the airplane.

3.6.2. STRUCTURAL APPLICATIONS OF FRP

FRP can be applied to strengthen the beams, columns and slabs in buildings. It is possible to increase strength of these structural members even after these have been severely damaged due to loading conditions. For strengthening beams, two techniques are adopted. First one is to paste FRP plates to the bottom (generally the tension face) of a beam. This increases the strength of beam, deflection capacity of beam and stiffness (load required to make unit deflection). Alternatively, FRP strips can be pasted in 'U' shape around the sides and bottom of a beam, resulting in higher shear resistance. Columns in building can be wrapped with FRP for achieving higher strength. This is called wrapping of columns. The technique works by restraining the lateral expansion of the column. Slabs may be strengthened by pasting FRP strips at their bottom (tension face). This will result in better performance, since the tensile resistance of slabs is supplemented by the tensile strength of FRP. In the case of beams and slabs, the effectiveness of FRP strengthening depends on the performance of the resin chosen for bonding [32].

3.6.3. GLASS FIBRE REINFORCED POLYMERS

Engine intake manifolds are made from glass fibre reinforced PA 66.

- Advantages this has over cast aluminum manifolds are:
- Up to a 60% reduction in weight
- Improved surface quality and aerodynamics
- Reduction in components by combining parts and forms into simpler molded shapes. Automotive gas and clutch pedals made from glass fibre reinforced PA 66 (DWP 12-13)
- Advantages over stamped aluminum are:
- Pedals can be molded as single units combining both pedals and mechanical linkages simplifying the production and operation of the design.
- Fibres can be oriented to reinforce against specific stresses, increasing the durability and safety.

3.6.4. DESIGN CONSIDERATIONS

FRP is used in designs that require a measure of strength or modulus of elasticity those non-reinforced plastics and other material choices are either ill suited for mechanically or economically. This means that the primary design consideration for using FRP is to ensure that the material is used economically and in a manner that takes advantage of its structural enhancements specifically. This is however not always the case, the orientation of fibres also creates a material weakness perpendicular to the fibres. Thus the use of fibre reinforcement and their orientation affects the strength, rigidity, and elasticity of a final form and hence the operation of the final product itself. Orienting the direction of fibres either, unidirectional, 2-dimensionally, or 3-dimensionally during production affects the degree of strength, flexibility, and elasticity of the final product. Fibres oriented in the direction of forces display greater resistance to distortion from these forces and vice versa, thus areas of a product that must withstand forces will be reinforced with fibres in the same direction, and areas that require flexibility, such as natural hinges, will use fibres in a perpendicular direction to forces. Using more dimensions avoids this either or scenario and creates objects that seek to avoid any specific weak points due to the unidirectional orientation of fibres. The properties of strength, flexibility and elasticity can also be magnified or diminished through the geometric shape and design of the final product. These include such design consideration such as ensuring proper wall thickness and creating multifunctional geometric shapes that can be molding as single pieces, creating shapes that have more material and structural integrity by reducing joints, connections, and hardware [30].

3.6.5. DISPOSAL AND RECYCLING CONCERNS

As a subset of plastic FR plastics are liable to a number of the issues and concerns in plastic waste disposal and recycling. Plastics pose a particular challenge in recycling processes because they are derived from polymers and monomers that often cannot be separated and returned to their virgin states, for this reason not all plastics can be recycled for re-use, in fact some estimates claim only 20% to 30% of plastics can be material recycled at all. Fibre reinforced plastics and their matrices share these disposal and environmental concerns. In addition to these concerns, the fact that the fibres themselves are difficult to remove from the matrix and preserve for re-use means FRP amplify these challenges. FRP are inherently difficult to separate into base a material that is into fibre and matrix, and the Fibre-reinforced plastic matrix into separate usable plastic, polymers, and monomers. These are all concerns for environmentally informed design today, but plastics often offer savings in energy and economic savings in comparison to other materials, also with the advent of new more environmentally friendly matrices such as bioplastics and UV-degradable plastics, FRP will similarly gain environmental sensitivity [29].

4. MECHANICAL PROPERTIES MEASUREMENTS

4.1. Strength

Strength is a mechanical property that you should be able to relate to, but you might not know exactly what we mean by the word "strong" when are talking about polymers. First, there is more than one kind of strength. There is tensile strength. A polymer has tensile strength if it is strong when one pulls on it. Tensile strength is important for a material that is going to be stretched or under tension. Fibers need good tensile strength.

Then there is compressional strength. A polymer sample has compressional strength if it is strong when one tries to compress it. Concrete is an example of a material with good compressional strength. Anything that has to support weight from underneath has to have good compressional strength [32]. There is also flexural strength. A polymer sample has flexural strength if it is strong when one tries to bend it.

There are other kinds of strength we could talk about. A sample torsional strength if it is strong when one tries to twist it. Then there is impact strength. A sample has impact strength if it is strong when one hits it sharply and suddenly, as with a hammer.

To measure the tensile strength of a polymer sample, we take the sample and we try to stretch. We usually stretch it with a machine for these studies. This machine simply has clamps on each end of the sample, then, when you turn it on it stretches the sample. While it is stretching the sample, it measures the amount of force (F) that it is exerting. When we know the force being exerted on the sample, we then divide that number by the cross-sectional area (A) of our sample. The answer is the stress that our sample is experiencing. Then, using our machine, we continue to increase the amount of force, and stress naturally, on the sample until it breaks. The stress needed to break the sample is the tensile strength of the material. Likewise, one can imagine similar tests for compressional or flexural strength. In all cases, the strength is the stress needed to break the sample. Since tensile stress is the force placed on the sample divided by the cross-sectional area of the sample, tensile stress, and tensile strength as well, are both measured in units of force divided by units of area, usually N/cm2. Stress and strength can also be measured in megapascals (MPa) or gigapascals (GPa). It is easy to convert between the different units, because 1 MPa = 100 N/cm^2, 1 GPa = 100,000 N/cm^2, and of course 1 GPa = 1,000 MPa. Other times, stress and strength are measured in the old English units of pounds per square inch, or psi. If you ever have to convert psi to N/cm^2, the conversion factor is 1 N/cm^2 = 1.45 psi.

4.2. Elongation

But there is more to understanding a polymer's mechanical properties than merely knowing how strong it is. All strength tells us is how much stress is needed to break something. It doesn't tell us anything about what happens to our sample while we're trying to break it. That's where it pays to study the elongation behavior of a polymer sample. Elongation is a type of deformation. Deformation is simply a change in shape that anything undergoes under stress. When we're talking about tensile stress, the sample deforms by stretching, becoming longer. We call this elongation, of course. Usually we talk about percent elongation, which is just the length the polymer sample is after it is stretched (L), divided by the original length of the sample (L_0), and then multiplied by 100.

There are a number of things we measure related to elongation. Which is most important depends on the type of material one is studying. Two important things we measure are ultimate elongation and elastic elongation. Ultimate elongation is important for any kind of material. It is nothing more than the amount you can stretch the sample before it breaks. Elastic elongation is the percent elongation you can reach without permanently deforming your sample. That is, how much can you stretch it, and still have the sample snap back to its original length once you release the stress on it. This is important if your material is an elastomer. Elastomers have to be able to stretch a long distance and still bounce back. Most of them can stretch from 500 to 1000 % elongation and return to their original lengths without any trouble [32].

4.3. Modulus

In the elastomers are need show the high elastic elongation. But for some other types of materials, like plastics, it usually they not stretch or deform so easily. If we want to know how well a material resists deformation, we measure something called modulus. To measure tensile modulus, we do the same thing as we did to measure strength and ultimate elongation. This time we measure the stress we're exerting on the material, just like we did when we were measuring tensile strength. First, is slowly increasing the amount of stress, and then we measure the elongation the sample undergoes at each stress level. We keep doing this until the sample breaks. This plot is called a stress-strain curve. (Strain is any kind of deformation, including elongation. Elongation is the word we use if we're talking specifically about tensile strain.) The height of the curve when the sample breaks is the tensile strength, of course, and the tensile modulus is the slope of this plot. If the slope is steep, the sample has a high tensile modulus, which means it resists deformation. If the slope is gentle, then the sample has a low tensile modulus, which means it is easily deformed. There are times when the stress-strain curve is not nice and straight, like we saw above. The slope isn't constant as stress increases. The slope, that is the modulus, is changing with stress. In a case like this we usually, the initial slope change as the modulus change [32].

In general, fibers have the highest tensile moduli, and elastomers have the lowest, and plastics have tensile moduli somewhere in between fibers and elastomers.

Modulus is measured by calculating stress and dividing by elongation, and would be measured in units of stress divided by units of elongation. But since elongation is dimensionless, it has no units by which we can divide. So modulus is expressed in the same units as strength, such as N/cm^2.

Intrinsic deformation is defined as the materials' true stress-strain response during homogeneous deformation. Since generally strain localization phenomena occur (like necking, shear banding, crazing and cracking), the measurement of the intrinsic materials' response requires a special experimental set-up, such as a video-controlled tensile or a uniaxial compression test. Although details of the intrinsic response differ per material, a general representation of the intrinsic deformation of polymers can be recognized [33], see Figure 1.

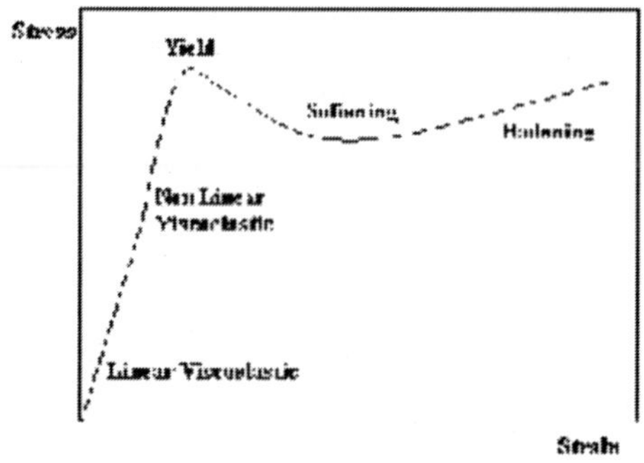

Figure 1. Schematic representation of the intrinsic deformation behavior of a polymer material [33].

4.4. Toughness

That plot of stress versus strain can give us another very valuable piece of information. If one measures the area underneath the stress-strain curve (figure 2), colored red in the graph below, the number you get is something we call toughness.

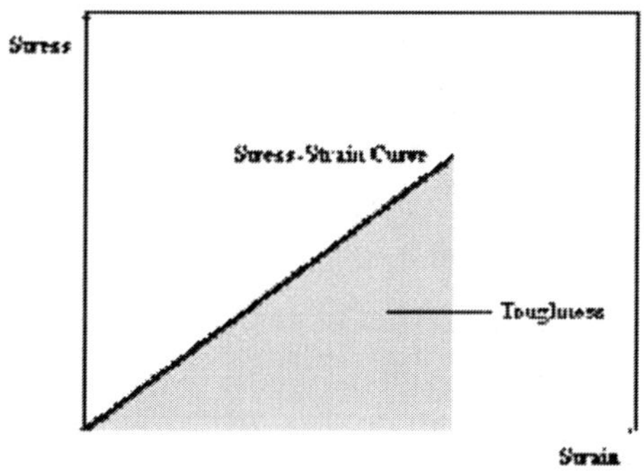

Figure 2. Plot of stress in function of strain.

Toughness is really a measure of the energy a sample can absorb before it breaks. Think about it, if the height of the triangle in the plot is strength, and the base of the triangle is strain, then the area is proportional to strength strain. Since strength is proportional to the force needed to break the sample, and strain is measured in units of distance (the distance the sample is stretched), then strength strain is proportional is force times distance, and as we remember from physics, force times distance is energy.

From a physics point of view the strength, is that strength tells how much force is needed to break a sample, and toughness tells how much energy is needed to break a sample. But that does not really tell you what the practical differences are. What is important knows that just because a material is strong, it isn't necessarily going to be tough as well [34-35].

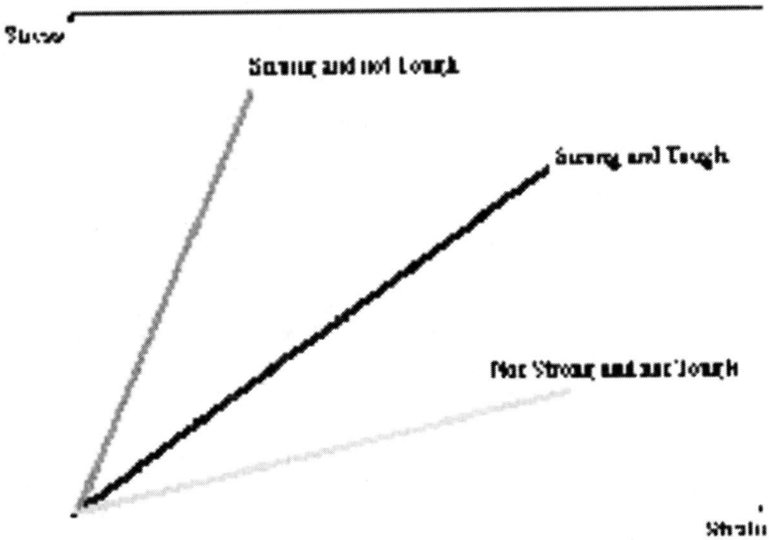

Figure 3. Plot of stress in function of strain, strong and tough concepts.

The gray plot is the stress-strain curve for a sample that is strong, but not tough (figure 3). As you can see, it takes a lot of force to break this sample. Likewise, this sample ca not stretch very much before it breaks. A material like this which is strong, but can not deform very much before it breaks is called brittle [36].

The gray plot is a stress-strain curve for a sample that is both strong and tough. This material is not as strong as the sample in the gray plot, but the area underneath its curve is a lot larger than the area under the gray sample's curve. So it can absorb a lot more energy than the gray-black sample plot.

The gray-black sample elongates a lot more before breaking than the gray sample does. You see, deformation allows a sample to dissipate energy. If a sample can't deform, the energy won't be dissipated, and will cause the sample to break [37].

In real life, we usually want materials to be tough and strong. Ideally, it would be nice to have a material that would not bend or break, but this is the real world. The gray-black sample has a much higher modulus than the red sample. While it is good for materials in a lot of applications to have high moduli and resist deformation, in the real world it is a lot better for a material to bend than to break, and if bending, stretching or deforming in some other way prevents the material from breaking, all the better. So when we design new polymers, or new composites, we often sacrifice a little bit of strength in order to make the material tougher.

4.5. Mechanical Properties of Real Polymers

The rigid plastics such as polystyrene, poly(methyl methacrylate or polycarbonate can withstand a good deal of stress, but they won't withstand much elongation before breaking. There is not much area under the stress-strain curve at all. So we say that materials like this are strong, but not very tough. Also, the slope of the plot is very steep, which means that it takes a lot of force to deform a rigid plastic. So it is easy to see that rigid plastics have high moduli. In short, rigid plastics tend to be strong, at resist deformation, but they tend not to be very tough, that is, they are brittle.

Flexible plastics like polyethylene and polypropylene are different from rigid plastics in that they don not resist deformation as well, but they tend not to break. The ability to deform is what keeps them from breaking. Initial modulus is high, that is it will resist deformation for awhile, but if enough stress is put on a flexible plastic, it will eventually deform. If you try to stretch it a plastic bag, it will be very hard at first, but once you have stretched it far enough it will give way and stretch easily. The bottom line is that flexible plastics may not be as strong as rigid ones, but they are a lot tougher.

It is possible to alter the stress-strain behavior of a plastic with additives called plasticizers. A plasticizer is a small molecule that makes plastics more flexible. For example, without plasticizers, poly(vinyl chloride), or PVC for short, is a rigid plastic. Rigid unplasticized PVC is used for water pipes. But with plasticizers, PVC can be made flexible enough to use to make things like hoses.

Fibers like Kevlar™, carbon fiber and nylon tend to have stress-strain curves like the aqua-colored plot in the graph above. Like the rigid plastics, they are more strong than tough, and do not deform very much under tensile stress. But when strength is what you need, fibers have plenty of it. They are much stronger than plastics, even the rigid ones, and some polymeric fibers, like Kevlar™, carbon fiber and ultra-high molecular weight polyethylene have better tensile strength than steel.

Elastomers like polyisoprene, polybutadiene and polyisobutylene have completely different mechanical behavior from the other types of materials. Take a look at the pink plot in the graph above. Elastomers have very low moduli. You can see this from the very gentle slope of the pink plot, but you probably knew this already. You already knew that it is easy to stretch or bend a piece of rubber [34]. If elastomers did not have low moduli, they would not be very good elastomers.

But it takes more than just low modulus to make a polymer an elastomer. Being easily stretched is not much use unless the material can bounce back to its original size and shape once the stress is released. Rubber bands would be useless if they just stretched and did not bounce back. Of course, elastomers do bounce back, and that is what makes them so amazing. They have not just high elongation, but high reversible elongation.

4.6. Tensile Properties

The discussion of which types of polymers have which mechanical properties has focused mostly on tensile properties. When we look at other properties, like compressional properties or flexural properties things can be completely different. For example, fibers have very high tensile strength and good flexural strength as well, but they usually have terrible compressional strength. They also only have good tensile strength in the direction of the fibers.

Some polymers are tough, while others are strong, and how one often has to make trade-offs when designing new materials; the design may have to sacrifice strength for toughness, but sometimes we can combine two polymers with different properties to get a new material with some of the properties of both. There are three main ways of doing this, and they are copolymerization, blending, and making composite materials.

The copolymer that combines the properties of two materials is spandex. It is a copolymer containing blocks of elastomeric polyoxyethylene and blocks of a rigid fiber-forming polyurethane. The result is a fiber that stretches. Spandex is used to make stretchy clothing like bicycle pants.

High-impact polystyrene, or HIPS for short, is an immiscible blend that combines the properties of two polymers, styrene and polybutadiene. Polystyrene is a rigid plastic. When mixed with polybutadiene, an elastomer, it forms a phase-separated mixture which has the strength of polystyrene along with toughness supplied by the polybutadiene. For this reason, HIPS is far less brittle than regular polystyrene [38].

In the case of a composite material, we are usually using a fiber to reinforce a thermoset. Thermosets are crosslinked materials whose stress-strain behavior is often similar to plastics. The fiber increases the tensile strength of the composite, while the thermoset gives it compressional strength and toughness.

5. CONCLUSIONS

This brief review of FRP has summarized the very broad range of unusual functionalities that these products bring (Polymers, Aramids, Composites, Carbon FRP, and Glass-FRP). While the chemistry plays an important role in defining the scope of applications for which these materials are suited, it is equally important that the final parts are designed to maximize the value of the inherent properties of these materials. Clearly exemplify the constant trade-off between functionality and processability that is an ongoing challenge with these advanced materials. The functionality that allows these materials to perform under extreme conditions has to be balanced against processability that allows them to be economically shaped into useful forms. The ability of a polymer material to deform is determined by the mobility of its molecules, characterized by specific molecular motions and relaxation mechanisms, which are accelerated by temperature and stress. Since these relaxation mechanisms are material specific and depend on the molecular structure, they are used here to establish the desired link with the intrinsic deformation behavior.

ACKNOWLEDGEMENT

The author would like to offer a special thanks to Universidad Nacional de San Luis, to Instituto de Física Aplicada, and to Consejo Nacional de Investigaciones Científicas y Técnicas for being generously support used in this research works.

REFERENCES

1. http://www.matter.org.uk/matscicdrom/manual/co.html.
2. Hinton M.J., Soden P.D., Kaddour A.S. Failure Criteria in Fibre-Reinforced-Polymer Composites: The World-Wide Failure Exercise. Elsevier 2004.
3. Tong L., Mouritz A.P., Bannister M. 3D Fibre Reinforced Polymer Composites. Elsevier 2002.
4. Ravi Jain, Luke lee. Fiber Reinforced Polymer (FRP) Composites for Infrastructure Applications. Focusing on Innovation, Technology Implementation and Sustainability. Springer 2012.
5. Kimg Hwee TAN. Fibre Reinforced Polymer. Reinforcement for Concrete Structures. Proceedings of the Sixth International Symposium on FRP Concrete Structures, volume 1-2 (FRPRCS-6). World Scientific 2003.
6. Erki M.A., and Rizkalla S.H. FRP Reinforcement for Concrete Structures. Concrete International (1993) 48-53.

7. Han E.H. Meijer, Govaert Leon E. Mechanical performance of polymer systems: The relation between structure and properties. Prog. Polym. Sci. 30 (2005) 915-938.

8. McGraw-Hill Science & Technology Encyclopedia

9. Entsiklopediia polimerov, vols. 1–3. Moscow, 1972-77. The Great Soviet Encyclopedia, 3rd Edition (1970-1979). 2010 The Gale Group, Inc.

10. Giancaspro James, Papakonstantinou Christos, Balaguru P. Mechanical behavior of fire-resistant biocomposite. Composites: Part B 40 (2009) 206-211.

11. Aubourg, P.F., Crall C., Hadley J., Kaverman R.D., and Miller D.M. Glass Fibers, Ceramics and Glasses. Engineered Materials Handbook, Vol. 4. ASM International, 1991, pp. 1027-31.

12. McLellan, G.W. and Shand E.B. Glass Engineering Handbook. McGraw-Hill, 1984.

13. Pfaender, H.G. Schott Guide To Glass. Van Nostrand Reinhold Company, 1983.

14. Tooley, F.V. "Fiberglass, Ceramics and Glasses", in Engineered Materials 14- 14-14- Handbook, Vol. 4. ASM International, 1991, pp. 402-08.

15. Hnat, J.G. "Recycling of Insulation Fiberglass Waste". Glass Production Technology International, Sterling Publications Ltd., pp. 81-84.

16. Webb, R.O. "Major Forces Impacting the Fiberglass Insulation Industry in the 1990s". Ceramic Engineering and Science Proceedings, 1991, pp. 426-31.

17. How fiberglass is made - material, used, processing, components, dimensions, composition, product, industry, machine, Raw Materials, The Manufacturing Process of fiberglass, Quality Control http://www.madehow.com/Volume-2/Fiberglass.html

18. Carbon fibers Seen as Having Big Long Term Growth Infrastructure is Next Big Trend Driver, "Advanced Materials & Composites" News, No. 3, 1999.

19. Structures, Advanced Materials & Composites News, No. 2, 1999.

20. New Company Launches Carbon Fiber Fabrics for Decorative Applications, Advanced Materials & Composites News, No. 8, 1998

21. Carbon fibers Electrical Conductivity Found to Offer New Uses, Composites News, No. 3, 1998, 9.

22. Donnet Jean-Baptiste, Roop Chand Bansal, "Carbon Fibers", published by Marcel Dekker Inc., 1990, p370.

23. Gullapalli Sravani; Wong Michael S. (2011). Nanotechnology: A Guide to Nano-Objects. Chemical Engineering Progress: 28–32.

24. Zhao Z., and Gou J. (2009). Improved fire retardancy of thermoset composites modified with carbon nanofibers. Sci. Technol. Adv. Mater. 10: 015005. doi:10.1088/1468-6996/10/1/015005.

25. Japan Carbon Fiber Manufacturers Association (English) (http://www.carbonfiber.gr.jp/ english/)

26. Handbook of Fiber Chemistry. International Fiber Science and Technology Series. Editor Menachem Lewin. Third Edition 2007 CRC Press, Taylor & Francis Group.

27. Jassal M., Ghosh S. Aramid Fibres – An Overview. Indian Journal of Fibre & Textile Research 27 (2002) 290-306.

28. José M. García, Félix C. García, Felipe Serna, José L. de la Peña. High-performance aromatic polyamides. Progress in Polymer Science 35 (2010) 623-686.

29. Smallman, R.E., and Bishop R.J.. Modern Physical Metallurgy and Materials Engineering. 6th ed. Oxford. Butterworth-Heinemann, 1999.

30. Erhard, Gunter. Designing with Plastics. Trans. Martin Thompson. Munich: Hanser Publishers, 2006.

31. Rosato Donald V., Rosato Dominick V., Murphy John. Reinforced Plastics Handbook. Elsevier; 2004; page 586

32. Composite moulding (http:/ / web. archive. org/ web/ 20080215010801/ http:/ / www. quatrocomposites. com/ comp101proc. htm).

33. Klompen, Edwin T.J. Mechanical properties of solid polymers. Constitutive modelling of long and short term behaviour. Eindhoven: Technische Universiteit Eindhoven, 2005.

34. FRP Library (http://www.compositepedia.com)

35. Odian, George. Principles of Polymerization. 3rd ed., J. Wiley, New York, 1991.

36. Jang, B.Z.; Advanced Polymer Composites: Principles and Applications, ASM International, Materials Park, OH, 1994.

37. Bhowmick, Anil. Mechanical Properties of Polymers. Volume 1, Material Sciences Engineering. Enciclopedia of Life Support Systems – UNESCO, 2011.

38. Hashim M.H.M., Sam A.R.M., Hussin M.W. The Future of External Application of Fibre Reinforced Polymer in Civil Infrastructure for Tropical Climates Region. International Journal of Mechanical and Materials Engineering (IJMME), 6, 2 (2011) 147-159.

Chapter 2

A Fiber Pull-Out Based Model for Synthetic Fiber Reinforced Concrete Beams under a Flexural Load

Alessandro Paglia

CE Department, Clarkson University, Potsdam, USA

ABSTRACT

This work is intended to be a simple contribution to building a model able to Implement theoretical results related to the random oriented fiber reinforced concrete in a procedure that could be used in structures analysis and design involving fiber reinforced elements. Here follows a short outline: In the introduction chapter the problem is presented together the work done. Section 2 develops some ancillary concepts of this material and its mechanical properties, while in Section 3, following the path of other researchers, the assumptions made to solve the problem are presented, together with the most relevant results related to presence of 3D randomly oriented fiber. In the following section a review of the mechanical process of fiber pull-out is done, and the results, mostly due to Victor Li researches, of a 3D randomly oriented synthetic fiber stress vs crack opening in a pull-out process from a cement matrix. In Section 5 the author, after making some assumptions about the configuration of the strain and crack geometry in the cross section where failure is assume to occur under flexural bending moment, the resultant stress is integrated to find the resultant internal moment vs increasing strain and crack width. In this analysis, the crack bridging law for synthetic fiber in FRC presented in the previous section is taken into account. In Section 6, a procedure to find a cross section configuration in equilibrium under external bending moment has been built. Under the assumption of a perfectly plastic collapse mechanism a numerical simulation is conducted on a specimen that undergoes a four-point bending test. A comparison with the trend of a similar test on a synthetic FRC sample has been done. The work is completed by the conclusions that could be inferred from this work.

Keywords: Fiber Reinforced Concrete; Concrete Sample Test; Synthetic Fiber Composite Materials

1. INTRODUCTION

This work deals with composite materials formed by cementitious matrix and inclusion of synthetic fibers, referring in particular to fiber mechanical parameters typical of polyethylene and polypropylene fibers. The goal of this work consists in finding a model, simple but efficient enough to describe the contribution of the inclusion of straight fibers, randomly distributed, to the toughness and ductility of the composite, after briefly illustrating the mechanics of FRC and in particular the fiber pull-out phenomenon, which is assumed, as we see later, to be the most important factor in determining the increased ductility of the material.

An analysis of a post-first crack behavior beam is conducted assuming a rigid-plastic model for a four-point bending test load picture. The assumptions on the strain diagrams and the crack opening profile, even though not directly based on fracture mechanics theory, allow us to define relations between an external moment after the first crack and related displacements. An iterative procedure has been set up to evaluate numerically beam configurations under bending moment and a subroutine has been programmed to implement this procedure.

2. FIBER REINFORCED CONCRETE

Mechanical Behavior of FRC Concrete: Ancillary Concepts

A composite material can be defined as a material formed by two or more elements mixed on a macroscale [1]. This new material has particular properties that none of its components, used alone, can provide. We know that cementitious materials generally have poor tensile strength and low fractural toughness. Fibers are inclusions for which one dimension, the length, is dominant over the other two. They can be aligned in one direction or randomly oriented in a plane or in space, and they can be made of different materials (i.e. steel glass, asbestos, carbon, nylon, polypropylene, Kevlar and others).

One of the most important limits to the fiber percentage is given by the workability and the bundle phenomenon. The first is well known in the concrete technology, and it regards the difficulty of pumping the mixture and casting it correctly into the molds due to an incorrect mix design or a too low water/cement ratio. The second limit is due to the trend of a group of fibers to be tightly bounded together, so that the total surface of the bundle in touch with the matrix is lower than the sum of fibers taken individually. To avoid this, it is necessary to reduce the percentage of fiber and to take care of the mixing phase technique. Currently, one of the materials most commonly used in FRC is steel fibers, which can be straight or hooked [1]. The other most common solution is given by synthetic fibers such as Polyethylene, Polypropylene, Nylon and Kevlar. Compared to steel fibers, synthetic fibers generally cannot provide neither the same stiffness increment nor the same peak value when tensile deformation of the composite reaches the ultimate strain of the matrix. However, [2] they present the advantages of a greater ductility and higher resistance to aggressive agents. Most importantly, unlike steel fibers, there are kinds of synthetic fibers, in which the interfacial shear stress during the fiber debonding shows a hardening behavior with slippage, due to surface abrasion (Victor Li et al., [3]). The main contribution of fibers to composite cementitious material consists in converting a brittle material to pseudo-ductile material. That is, independently of the behavior of composite after its first crack (hardening of softening), the ultimate FRC strain is generally much greater with respect to the plain matrix; In general, although the failure mechanisms are more complicated, we can divide the failure mechanisms related to cementitious matrix composites into the following modes:

1) fiber pull-out (the composite failure occurs when the fibers are completely debonded).

2) fiber failure (fiber fracture).

According Kelly and Lilholt (1973) (cfr [1]) the composite tensile strength for the two specific modes can be predicted as follows:

Fiber pull-out:

$$\sigma_c = \left(1 - V_f\right)\sigma_m \Phi_i V_f s_f \tag{1}$$

Fiber failure:

$$\sigma_c = \left(1 - V_f\right) E_m / E_f \, \sigma_f \, \Phi_i V_f \sigma_f \tag{2}$$

where the coefficient Φ_i assumes values around 1/6. V_f is fibers volume/matrix volume ratio σ_m and σ_f are the failure stress for matrix and fibers respectively E_m, E_f are the Young moduli for matrix and fibers respectively.

In cement based composites the fiber failure strain is usually much greater than that of the matrix

$$\left(\varepsilon_{fu} \gg \varepsilon_{mu}\right).$$

That means fibers carry an additional load after matrix failure. The minimum amount of fiber, related to the unit volume that ensures fibers can entirely withstand this additional load after matrix failure occurs is defined as the Critical Fiber Volume Fraction.

This value has been derived by Hannant, D. J. (1978) [4] as

$$V_{cr} = \frac{\sigma_{mu}}{\sigma_{mu} + \left(\sigma_{fu} - \sigma'_f\right)} \quad (3)$$

where σ_{fu} and σ_{mu} are the ultimate stress of fiber and matrix respectively, and σ'_f is the fiber stress at matrix failure. The critical fiber volume has been estimated at around 0.31% for steel fibers and 0.75% for polypropylene fiber.

3. PRINCIPAL ASSUMPTIONS FOR THE MODEL IN THE PRE-CRACKING ANALYSIS

Before you begin to format your paper, first write and save the content as a separate text file. Keep your text and graphic files separate until after the text has been formatted and styled. Do not use hard tabs, and limit use of hard returns to only one return at the end of a paragraph. Do not add any kind of pagination anywhere in the paper. Do not number text heads—the template will do that for you.

Finally, complete content and organizational editing before formatting. Please take note of the following items when proofreading spelling and grammar:

3.1. Model Behavior in the Pre-Cracking Phase

The principal assumptions made related to this phase are the following:

1) the mod opening depends only on the tensile principal stress (the biggest if both principal stresses are tensile) and the crack direction is orthogonal to this stress.
2) there are no fiber bundles to consider, as the fiber volume Vs cement volume ratio is below 3%, so the fibers are expected to work independently of one another.
3) the fibers are randomly oriented with respect to the crack direction, with somewhat a uniform 3D distribution.
4) the strength of every fiber is high enough to avoid the fiber failure.
5) the behavior of both concrete and fibers remains linear elastic.
6) the axial stiffness is the only relevant stiffness in synthetic fibers.
7) in the pre-cracking phase there is no slippage between the fibers and the matrix, so the fiber strain is equal to the matrix strain till the matrix tensile strain reaches the ultimate strain, that is

$$\varepsilon_c = \varepsilon_{cu}$$

These assumptions are somewhat similar to the one on which the reinforced concrete theory is based. We assume the fibers contribute to the composite strength just as the steel rebar contributes to the reinforced concrete strength. The main difference is due to random orientation, and in this work we assume the fibers were cast randomly and are distributed uniformly in space (3D) inside the matrix.

3.2. Stress vs Deformation before the First Cracking, with Fibers Randomly Oriented in 3D

The approach to the 3D model necessary leads to a probabilistic approach [4,5]. We consider a rectangular specimen as shown in **Figure 1**:

The position of a fiber and the incident angle formed with the normal to a generic cross section can be determined by the following parameters (see **Figure 2**) x' : the distance between the fiber centroid from this cross section, and ψ the angle in the space formed by the fiber axis with the x direction, corresponding to the stress direction. We need to obtain the probability function for a fiber to have its position determined by x' and ψ with respect to any cross section normal to x direction (see **Figure 2**).

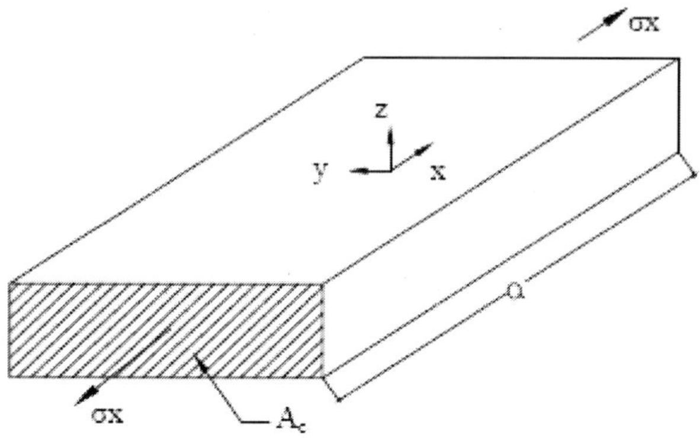

Figure 1. Specimen geometry.

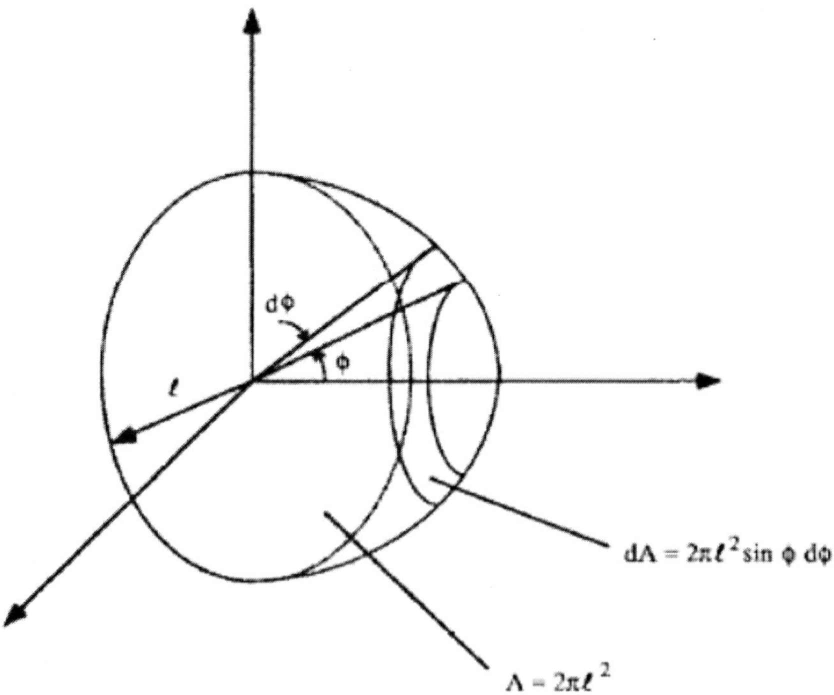

Figure 2. 3D randomly oriented fiber with embedded length l and its end over the hemispherical surface, [4].

If we define E1 as the probability that the centroid is inside a range $(x', x' + dx')$ and E2 as the probability that the fiber has an incident angle between ψ and $\psi + d\psi$, the probability density is given by P[E1]*P[E2], according Bayes's theory for independent events. (or above version).

However, the only values of x' are those that intercept the cross section plane, i.e. $x' \in (-L/2\cos\psi, L/2\cos\psi)$.

We assume the fiber centroid is uniformly distributed so that P[E1] =1/a, but if we refer only to a semi-domain $(0, L/2\cos\psi)$ as in **Figure 3**, we will obtain the same result multiplying the probability by 2, i.e. P[E1'] =2/a. The probability function P[E2] for uniformly spherical is equal to:

$$P[E2] = f_\psi \sin(\psi) = \sin(\psi) \tag{4}$$

that means the number of fibers dN crossing the cross section plane with $x' \in (x', x' + dx')$ and

$\psi \in (\psi, \psi + d\psi)$ is given by

$$dN = \frac{2}{a} N_{tot,f} \sin\psi d\psi dx' \tag{5}$$

and dN = 0 elsewhere, while where $N_{TOT.F}$ is the total number of fibers in the specimen, equal to:

$$N_{tot.f} = \frac{A_c a V_f}{A_f L_f} \tag{6}$$

The contribution of any fibers to the stress in x direction is equal to

$$T_f = E_f A_f \varepsilon_f \cos\psi \tag{7}$$

where ε_f, the fiber strain is equal to

$$\varepsilon_f = \varepsilon_x \cos^2\psi \tag{8}$$

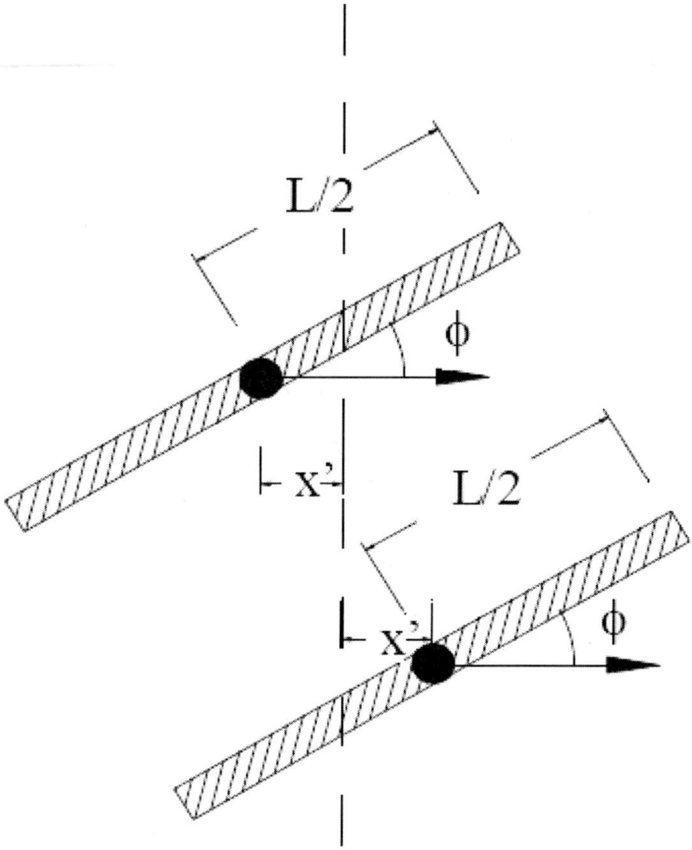

Figure 3. A fiber crossing a matrix crack and its centroid position.

so the axial resultant due to the fiber elastic deformation

$$T_f = N_{tot.f} \frac{E_f A_f \varepsilon_x L_f}{5a} \tag{9}$$

Leads to:

$$T_f = 0.2 E_f A_C \varepsilon_x V_f \tag{10}$$

The total tensile force due a strain $0 \le \varepsilon_x \le \varepsilon_{cu}$ is equal to the total tensile force due a strain is equal to

$$T = T_c + T_f = E_{ct}\varepsilon_x A_c\left[\left(1-0.2V_f\right)+0.2V_f\frac{E_f}{E_{ct}}\right] \tag{11}$$

It is possible to define an equivalent Young modulus equal to:

$$E_{eq.} = E_{ct}\left[\left(1-0.2V_f\right)+0.2V_f\frac{E_f}{E_{ct}}\right]$$

In many cases, synthetic fibers generally have a Young's modulus not much higher than mortar and concrete's modulus in the elastic phase under traction, and the percentage used in FRC composites is below 3%, thus we can assume reasonably for our goals:

$$E_{eq} = E_{ct} \tag{12}$$

4. PULL-OUT THEORY AND MODELING FOR SYNTHETIC FRC

4.1. Single Fiber Pull-Out Theory

The analysis of the single fiber embedded in a matrix is not simple as it depends in general on many parameters; thus, it is necessary to make some assumptions to obtain an analytical solution that can be used with a reasonable number of parameters and produces results generally in agreement with experimental tests. This section has been developed in according with the articles written by other researcher, in particular [5-7] We confirm the assumption from 1) to 6) but they are not enough to work in this scenario, and 7) no longer remains valid for every point of the fiber anymore. Four further important assumptions are necessary:

7') The fiber that remains in an elastic field (this can't be always true) is frictionally bonded to the matrix; that is, interfacial stress depends only on the elastic stress at the tip of the debonded zone and on the fiber-matrix relative slippage.

8) The effect of elastic interfacial stress is negligible in the bonded area as it is assumed to decay quickly.

9) After the debonding has occurred for the whole embedded length, the further fiber axial deformation is not considered, so the fiber moves a rigid motion.

10) The effect of the angle ψ formed between the fiber axis and the normal to the crack plane can be taken into account considering the fiber as a frictional rope slipping on a pulley. As a result,

$$P(l,\psi) = P(l)_{\psi=0}\, e^{f\psi} \tag{4}$$

for a single fiber.

11) Before the crack opens, the matrix strain is assumed to be the same as the fiber strain, and no slippage occurs. After a crack opens, the fiber undergoes a jump in tensile stress as the capacity of the cracked matrix to carry loads vanishes or rapidly decreases with the crack mod. The pull-out of the single fiber and the analysis of fiber matrix interaction after the first crack appears, can be divided in two phases, as shown in **Figure 4** consistenly with the assumptions made in 7'), 8) and 9);

a) Fiber debonding: the breakage of the initial cohesion (mechanical or chemical) existing between the fiber and the matrix. It begins where the elastic interfacial stress is supposed to be the highest, in a homogeneous matrix at the crack tip, and it ends with the complete breakage of the cohesion all along the fiber length, i.e. the debonded length is equal to the embedded length.

b) Fiber pull-out: which occurs after the complete debonding.

The fiber experiences a complete slippage with respect to the matrix. Generally, there still exists an interfacial shear stress that depends on the physical characteristics of the fiber surface and varies with the fiber slippage. This phase ends when the fiber is completely pulled out from the matrix.

An important condition to ensure consistency with assumption 4) must be established: the embedded fiber length l must be shorter than the critical length, L_c, defined as the embedded length at which fiber failure can occur.

$$L_C = \frac{d_f \sigma_u}{4\tau^*} \tag{14}$$

4.2. Interfacial Shear Stress vs Slippage Law

According to [5] the hardening behavior is relevant for synthetic fibers: in this work, the linear hardening model proposed by Bao and Song [8] has been used, thus the

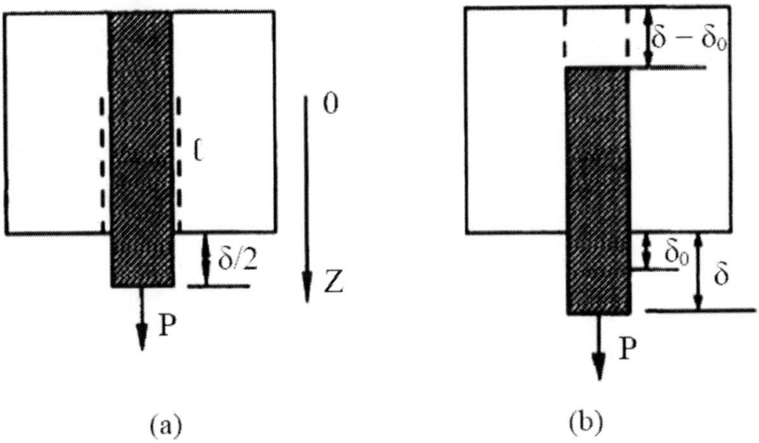

Figure 4. Fiber debonding and fiber pull-out [3]. (a) Fiber deboning; (b) Fiber pull-out.

relation between the shear interface stress and the fiber slippage is the following:

$$\tau = \tau_0 \left(1 + \frac{\beta S}{d_f}\right) \tag{15}$$

where S is slippage, d_f is fiber diameter, Young's modulus not much higher than mortar and concrete's modulus in the elastic phase The values of τ_0 and β must be determined from experimental tests.

Phase a) $l \leq L$ (debonding) where l is the debonded length and L is the embedded length. We obtain, after setting $\eta = \frac{V_f E_f}{V_m E_m}$ $\omega^2 = \frac{4\beta\tau_0(1+\eta)}{V_m E_m}$ the governing equation in the form [5]:

$$\frac{\partial^2 S(z)}{\partial z^2} - \frac{\omega^2}{d_f} S(z) = \frac{\omega^2}{\beta d_f} \tag{16}$$

and, applying the boundary conditions

1) $S(0) = 0$

2) $\varepsilon_f = \varepsilon_m \Rightarrow \frac{\partial S(z)}{\partial z} = 0$ for z = 0 it is possible to obtain the solution [2].

$$S(z) = \frac{d_f}{\beta}\left[\cosh\left(\frac{\omega z}{d_f}\right) - 1\right] \tag{17}$$

thus the load at the free end of the fiber is equal to

$$P(l) = \tau_0 \sinh\left(\frac{\omega l}{d_f}\right)\pi\frac{d_f^2}{\omega}(1+\eta) \tag{18}$$

and at the end of Phase 1) l = L

$$P(L) = \tau_0 \sinh\left(\frac{\omega L}{d_f}\right)\pi\frac{d_f^2}{\omega}(1+\eta) \tag{19}$$

As this phase is bilateral and symmetric, we can set the crack opening $\delta(z) = 2u(z)$.

$$\delta_0 = \frac{2d_f}{\beta}\left[\cosh\left(\frac{\omega L}{d_f}\right) - 1\right] \quad \text{for } l = L \tag{20}$$

and the load is equal to:

$$P(L) = \tau_0 \pi \frac{d_f^2}{\omega}(1+\eta)\sqrt{\left(1+\frac{\beta\delta}{2d_f}\right)^2 - 1} \quad \text{for } \delta < \delta_0 \; . \tag{21}$$

Phase 2) (the fiber has been completely debonded).

In this phase, ignoring the further axial deformation according to 9) the slippage is equal to:

$$S(z) = \frac{d_f}{\beta}\left[\cosh\left(\frac{\omega z}{d_f}\right) - 1\right] + (\delta - \delta_0) \tag{22}$$

and the load to:

$$P(L)=\tau_0\left[\sinh\left(\frac{\omega L}{d_f}\right)-\sinh\left(\omega\frac{\delta-\delta_0}{d_f}\right)\right]\pi\frac{d_f^2}{\omega}(1+\eta)$$
$$+\pi\tau_0\beta(\delta-\delta_0)(1+\eta)(L-(\delta-\delta_0))$$
.(23)

This value of P (L, δ) is valid only for one side pullout behavior (see **Figure 4**). However, it has been demonstrated [3] that fibers that can undergo a double side for most actual situations must have the shorter embedded length close to L/2 (see **Figure 5** for details) so their relative percentage is low and they can be disregarded. If we had had a constant interfacial shear stress (i.e. $\tau = \tau_0$) we would have obtained the following results for $\delta_0 < \delta < L$ (linear decreasing).

$$P = \pi\tau_0 L d_f\left(1-\frac{\delta-\delta_0}{L}\right)$$
(24)

4.3. 3D Uniformly Distributed Fibers Pull-Out Theory

The theory developed so far must be extended to a great number of fibers whose position is random to achieve useful results in terms of stress and crack opening. We maintain the assumptions stated before. Thus it is necessary to evaluate the probability that, depending on a given crack opening a fiber can be 1) still embedded in both sides but in a debonding phase;

2) completely debonded and in the phase of slipping in the shorter side and in the debonding phase in the longer side but still intersecting both crack faces.

3) completely pull-out from the shorter embedded side and then removed from the model as providing no contribution to the tensile strength. Thus the bridging stress is according to 9) the slippage is equal to:

$$\sigma_B = \frac{4V_f}{\pi d_f^2}\int_0^{\frac{\pi}{2}}\int_{z=0}^{\frac{L_f}{2}\cos\varphi} P(\delta)\,p(\varphi)\,\mathrm{d}z\mathrm{d}\varphi$$
(25)

that gives the following results [6] expressed in non-dimensional terms for $0 \le \hat{\delta} \le \hat{\delta}^*$ and the load to: (please see Equation (26) below)

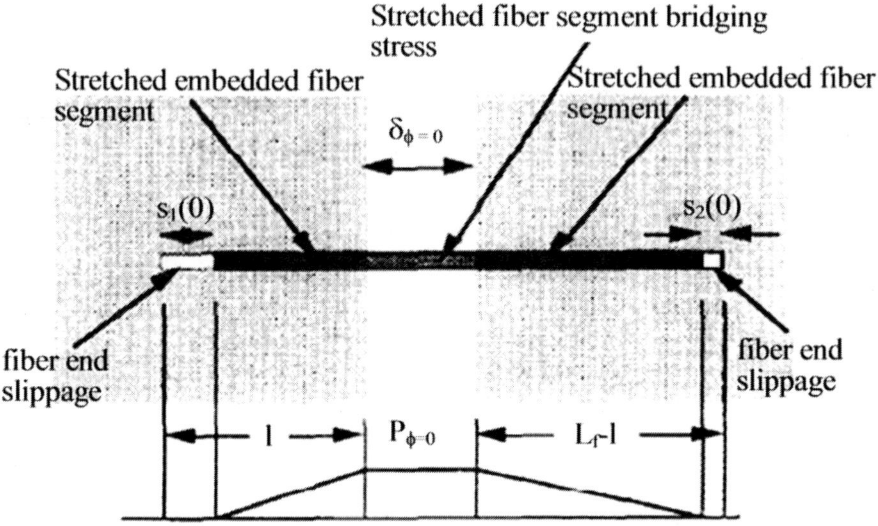

Figure 5. Bilateral pull-out [2].

and for $\hat{\delta}^* \leq \hat{\delta} \leq 1$

$$\tilde{\sigma}_B = \frac{\sigma_B}{\sigma_0} = \left(1 + \frac{\beta L_f}{2d_f}\hat{\delta}\right)\left(1-\hat{\delta}\right)^2 \tag{27}$$

While, for $\hat{\delta} \geq 1$

$$\tilde{\sigma}_B = \frac{\sigma_B}{\sigma_0} = 0 \tag{28}$$

$k = \frac{\omega L_f}{2d_f}$, $\lambda = \cosh(k) - 1$, $\hat{\delta} = \frac{\delta}{\left(\frac{L_f}{2}\right)}$,

$\hat{\delta}^* = \frac{4d_f}{\beta}\frac{\left[\cosh(k)-1\right]}{L_f}$ is the normalized crack opening at which all fibers have completed debonding, and

$\sigma_0 = \frac{1}{2}g\tau_0 V_f(1+\eta)\frac{L_f}{d_f}$ is the normalizing stress and g is the snubbing factor related to snubbing coefficient f with the relation for uniform fiber distribution in 3D. [1] the influence of slip-hardening parameter β in the debonding stage $\left(0 \le \hat{\delta} \le \hat{\delta}^*\right)$ is practically negligible, because k is generally around 0.1 for most synthetic fibers/ cement matrices The maximum stress in the post debonding phase can be found using the condition:

$$\frac{\partial \tilde{\sigma}_B}{\partial \hat{\delta}} = 0$$

which occurs that occurs when

$$\hat{\delta} = \frac{c-2}{3c} = \hat{\delta}_1$$

with the condition $\hat{\delta}_1 \ge \hat{\delta}^*$ and we obtain

$$\frac{\beta L_f}{2d_f} = \frac{2}{1-3\hat{\delta}^*} \cong 2 \tag{29}$$

as $\hat{\delta}^* \ll 1$. That means a rising value of the normalized bridging stress is possible in the pull-out phase only if [6].

$\frac{\beta L_f}{2d_f} \ge 2$ with an interfacial shear stress constant (i.e. $\tau = \tau_0$), the result would have been:

$$\tilde{\sigma}_B = \frac{\sigma_B}{\sigma_1} = g\left[2\left(\frac{\hat{\delta}}{\hat{\delta}^*}\right)^{\frac{1}{2}} - \frac{\hat{\delta}}{\hat{\delta}^*}\right] \quad \text{for } 0 \le \hat{\delta} \le \hat{\delta}^* \tag{30}$$

where g is the snubbing coefficient and

$$\sigma_1 = \frac{1}{2} V_f \tau \left(\frac{L_f}{d_f}\right) \tag{31}$$

This relation will be used in our model, as the influence of the hardening parameter β is generally negligible in the debonding phase.

4.4. Composite Ultimate Tensile Stress

For what we pointed out above, there is a value of the parameter beyond which the bridging stress vs crack opening shows a rising branch whose value is around 2. [5]. Thus, if the parameter $(bL_f)/2d_f$ is smaller than 2, the maximum bridging stress occurs at $\hat{\delta} = \hat{\delta}^*$ that is equivalent to state the following:

$$\sigma_{cu} = \frac{2\lambda}{k^2} \tag{32}$$

for the model with linear hardening interfacial shear stress, or

$$\sigma_{cu} = \sigma_1 \tag{33}$$

for the model with constant interfacial shear stress, while for $(bL_f)/2d_f > 2$ equivalent to state the following:

$$\sigma_{cu} = \sigma_1\left(1 + \frac{\beta L_f}{2d_f}\hat{\delta}_1\right)\left(1 - \hat{\delta}_1\right)^2 \tag{34}$$

for the model with linear hardening.

5. ANALYSIS OF A FRC CROSS SECTION UNDER A BENDING MOMENT AFTER CRACK OPENING

In the previous section we faced the problem of randomly oriented fibers debonding and pull-out from a cementitious matrix. We also found the stresses transmitted by the fiber after matrix cracking, relating to the various phases of crack opening. However, we have not yet dealt with the process that leads to define the crack width and its opening as forces acting on the cross section increase. For this purpose, a post crack analysis will be conducted on a specimen that undergoes a four-point bending test comparing the theoretical results relative to load vs deflection and moment vs curvature will be compared to the test data. In this analysis, the crack bridging law for synthetic fiber in FRC presented in the last chapter will be considered.

$$\tilde{\sigma}_B = \frac{\sigma_B}{\sigma_0} = \frac{2}{k}\left\{\left[1 - \frac{1}{k}\cosh^{-1}\left(1 + \lambda\frac{\hat{\delta}}{\hat{\delta}^*}\right)\right] \times \sqrt{\left(1 + \lambda\frac{\hat{\delta}}{\hat{\delta}^*}\right)^2 - 1} + \frac{\lambda}{k}\frac{\hat{\delta}}{\hat{\delta}^*}\right\} \quad (26)$$

5.1. Models for Post Cracking Behavior and Preliminary Assumptions for the Cracked Beam Analysis

We consider a beam under an increasing bending moment M; until the moment reaches a critical value MCR, the displacements increase nearly linearly with the load and no macro crack has yet occurred. Suddenly there is a jump; that is, without significant load increases, a crack opens and displacements show a considerable increase. In this chapter we consider a cracked realistic configuretion for strains and crack opening, see **Figure 6** trying only to study the equilibrium between the stress on the concrete and the stress given by the bridging effect of the fiber. We are going to analyze the model of a beam under a condition of four-point bending. Our goal is to include the fiber bridging stress in the cracked cross section equilibrium, and to estimate this contribution to beam toughness and ductility.

For this purpose the following assumptions are made:

1) The behavior of the beam before cracking is linear elastic; after its cracking for its particular load pattern it is rigid-plastic [9], that is, all the further deformations beyond the elastic limit are only related to the deformation of the cracked cross section.

2) The position crack tip is located in the tensile stress area where the matrix strain reaches the value of $\varepsilon_{CR} = \frac{f_{CR}}{E_c}$: the strain and the displacements related to the crack opening are both linear along the depth of the cross section. In particular, the crack opening is assumed linear from the crack tip to the value of C_{mod} at the lowest fiber of the cross section.

3) The stress vs strain law of concrete in compression is theparabola rectangle EC2 [9] shown in**Figure 7** then

$$f_c = f_c'\left[2\frac{\varepsilon_c}{\varepsilon_0} - \left(\frac{\varepsilon_c}{\varepsilon_0}\right)^2\right] \quad \text{for } 0 \le \varepsilon_C \le \varepsilon_0 \quad (35)$$

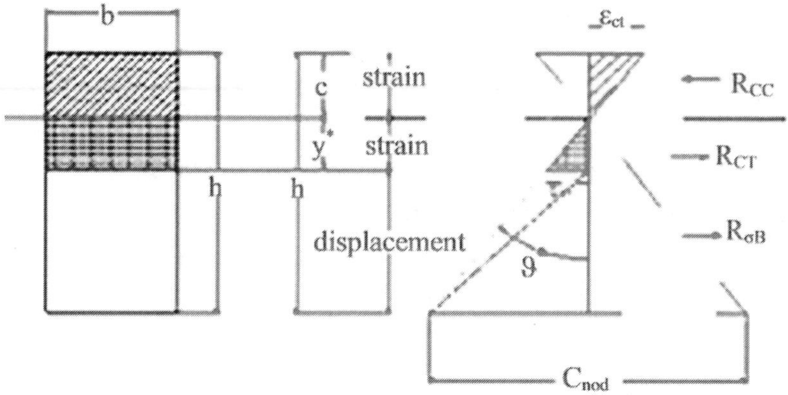

Figure 6. Assumed strains and displacements diagrams in the cracked cross section.

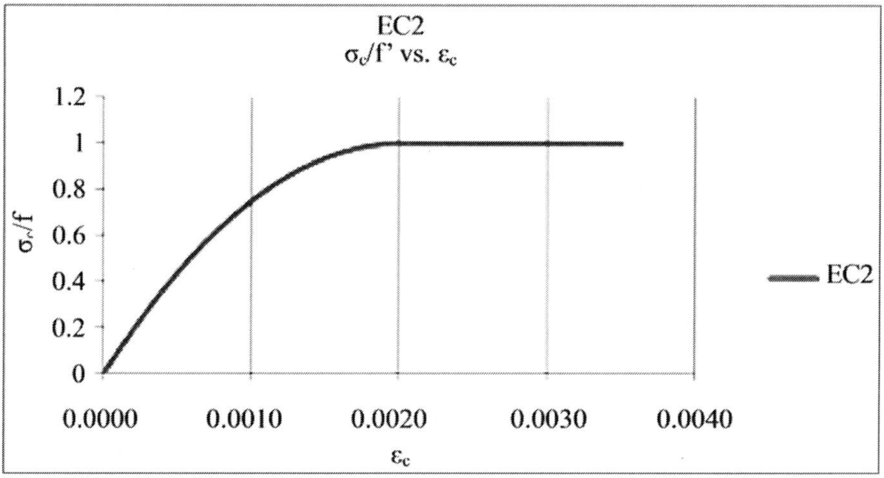

Figure 7. Compressive stress vs strain according EC2 [9].

$$f_c = f_c' \quad \text{for } \varepsilon_0 \leq \varepsilon_C \leq \varepsilon_{CU} \tag{35 bis}$$

where we assumed $\varepsilon_0 = \frac{2f_C'}{E_t} \cong 0.002$ and $\varepsilon_{CU} = 0.0035$

For concrete tensile stress vs strain, we can assume [8] the relations law is linear

$$\sigma_{CT} = E\varepsilon_{CT} \quad \text{for } 0 \le \varepsilon_{CT} \le \varepsilon_{CR} \tag{36}$$

$$\sigma_{CT} = \frac{f_{CR}}{1+\sqrt{200\varepsilon}} \quad \text{for } \varepsilon_{CT} \ge \varepsilon_{CR} \tag{36 bis}$$

where E, the Young modulus (tangent to the curve at the 0,0 point) is equal to $E = 5.500\sqrt{f_C'}\ (f_C' \text{ in MPa})$ and $f_{CR} = E\varepsilon_{CR}$.

In this chapter, however, we consider the concrete cracked zone unable to transmit tensile stress, so the post-peak branch is ignored.

3) The process of crack front proceeding can be considered quasi-static, i.e. no dynamic effects are considered.

5.2. Compatibility and Equilibrium Equations for the Cracked Cross Section

If we define the strain at the upper surface of the beam over the neutral axis as ε_{cx} and its maximum value as ε_{cf}, the displacement of the extreme fiber for this load picture is given by

$$\Delta_0 = \int_0^L \varepsilon_{cx} \mathrm{d}x = \frac{2}{3}\varepsilon_{cf} L \tag{37}$$

As Δ_0 is equal to the "virtual" overlapping upper cross section fiber in the rigid body motion, [10] we can define the slope semi-angle formed by the crack opening equal to:

$$\Delta\vartheta = \frac{\Delta_0}{2c} \tag{38}$$

where c is the depth of the neutral axis (see **Figure 8**) and the settlement of the mid-span point is equal to

$$d\delta = d\vartheta\frac{l}{2} \tag{39}$$

plus the elastic deflection, which, according to the EulerBernoulli's beam theory, is given by

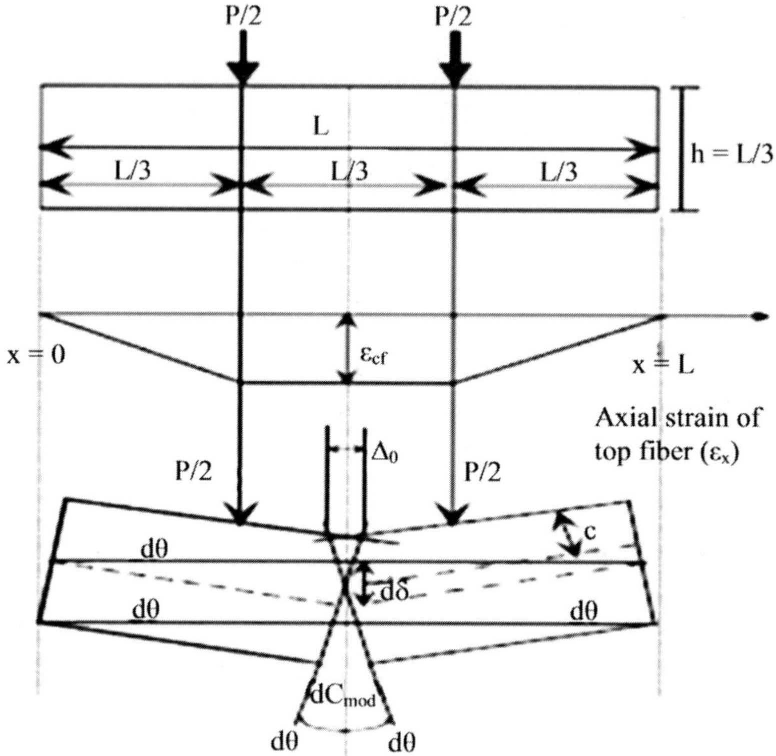

Figure 8. A kinematic mechanism of a fai1ure [10].

$$\delta_e = \frac{23PL^3}{1296EI} = \frac{23ML^2}{216EI} \tag{40}$$

The crack opening is given by (see **Figure 8**)

$$dC\,\text{mod} = 2d\vartheta(h-c) \tag{41}$$

The translation equilibrium equation is the following:

$$\int_0^c \sigma_C(y)b\text{d}y + \int_{y^*}^0 \sigma_{CT}(y)b\text{d}y + \int_{h-c}^{y^*} \sigma_B(y)b\text{d}y = 0 \tag{42}$$

while the rotation equilibrium equation is the following

$$\int_{0}^{c}\sigma_{C}(y)\,by\,dy+\int_{y^*}^{0}\sigma_{CT}(y)\,by\,dy+\int_{h-c}^{y^*}\sigma_{B}(y)\,by\,dy=M$$

(43)

As $\varepsilon_{cx}=\dfrac{\varepsilon_{cf}}{c}y$ in compression and $\Delta_X=-2d\vartheta y$ in the cracked area we can rewrite the for a rectangular section so

$$R_{CC}=f_C'bc\left(\frac{\varepsilon_{cf}}{\varepsilon_0}-\frac{\varepsilon_{cf}^2}{3\varepsilon_0^2}\right)\quad \text{for } \varepsilon_{cf}\le\varepsilon_0$$

(44)

$$R_{CC}=f_C'bc\left(\frac{2\varepsilon_0}{3\varepsilon_{cf}}+\frac{\left(\varepsilon_{cf}-\varepsilon_0\right)}{\varepsilon_{cf}}\right)\quad \text{for } \varepsilon_0\le\varepsilon_{cf}\le\varepsilon_{cu}$$

(44 bis)

while the resultant moment of the compressed concrete with respect to the neutral axis, is equal to

$$MR_{CC}=f_C'bc^2\left(\frac{2}{3}\frac{\varepsilon_{cf}}{\varepsilon_0}-\frac{1}{4}\frac{\varepsilon_{cf}^2}{\varepsilon_0^2}\right)\quad \text{for } \varepsilon_{cf}\le\varepsilon_0$$

, (45a)

$$MR_{CC}=f_C'bc^2\left(\frac{1}{2}-\frac{1}{12}\frac{\varepsilon_0^2}{\varepsilon_{cf}^2}\right)\quad \text{for } \varepsilon_0\le\varepsilon_{cf}\le\varepsilon_{cu}$$

(45b)

As we assume in c), the concrete is able to transmit tensile stress to where the cross section is not cracked. We can evaluate the tensile concrete stress resultant then

$$R_{CT}=\frac{E_T cb\hat{\varepsilon}_{CT}^2}{2\varepsilon_{CF}}\quad \text{for } 0\le\hat{\varepsilon}_{CT}\le\varepsilon_{CR}$$

(46a)

Where $\hat{\varepsilon}_{CT}=\dfrac{\varepsilon_{CF}(h-c)}{c}$ is the concrete strain at lowest fiber, and

$$R_{CT} = \frac{E_T c b \varepsilon_{CR}^2}{2\varepsilon_{CF}} \quad \text{for } \hat{\varepsilon}_{CT} \geq \varepsilon_{CR}$$, (46b)

The moment of R_{CT} is equal to

$$MR_{CT} = \frac{E_t c^2 b \hat{\varepsilon}_{CR}^3}{3\varepsilon_{CF}^2} \quad \text{for } 0 \leq \hat{\varepsilon}_{CT} \leq \varepsilon_{CR}$$ (47a)

$$MR_{CT} = \frac{E_t c^2 b \varepsilon_{CR}^3}{3\varepsilon_{CF}^2} \quad \text{for } \hat{\varepsilon}_{CT} \geq \varepsilon_{CR}$$ (47b)

5.3. Resultant of Fiber Bridging Stress

The resultant of fiber bridging stress and its position depends, as we indicate in (3, 27) on crack opening profile. As we can see from **Figure 9**, three situation can be verified:

1) C_{mod} corresponds to a normalized crack opening $0 \leq \hat{C}_{mod} \leq \hat{\delta}^*$.

2) C_{mod} corresponds to a normalized crack opening $\hat{\delta}^* \leq \hat{C}_{mod} \leq 1$.

3) C_{mod} corresponds to a normalized crack opening $\hat{C}_{mod} > 1$.

where $\hat{C}_{mod} = \frac{2C_{mod}}{L_f}$ the normalized maximum crack opening.

In 1) the stress resultant is given by:

$$R_{\sigma b}^a = \sigma_1 g b \left(h - c - y^*\right)\left(\frac{4}{3}\sqrt{\frac{\hat{C}_{mod}}{\hat{\delta}^*}} - \frac{1}{2}\frac{\hat{C}_{mod}}{\hat{\delta}^*}\right)$$ (48)

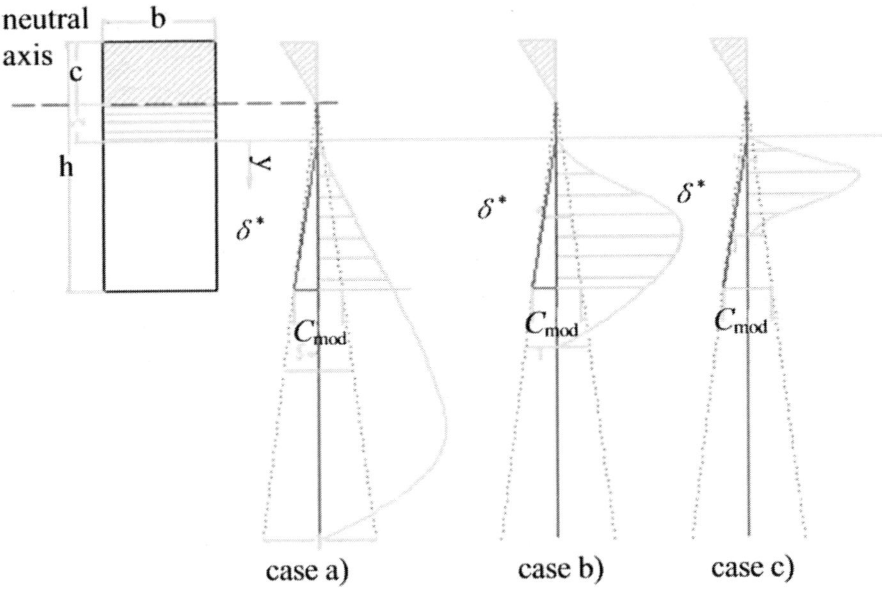

Figure 9. Possible bridging stress diagrams in function of the normalized C_{mod}.

Then In 2) the stress resultant is made of two terms as the bridging law changes for $\hat{\delta}^* \leq \hat{\delta} \leq 1$.

Then

$$R^b_{\sigma B} = R^b_1 + R^b_2 \tag{49}$$

where

$$R^b_1 = \sigma_1 g b \left(h - c - y^*\right) \frac{5}{6} \frac{\hat{\delta}^*}{\hat{C}_{mod}} \tag{50}$$

$$R^b_2 = \sigma_0 b \frac{\left(h-c-y^*\right)}{\hat{C}_{mod}} \left[\hat{C}_{mod} - \hat{\delta}^* + \frac{1}{2}(\gamma - 2)\left(\hat{C}^2_{mod} - \hat{\delta}^{*2}\right) + \frac{1}{3}(1-2\gamma)\left(\hat{C}^3_{mod} - \hat{\delta}^{*3}\right) + \frac{1}{4}\gamma\left(\hat{C}^4_{mod} - \hat{\delta}^{*4}\right)\right] \tag{51}$$

where $\gamma = \frac{\beta L_f}{2d_f}$

In 3) $R_{\sigma B}^c = R_1^c + R_2^c$ where $R_1^c = R_1^b$ remains the same, while

$$R_2^c = \sigma_0 b \frac{(h-c-y^*)}{\hat{C}_{mod}} \left[1-\hat{\delta}^* + \frac{1}{2}(\gamma-2)\left(1-\hat{\delta}^{*2}\right) + \frac{1}{3}(1-2\gamma)\left(1-\hat{\delta}^{*3}\right) + \frac{1}{4}\gamma\left(1-\hat{\delta}^{*4}\right) \right] \quad (52)$$

It is not easy to find the value of C_{mod} for which the bridging stress reaches its maximum value. However it is clear that for $C_{mod} > \frac{L_f}{2}$, i.e. $\hat{C}_{mod} > 1$ for a given crack length the contribution of $R_{\sigma B}$begins to decrease, so does its moment as the arm decreases too.

This means that unless there are concentrated reinforcements (e.g. steel bar), it is likely the crack will begin to propagate rather than the C_{mod} increasing beyond this value.

The resultant moment with respect to the neutral is given, referring to the same situation 1), 2), 3), by the following relations:

1) $0 \le \hat{C}_{mod} \le \hat{\delta}^*$

$$M\sigma_{B1}^a = \sigma_1 g b \left[(h-c-y^*)^2 \left(\frac{4}{5}\sqrt{\frac{\hat{C}_{mod}}{\hat{\delta}^*}} - \frac{1}{3}\frac{\hat{C}_{mod}}{\hat{\delta}^*} \right) + y^*(h-c-y^*)\left(\frac{4}{3}\sqrt{\frac{\hat{C}_{mod}}{\hat{\delta}^*}} - \frac{1}{2}\frac{\hat{C}_{mod}}{\hat{\delta}^*} \right) \right] \quad (53)$$

2) $\hat{\delta}^* \le \hat{C}_{mod} \le 1$ there are two components $M_B^b = M_{B1}^b + M_{B2}^b$ (54a)

where

$$M_{B1}^{b} = \sigma_1 g b \frac{7}{15} \frac{\left(h-c-y^*\right)^2 \hat{\delta}^{*2}}{\hat{C}_{mod}^2} + R_1 y^* \tag{54b}$$

$$M_{B2}^{b} = \sigma_0 b \frac{\left(h-c-y^*\right)^2}{\hat{C}_{mod}^2}\left[\frac{1}{2}\left(\hat{C}_{mod}^2 - \hat{\delta}^{*2}\right)\right.$$
$$+\frac{1}{3}(\gamma-2)\left(\hat{C}_{mod}^3 - \hat{\delta}^{*3}\right) + \frac{1}{4}(1-2\gamma)\left(\hat{C}_{mod}^4 - \hat{\delta}^{*4}\right)$$
$$\left.+\frac{1}{5}\gamma\left(\hat{C}_{mod}^5 - \hat{\delta}^{*5}\right)\right] + R_2 y^* \tag{54c}$$

3) $\hat{C}_{mod} > 1$

$$M_B^c = M_{B1}^c + M_{B2}^c \tag{55a}$$

where $M_{B1}^c = M_{B1}^b$ and:

$$M_{B2}^{c} = \sigma_0 b \frac{\left(h-c-y^*\right)^2}{\hat{C}_{mod}^2}\left[\frac{1}{2}\left(1-\hat{\delta}^{*2}\right) + \frac{1}{3}(\gamma-2)\left(1-\hat{\delta}^{*3}\right)\right.$$
$$\left.+\frac{1}{4}(1-2\gamma)\left(1-\hat{\delta}^{*4}\right) + \frac{1}{5}\gamma\left(1-\hat{\delta}^{*5}\right)\right] + R_2 y^* \tag{55b}$$

6. PROCEDURE TO FIND EXTERNAL MOMENT VS CURVATURE RELATION IN THE CRACKED CROSS SECTION

In the last paragraph, we defined for a given cross section configuration the stresses resultant and their moment resultant with respect to the neutral axis once we know the upper fiber strain e_{cf}, the neutral axis position c and the crack opening C_{mod} (cfr. **Figure 8**).

The parameter C_{mod} is not independent from the others. In fact, as we showed before, using the rigid plastic model we can define Δ_0 and $\Delta\vartheta$ [10] and then

$$C_{\mathrm{mod}} = \Delta_0 \frac{(h-c)}{c} = \frac{2}{3}\varepsilon_{cf} L \frac{(h-c)}{c} \tag{56}$$

So the equation system is defined by two equation in two unknown

$$R_{CC} + R_{CT} + R_{\sigma b} = N = 0 \tag{57}$$

$$M_{RCC} + M_{RCT} + M_{R\sigma b} = P \times \frac{L}{6} \tag{58}$$

and the unknowns are e_{cf} and c. Once we know this parameter, we are able to obtain C_{mod} and the rigid plastic deflection.

6.1. Equation in the Non-Dimensional Form

The two equilibrium equations can be written using nondimensional terms, avoiding reference to the cross section dimensions. If we divide the first by $f_c'bh$ and the second by $f_c'bh^2$ setting $\xi = \frac{c}{h}$, $\bar{f}_{CR} = \frac{f_C'}{f_{CR}'}$ we obtain

$$\bar{\upsilon} = \frac{N}{f_C'bh} = \frac{R_{CC} + R_{CT} + R_{\sigma B}}{f_C'bh} = \upsilon_{CC} + \upsilon_{CT} + \upsilon_{\sigma B} \tag{60}$$

where

$$\upsilon_{CC} = \frac{R_{CC}}{f_C'bh} = \xi A(\varepsilon_{CF}) \quad , \quad \upsilon_{CT} = \frac{R_{CT}}{f_C'bh} = \xi L(\varepsilon_{CT}, \varepsilon_{CF})$$

$\upsilon_B = \upsilon_{B1} + \upsilon_{B2}$ with

$$\upsilon_{B2} = \frac{\sigma_0}{f_C'}\left(1 - \xi - \xi\frac{\varepsilon_{CR}}{\varepsilon_{CF}}\right) D(C_{MOD})$$

$$\upsilon_{B1} = \frac{\sigma_1 g}{f_C'}\left(1 - \xi - \xi\frac{\varepsilon_{CR}}{\varepsilon_{CF}}\right) C(C_{MOD})$$

where the coefficients A, C, D, L, are expressed in Appendix 3 and if no external force is applied to the beam we have

$$\overline{\upsilon} = 0$$

The non-dimensional expression for the moment can be put in the following form;

$$\overline{\mu} = \frac{M}{f_C' b h^2} = \mu_{CC} + \mu_{CT} + \mu_B \tag{61}$$

where $\mu_{CC} = \xi^2 B(\varepsilon_{CF})$ is the. moment due to compressed concrete' $\mu_{CC} = \xi^2 L(\varepsilon_{CF}, \varepsilon_{CT})$ is the moment due to concrete in tension $\mu_B = \mu_{B1} + \mu_{B2}$ is the moment due to bridging fiber stress where

$$\mu_{B1} = \frac{\sigma_1 g}{f_c'}\left(1 - \xi - \xi\frac{\varepsilon_{CR}}{\varepsilon_{CF}}\right) \times\left[\left(1 - \xi - \xi\frac{\varepsilon_{CR}}{\varepsilon_{CF}}\right)F\left(\hat{C}_{\text{mod}}\right) + \xi\frac{\varepsilon_{CR}}{\varepsilon_{CF}}C\left(\hat{C}_{\text{mod}}\right)\right] \tag{62}$$

$$\mu_{B2} = \frac{\sigma_0}{f_c'}\left(1 - \xi - \xi\frac{\varepsilon_{CR}}{\varepsilon_{CF}}\right) \times\left[\left(1 - \xi - \xi\frac{\varepsilon_{CR}}{\varepsilon_{CF}}\right)H\left(\hat{C}_{\text{mod}}\right) + \xi\frac{\varepsilon_{CR}}{\varepsilon_{CF}}D\left(\hat{C}_{\text{mod}}\right)\right] \tag{63}$$

The coefficients A, B, C, D, F, H, L, are expressed in Appendix 3.

6.2. Iterative Procedure

The goal of this procedure is to find a link between external moment vs curvature and vs crack opening in the post cracking phase i.e. just after the first crack appears. This procedure operates on equilibrium equation, disregarding the problems related to the crack stability and propagation. The flow-chart is shown in **Figure 10**. For a given concrete strain in compression at the upper fiber e_{cf} going from 0 to e_{cu} (set equal to −0.0035) with a defined step, a trial position of the neutral axis is assumed (e.g.

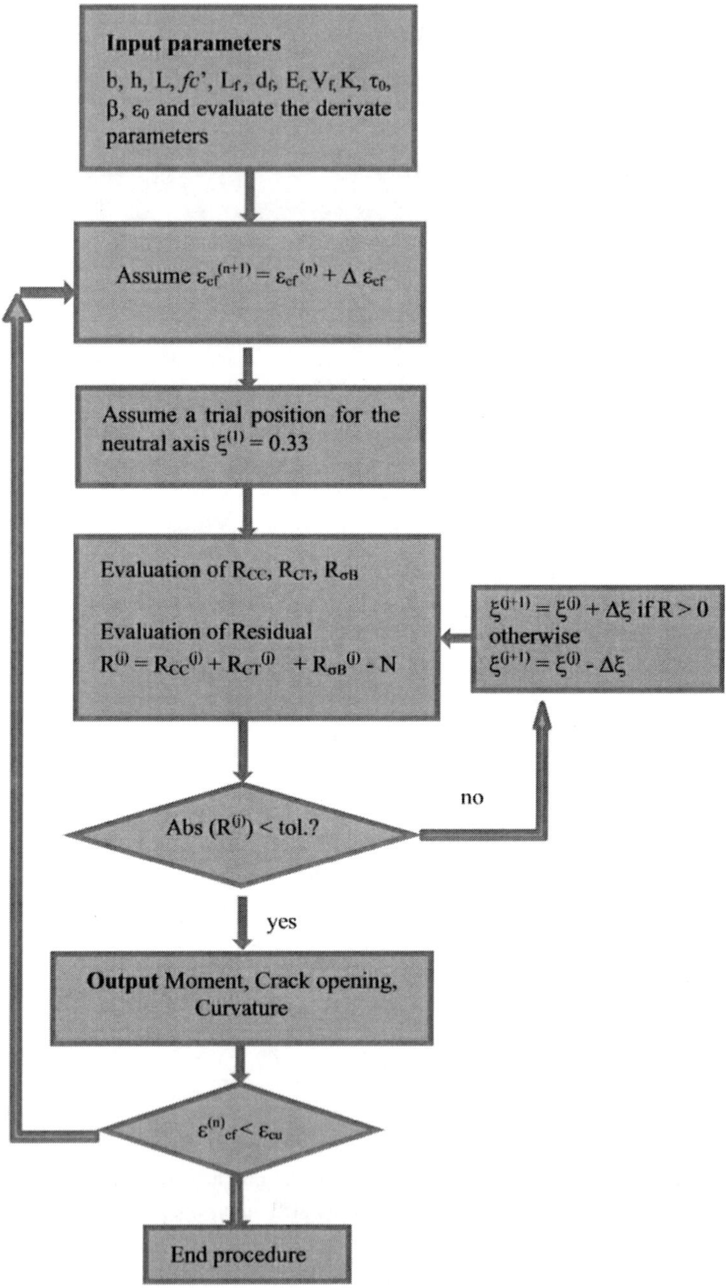

Figure 10. The procedure flow chart.

$\xi_{trial}^{(1)} = 0.33$). Then it is possible to define all the nondimensional resultant and check the equilibrium; if

$\sum R^{trial(1)} > 0$ then tensile components overcome then it is necessary to increase the neutral axis parameter

$\xi_{trial}^{(2)} = \xi_{trial}^{(1)} + \Delta\xi$ otherwise if $\sum R^{trial(1)} < 0$ compression components are prevalent so neutral axis parameter must be decreased then $\xi_{trial}^{(2)} = \xi_{trial}^{(1)} - \Delta\xi$. The process can be repeated until a satisfactory convergence is achieved.

The convergence criterion can be focused on the resultant residual i.e.

$R^{(n)} = \sum R^{trial(n)} < \text{tol.}$ where tol. is a tolerance value

(i.e. tol. = 0.01).

If the convergence is achieved it is possible to define an external moment, the crack opening from the Equation (10), and the curvature from the relation

$$K = \frac{1}{\rho} = \frac{\varepsilon_{cf}}{c} \tag{64}$$

Then for a given fiber reinforced concrete beam with given mechanical and geometric properties is possible to obtain a series of data related. Similarly to a reinforced concrete cross section, the ultimate strength and deformation field are defined by geometrical and mechanical parameters, which have to be investigated for every situation depending on fibers length, strength, Young modulus, interfacial stress and volume percentage above all.

The default tolerance is set, for the normalized residual, equal to 10^{-3}, while the incremental step for e_{cf} can be set to $\Delta e_{cf} = 0.0001$. An important influence in the convergence has the incremental step for ξ, that can be initially set to $\Delta\xi = 0.01$ but in case of switching sign of the residual R, it could need being reduced (i.e. using bisection method). A simple flow chart is represented in the next page (**Figure 10**).

6.3. A Numerical Example

A first numerical example will be performed with the model SPECTRA 900 ECC (randomly distributed Spectra fibers in a cement matrix) [3].

According to this article the properties of fiber and matrix are the following

$\sigma_{mu} = 2.2\ \text{MPa}$ $E_m = 25\ \text{GPa}$ g = 2 (where g is the snubbing coefficient).

Polyethylene fibers can be untreated and plasma treated; in the second kind the elastic interfacial shear strength is higher (0.8 MPa vs 0.47 MPa). In this numerical test we assume fiber percentage to be 0.75%, that is according to (2) enough to be above the critical percentage.

The other mechanical parameter are summarized in **Table 1**. The cylindrical compressive concrete strength and the peak tensile concrete strength are obtainable by the following relations

$$f_C' = \frac{E_m^2}{5.500^2} = 20.66\ \text{MPa}$$

$$f_{CR} = \frac{1}{10} f_C' = 2.066\ \text{MPa}$$

The result in italic font are related to the first crack, assumed to occur, as we mentioned before, when the elastic stress at the lower fiber of the cross section reaches the characteristic tensile strength, assumed, for the cement and concrete matrix, to be around 1/10 of the characteristic cylindrical compression strength. Moment vs crack opening has been plotted for two series, the first is made up of the data related to hardening parameter β = 0.00125 and the second to β = 0. These diagrams will be compared to the similar given by [5].

The numerical simulation is conducted assuming a four point bending test on prismatic specimen 100 × 100 × 350 assuming a span length of 300 mm. The results can be summarized in the Tables 2 and 3.

The result in italic font are related to the first crack, assumed to occur, as we mentioned before, when the elastic stress at the lower fiber of the cross section reaches the characteristic tensile strength, assumed, for the cement and concrete matrix, to be around 1/10 of the characteristic cylindrical compression strength.

Table 1. SPECTRA 900 ECC parameters [2].

Fiber	*l*. (mm.)	*d*. (μm)	*E* (Gpa)	β	τ_0 (MPa)	$\beta L_f/(2d_f)$
Untr.	12.7	38	117	0.0125	0.48	2.09
Plasma tr.	12.7	38	117	0.0128	0.8	2.14

Table 2. Numerical results for SPECTRA 900 untreated τ_0 = 0.48 MPa b = 0.0125.

ε_{cf}	ξ_i	μ_υ	C_{mod} (mm.)	Total displ.	P (kN)	Crack length
10^{-3}		**10^{-3}**		(mm.)		(mm.)
0.00		0	0	0.000	0.00	0
0.05		5.4		0.004	2.23	0
0.083		**13.8**		**0.011**	**5.69**	**0**
0.100	0.471	14.7	0.022	0.033	6.07	13.99
0.200	0.355	20.1	0.072	0.068	8.30	49.83
0.300	0.283	22.7	0.152	0.117	9.38	63.90
0.400	0.235	24.0	0.260	0.181	9.96	71.64
0.500	0.202	25.0	0.395	0.259	10.33	76.46
0.600	0.177	25.5	0.558	0.350	10.57	79.86
0.700	0.158	25.8	0.751	0.454	10.66	82.33
0. 80	0.143	25.8	0.96	0.571	10.66	84.22
0. 90	0.13	25.7	1.20	0.704	10.61	85.81
1.00	0.119	25.3	1.48	0.852	10.45	87.12
1.10	0.110	24.7	1.78	1.016	10.20	88.22
1.20	0.101	23.9	2.15	1.205	9.87	89.25
1.30	0.092	22.6	2.57	1.424	9.33	90.21
1.40	0.083	20.8	3.09	1.698	8.59	91.21
1.50	0.073	18.0	3.81	2.066	7.43	92.30
1.60	0.051	9.9	5.95	3.148	4.09	94.63
1.70	0.001	0	-			

Table 3. Results for SPECTRA 900 ECC plasma treated τ_0 = 0.8 MPa b = 0.0128.

ε_{cf}	ς_i	μ_u	C_{mod}	displ.	P (kN)	Crack
			(mm.)	mm.		mm.
0.00E+00		0.000	0	0.000	0	0.00
5.00E−05		0.005		0.004	2.23	0.00
8.26E−05		**0.014**		**0.011**	**5.69**	**0.00**
1.00E−04	0.492	0.016	0.0207	0.032	6.57	10.16
2.00E−04	0.432	0.026	0.0526	0.058	10.87	38.96
3.00E−04	0.373	0.032	0.1009	0.092	13.18	52.43
4.00E−04	0.326	0.035	0.1654	0.134	14.59	60.67
5.00E−04	0.289	0.038	0.2466	0.185	15.5	66.38
6.00E−04	0.260	0.039	0.3424	0.243	16.11	70.48
7.00E−04	0.236	0.040	0.4532	0.308	16.53	73.62
8.00E−04	0.217	0.041	0.5782	0.380	16.86	76.09
9.00E−04	0.201	0.041	0.7155	0.459	17.07	78.06
1.00E−03	0.187	0.042	0.8681	0.545	17.15	79.73
1.10E−03	0.176	0.042	1.0336	0.638	17.19	81.13
1.20E−03	0.166	0.042	1.2102	0.736	17.15	82.31
1.30E−03	0.156	0.041	1.404	0.843	17.07	83.39
1.40E−03	0.148	0.041	1.6119	0.957	16.86	84.33
1.50E−03	0.141	0.040	1.8352	1.079	16.57	85.18
1.60E−03	0.133	0.039	2.0815	1.212	16.2	85.99
1.70E−03	0.127	0.038	2.3477	1.355	15.74	86.74
1.80E−03	0.120	0.037	2.6526	1.518	15.12	87.50
1.90E−03	0.113	0.035	2.9978	1.700	14.34	88.26
2.00E−03	0.106	0.032	3.3915	1.907	13.39	89.01

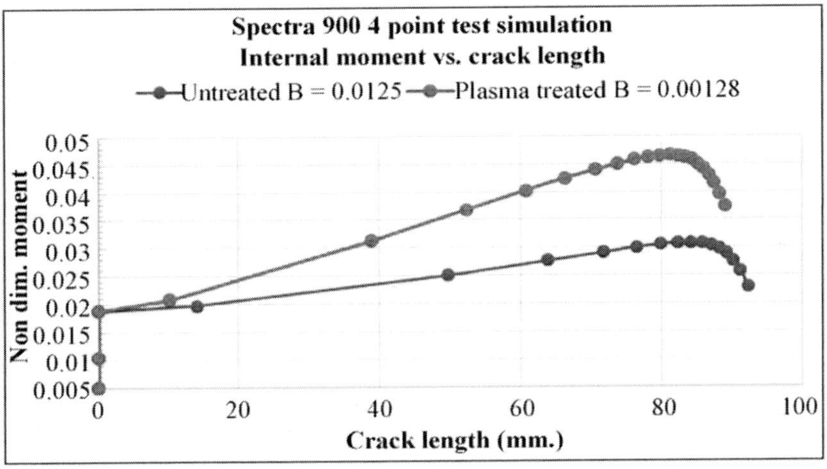

Figure 11. Diagrams moments vs crack length for SPECTRA 900 ECC according the built model.

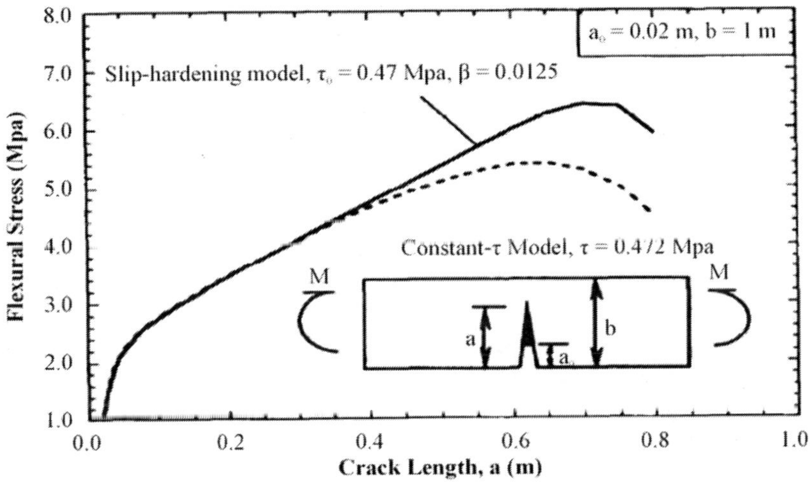

Figure 12. Flexural stress vs. crack length relation of a precracked beam [5].

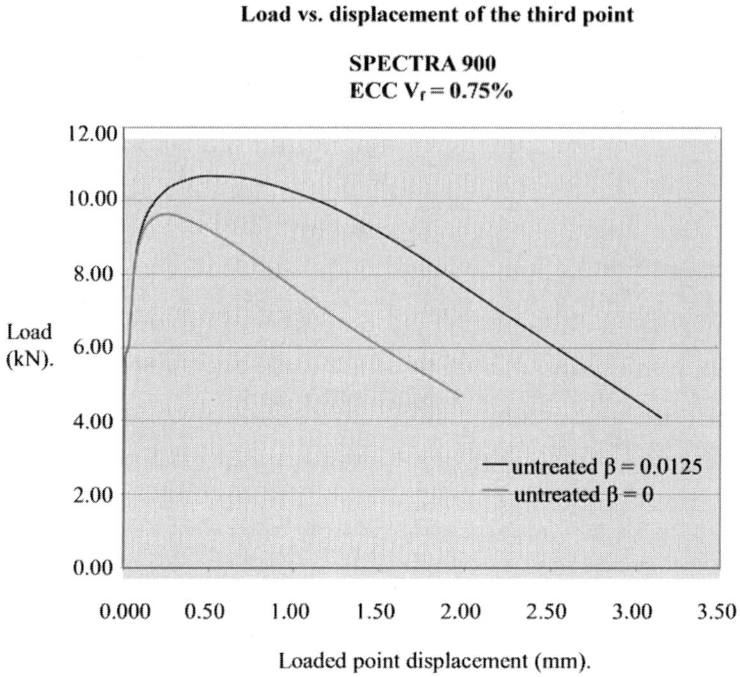

Figure 13. Load vs displacement of the third points for b = 0 and b = 0.0125.

Moment vs crack opening has been plotted for two series, as shown in **Figure 11** the first is made up of the data related to hardening parameter β = 0.00125 and the second to β = 0.

These diagrams will be compared to the similar given by [5] (see **Figure 12**). As we can see comparing the two diagrams above, even though they refer to little different picture but in any case to the same mechanical properties of the material (the numerical simulation related to Z.

Lin and V. C. Li work is related to a beam with an initial little crack while in our example we assume the central cross section uncracked, in addition the heights are different), the trend external moment vs crack opening is confirmed as the trend of different behavior between the configuration fiber with hardening parameter with respect to the one with fiber the constant shear interfacial stress. In our simulation the initial moment has been obtained assuming linear elastic model until the moment reaches the value of the first crack moment and setting the crack lengths equal to zero.

Even the diagrams load vs displacement of the loaded points (see **Figure 13**), confirms that, despite the simplicity of assumptions on which this analysis is based on, it is possible to implement the FRC model based on the fiber pull-out in a procedure able to catch the main trends of mechanical behavior in a beam synthetic fiber reinforced cementitious matrix composites element that undergoes an external moment.

7. CONCLUSIONS

In this paper, an attempt to link the behavior of a FRC rectangular beam under a four point bending test, has been made. The assumption about a failure configuration where a strain pattern for the cross section is consistent with the opening of crack after the maximum tensile stress overcoming reaches the nominal concrete tensile strength allows us to simulate the behavior of a sample of a synthetic FRC sample under increasing flexural load. Despite its simplicity the procedure used in this numerical performance in the previous paragraph is able, for the numerical performance analyzed so far, to catch the trend of the load test behavior as showed in [5].

This present work is not to be considered more than a contribution to developing a model based on the fibers pull-out theories for a better comprehension of fiber reinforced concrete behavior. Obviously further studies must be developed and experimental test must be conducted to confirm the general reliability of this model. In case of a confirmation of the cracking model it can be used to evaluate not only the ultimate strength of a synthetic FRC beam, but also the effect of the main parameters of fibers, like fiber volume fraction, aspect ratio, interfacial shear stress on the toughness and ductility.

8. ACKNOWLEDGEMENTS

An important contribution to this work has been given by Ph. D. Prechaporn Suwatnodom who gave me most of the needed papers about the FRC topics and for his explanations about the basic concepts. I also want to thank Prof. J. W. Ju for his final review of my Master Thesis, from which most of this work comes from, and Prof. Francesco Federico, always available to give me important suggestions to go on in the research world.

REFERENCES

1. P. Suwatnodom, "3D Micromechanical Damage Models, Fiber Pull-Out Models and Fracture Toughness of Discontinuous Steel Fiber Reinforced Cementitious Composites," UCLA Ph.D. Thesis, 2007.

2. N. Banthia, "Micro-Fiber Reinforced Cement Composites," Indian Concrete Journal, pp. 533-542.

3. D. J. Hannant, "Fibre Cements and Fibre Concrete," John Wiley and Sons, New York, 1978, pp. 219-231.

4. L. K. Jain and R. C. Wetherhold, "Effect of Fiber Orientation Composite," Acta Metallurgica et Materiala, Vol. 40, No. 6, 1992, pp. 1135-1140.

5. Z. Lin and V. C. Li, "Crack Bridging in Fiber Reinforced Cementitious Composites with Slip-Hardening Interface," Journal of the Mechanics and Physic, Vol. 45, No. 5, 1996, pp. 763-787.

6. Y. Wang, V. C. Li and S. Backer, "Modeling of Fiber Pull-Out from a Cement Matrix," Journal of Composite, 1988.

7. V. C. Li, Y. Wang and S. Backer, "A Micromechanical Model of Tension-Softening and Bridging Toughening of Short Random Fiber Reinforced Brittle Matrix Composites," Journal of the Mechanics and Physic, Vol. 39, No. 5, 1991, pp. 607-625.

8. G. Bao and Y. Song, "Crack Bridging Models for Fiber Composites with Slip-Dependent Interfaces," Solids, Vol. 41, No. 9, 1993, pp. 1425-1444.

9. Vecchio and Collins, "The Modified Compression-Field Theory for Reinforced Concrete Elements Subjected to Shear," ACI, Vol. 83, No. 22, 1986, pp. 219-231.

10. B. H. Oh, D. G. Park, J. C. Kim and Y.C. Choi, "Experimental and Theoretical Investigation on the Post-Cracking Inelastic Behavior of Synthetic Fiber Reinforced Concrete Beam," Cement and Concrete Research, Nutritional Aspects of Trace Elements, Vol. 35, No. 2, 2005, pp. 384- 392.

Chapter 3

Seismic Behavior of Substandard RC Columns Retrofitted with Embedded Aramid Fiber Reinforced Polymer (AFRP) Reinforcement

Engin C. Seyhan, Caglar Goksu *, Ahmet Uzunhasanoglu and Alper Ilki

Department of Civil Engineering, Civil Engineering Faculty, Istanbul Technical University, Ayazaga, Sariyer, Istanbul 34469, Turkey

ABSTRACT

Many existing reinforced concrete structures were constructed with substandard characteristics. Low quality concrete, poor transverse reinforcement details and insufficient flexural strength are among the most common deficiencies. While substandard structures are in need of retrofitting, particularly in seismic areas, problems such as high costs and disturbance to occupants are major obstacles for retrofit interventions. Fiber reinforced polymers can provide feasible retrofit solutions with minimum disturbance to occupants. In this study, the basic aim is to investigate the flexural seismic performance of substandard reinforced concrete columns retrofitted with embedded longitudinal fiber reinforced polymer reinforcement without increasing the original dimensions of the columns. In the experimental study, the reference and retrofitted columns were tested under constant vertical and reversed cyclic lateral loads. Three different connection methods of aramid fiber reinforced polymer reinforcement to the footing were investigated experimentally. A significant enhancement was obtained in lateral flexural strength through the proposed retrofitting method. Furthermore, it was observed that the cyclic lateral drift capacities of the retrofitted columns were as high as 3%, which can be deemed as quite satisfactory against seismic actions. The comparison of the experimental data with analytical calculations revealed that a conventional design approach assuming composite action between concrete and fiber reinforced polymer reinforcement can be used for flexural retrofit design. Experimental results also demonstrated that strain limit for longitudinal fiber reinforced polymer (FRP) reinforcement should be remarkably lower in case of reversed cyclic loading conditions.

Keywords: aramid; flexure; FRP reinforcement; reinforced concrete; retrofitting; seismic

1. INTRODUCTION

Many substandard reinforced concrete (RC) structures should be seismically retrofitted to reduce their vulnerabilities against seismic actions. Financial constraints, disturbance to the occupants and disruption of functions of the structures are the main obstacles for proper seismic retrofitting of these substandard existing structures. In recent years, use of fiber reinforced polymers (FRP) in construction industry has become quite common. They offer feasible and innovative solutions for seismic retrofitting due to their lightweight, high tensile strength and noncorrosive character [1,2,3,4,5,6]. FRPs are also utilized as external tendons for strengthening concrete structures [7,8,9]. While there are many studies on external confinement of columns or joints [10,11,12,13,14,15,16,17,18,19,20,21,22,23,24,25,26], and on flexural retrofit with near surface mounted (NSM) FRP rods or pultruded strips under monotonic loading conditions [27,28,29, 30,31, 32,33,34, 35,36,37, 38,39, 40,41, 42], only few studies are available on flexural retrofitting using FRPs under repeated loading conditions [43,44,45]. According to the best knowledge of the authors', reversed cyclic flexural behavior of RC members strengthened with longitudinal FRP reinforcement was studied for the first time by Ilki and Kumbasar [46]. More recently, Bournas and Triantafillou [47], Goksu *et al.* [48], Fahmy and Wu [49], Vrettos *et al.* [50], Li *et al.* [51], Goksu *et al.*[52] and Faustino and Chastre [53] also studied this type of behavior. However, in all cases, except the studies of Goksu *et al.* [48,52], it is difficult to perceive the performance of the NSM FRP system and anchoring of NSM reinforcement into the footing in case of low strength concrete. In the study of Goksu *et al.* [48], which was the precursor of this study, the possibility of using carbon FRP (CFRP) longitudinal (rod, laminate, sheet) and transverse (sheet) reinforcement for the flexural seismic retrofit of low strength RC members under reversed cyclic loading conditions was investigated. The aim of flexural retrofitting by using FRP reinforcement, mainly in the longitudinal direction, was to obtain flexural strength enhancement under reversed cyclic loading, since many existing structures, among other deficiencies, suffer from lack of sufficient flexural strength in case of seismic loading. Furthermore, it was targeted not to sacrifice from drift capacity, since drift capacity is as important as strength in case of seismic actions. In the study of Goksu *et al.* [48], an enhancement in flexural capacity was obtained until large drift ratios (approximately 6% drift ratio) through flexural FRP retrofit. Moreover, after testing of performance of several anchorage types for connection of FRP reinforcement to the footing, the most effective anchorage detail was obtained, and was also utilized in the current testing program. The major differences between the current testing program and the work reported by Goksu *et al.* [48] are (i) the presence of axial load (20% of axial load capacity of the columns, whereas no axial load was applied on the specimens in the study of Goksu *et al.* [48]), (ii) the different type of FRP reinforcement utilized in the current study, aramid FRP pultruded laminates (AFRP), and (iii) intentional partial debonding of FRP reinforcement in the anchorage hole to allow more distributed FRP deformations at the critical interface between the column and footing in the current study. The reason of utilizing AFRP reinforcement rather than CFRP or GFRP in the current study is the expectation of potentially better performance of AFRP due to its better toughness characteristics [54].

In this study, four cantilever RC columns were constructed using low strength concrete and plain reinforcing bars for representing relatively old substandard structures. According to the compressive tests of core specimens extracted from many different existing buildings constructed before 1990s in Turkey, the concrete compressive strengths are found to be between 5 and 13 MPa. The microstructural analyses of these core specimens revealed that the concretes were very porous and the water/cement ratio used for making them was excessively high [55]. The columns were then tested under reversed cyclic lateral and constant axial loads before and after retrofitting. Three different anchorage details were designed for embedding AFRP laminates to the existing footing. These different details constitute the main test variable. While the applied retrofitting technique is similar to NSM technique in terms of mechanical contribution of FRP reinforcement to the flexural strength, application technique of retrofitting is different than NSM technique, as has been introduced by Goksu *et al.* [48] before. As is well known, concrete cover is generally deteriorated due to corrosion of internal steel reinforcing bars of substandard structures constructed with low strength concrete. Typical appearances of columns of existing substandard RC frame buildings with corroded reinforcement are shown in Figure 1.

In the applied retrofitting technique, firstly, deteriorated concrete cover is removed. Secondly, a layer of high strength structural repair mortar (SRM) is applied to obtain a flat surface for bonding AFRP reinforcement. After hardening of SRM, AFRP longitudinal reinforcement is bonded to the substrate by using epoxy adhesive. Then, a second layer of SRM is applied to form the new concrete cover. As the final stage of seismic retrofit, the strengthened column is externally jacketed with CFRP sheets in transverse direction. Consequently, by this technique, all retrofitting application was conducted within the thickness of the original concrete cover, which may be deemed as an advantage with respect to the conventional retrofitting techniques. Clearly, AFRP reinforcement, which is embedded into SRM, is expected to perform better than NSM application in case of poor quality concrete in terms of structural integrity.

Figure 1. The appearances of existing substandard columns of RC frame structures.

Test results, supported by the analytical calculations, revealed that the cyclic flexural capacity of the substandard RC members can be increased significantly by the use of AFRP pultruded laminates through the applied retrofitting method. Current technical guidelines/documents do not allow use of longitudinal FRP reinforcement for flexural retrofit under cyclic loading in potential plastic hinge regions of RC members [5,56,57]. This is basically due to concerns related with ductility. In this study, however, it was observed that the deformation capacity of the retrofitted columns was also acceptable for a satisfactory seismic performance.

It was observed from the seismic tests that the measured rupture strain of AFRP reinforcement, which was subjected to cyclic loading, was approximately 43% of design rupture strain of FRP reinforcement (design rupture strain of FRP reinforcement = environmental reduction factor × manufacturer declared ultimate rupture strain). For NSM applications under monotonic loading conditions, ACI 440-2R-08 [5] recommends to consider 70% of ultimate strain of FRP. In this study, it is demonstrated clearly that allowable strain limit for longitudinal FRP reinforcement should be lower in case of reversed cyclic loading conditions (only 43% of manufacturer declared ultimate strain of FRP reinforcement). It is also shown that the increased flexural capacity as well as nonlinear inelastic displacement response of columns after strengthening could be predicted satisfactorily using conventional RC theory.

2. EXPERIMENTAL PROGRAM

The aim of the applied retrofitting technique is the enhancement of column flexural capacity using AFRP reinforcement under cyclic lateral loading in the presence of constant axial load without a significant sacrifice from deformation capacity. On the other hand, like many existing substandard RC columns, the spacing of the transverse reinforcement of the columns was insufficient. This might have caused deficiencies in terms of ductility and shear strength after flexural strength enhancement through FRP retrofitting. Therefore, after the intervention made for flexural strength enhancement through longitudinal AFRP reinforcement, the columns were also jacketed externally with CFRP sheets in transverse direction. As known, while external FRP confinement has a remarkable contribution to deformability and shear strength, its contribution to flexural strength (which is the main issue investigated in this study) is marginal with respect to the contribution of longitudinal AFRP reinforcement. Nevertheless, both contributions were taken into account during the analytical calculations presented in this paper.

2.1. Description of Test Specimens

Four substandard symmetrically reinforced cantilever RC columns were tested under reversed cyclic lateral and constant axial load. The main features of the specimens are shown in Table 1. While lateral loads were applied through displacement cycles in pushing and pulling directions with increasing amplitudes, axial load was kept constant. All columns were identical (before retrofitting) and flexure-critical.

Table 1. Main Features of the Specimens.

Specimen	Main longitudinal AFRP [a] reinforcement	Additional anchorage reinforcement [a]	Anchorage type [a]	Transverse CFRP [b] reinforcement
REF	–	–	–	–
LAM	2 × 2 (each 42 mm x 1.4 mm)	–	Fully bonded	2 plies × 0.166 mm
LAM-PB	2 × 2 (each 42 mm x 1.4 mm)	–	Partially bonded	2 plies × 0.166 mm
LAM-LAM	2 × 2 (each 42 mm × 1.4 mm)	2 × 2 (each 42 mm × 1.4 mm)	Fully bonded	2 plies × 0.166 mm

[a] Aramide based, pultruded laminates; [b] Carbon based, unidirectional sheets fully wrapped around the retrofitted specimens. REF: reference specimen; LAM, LAM-PB, LAM-LAM: retrofitted specimens.

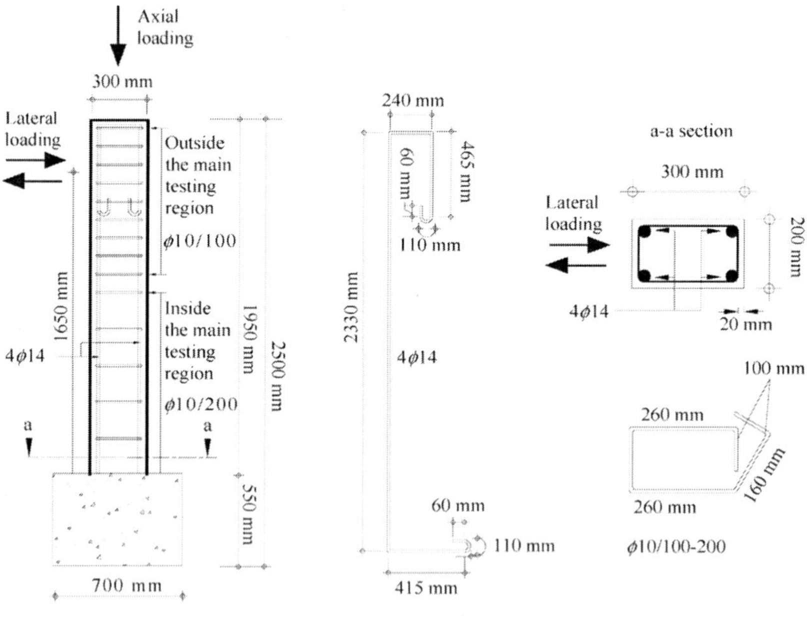

Figure 2. The geometry and reinforcement details of the columns.

The geometry and the reinforcement details of the columns are presented in Figure 2. In this figure, it can be observed that the longitudinal bars were

(4ϕ14) continuous from the bottom of the footing to the top of the column. The spacing of transverse bars was 200 mm in the testing zone (ϕ10/200). All internal reinforcing bars were plain round bars, which had been used commonly until 2000s in Turkey.

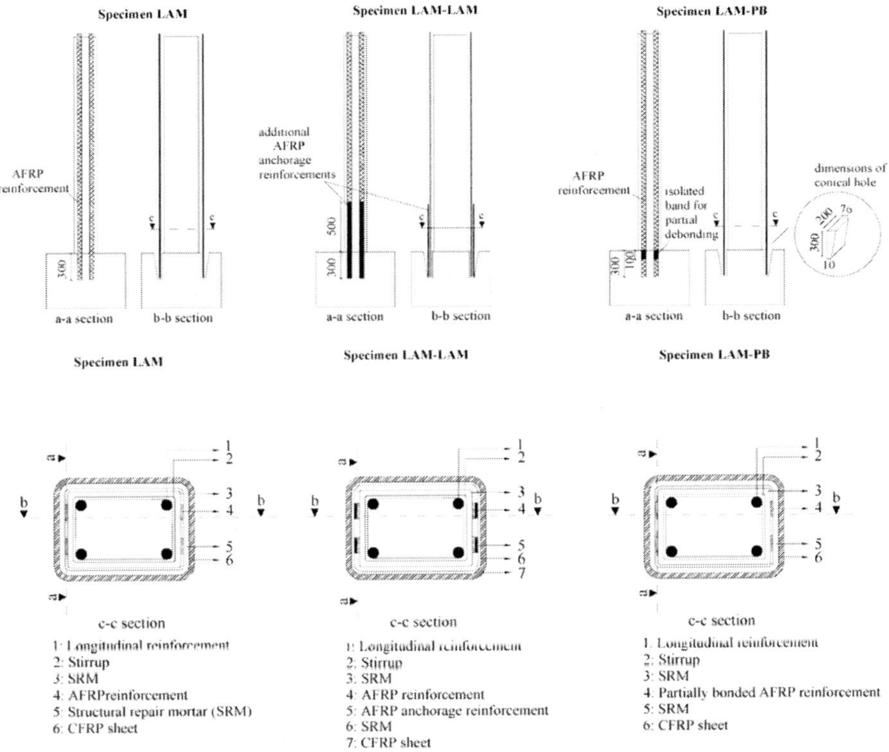

Figure 3. The retrofit schemes of the specimens (Note: Dimensions in mm).

The reference specimen, denoted as REF, was tested without any retrofit. The other columns (LAM, LAM-LAM and LAM-PB) were tested after they were retrofitted with AFRP reinforcement in longitudinal and CFRP reinforcement in transverse directions. The longitudinal AFRP reinforcement embedded in the columns was anchored directly to the footing in all retrofitted specimens, while additional AFRP anchorage reinforcement was used for the specimen LAM-LAM to achieve a better performance by avoiding potential failures of FRP reinforcement at the interface of the column and the footing. Contrary to the fully bonded anchorages of longitudinal FRP reinforcement in case of the specimens LAM and LAM-LAM, the anchorage of the FRP reinforcement in case of the specimen LAM-PB was intentionally partly debonded from concrete using insulating tape.

Partial debonding of AFRP reinforcement in the anchorage zone was for preventing the localized damage of the AFRP reinforcement at the interface of column and footing by avoiding localization of stresses at this critical zone. Details on each retrofitting scheme are explained below and presented in Figure 3.

2.2. Characteristics of Materials

Low strength concrete and plain reinforcing bars were used for the construction of the columns for representing relatively old, existing substandard structures. All RC columns were cast at the same day, from the same concrete production batch. The 28th-day average compression strength was measured as 10.3 MPa. An additional compressive strength test, performed on the 750th day, yielded an average compressive strength of 15.2 MPa. The compressive strength, at the time of testing of columns (between 57th and 64th days), was calculated as 12.0 MPa, according to the procedure described in Eurocode 2-Part 1 (2004) [58]. For the numerical analysis, this value was multiplied with 0.85 to account for the size effect in concrete columns. The concrete mix-proportions of cement: aggregate (0–4 mm): aggregate (5–12 mm): sand: water were 1:3.8:2.2:4.7:1 (by weight). The yield stress (f_y) and tensile strength (f_{max}) of longitudinal bars were 296 and 400 MPa, while those of transverse bars were 315, and 400 MPa, respectively. The compression tests of concrete specimens and tension tests of steel bars were carried out in accordance with TS-EN Turkish Standards 12390-3 [59] and TS 708 [60], respectively. Two different types of FRP reinforcement were used in retrofitting; AFRP pultruded laminates in longitudinal and CFRP sheets in transverse direction. The appearance and geometrical and mechanical properties of FRP reinforcement are presented in Figure 4 and Table 2, respectively. In this table, t_f, w_f, E_f and ε_{fu}* are the effective thickness, the effective width, the tensile elastic modulus and the ultimate rupture strain of FRP reinforcement declared by the manufacturer (BASF Construction Chemicals, Augsburg, Germany). The compressive strengths of SRM (cement based structural repair mortar, EN-1504-3 [61]) used for forming the new concrete cover, the epoxy adhesive used for bonding the AFRP reinforcement to the column surface (EN-1504-4 [62]), the epoxy adhesive used for bonding CFRP sheets in transverse direction to the column surface, and the epoxy grout used for anchoring AFRP reinforcement in the footing were 50, 75, 60 and 80 MPa (obtained from 40 mm× 40 mm × 160 mm prism test results at 7 days of age), respectively. The compressive characteristics of all materials were determined according to TS-EN 196-1 [63], and the mechanical characteristics of CFRP sheets were determined according to ASTM D3039 [64].

Table 2. Properties of FRP reinforcement.

FRP reinforcement	E_f (N/mm^2)	t_f (mm)	w_f (mm)	ε_{fu}*
AFRP (laminate)	60,000	1.4	42	0.023
CFRP (sheet)	230,000	0.166	500	0.015

FRP, fiber reinforced polymers; AFRP, aramide reinforced polymers; CFRP, carbon reinforced polymers; E_f, tensile elastic modulus of FRP reinforcement; t_f, effective thickness of FRP reinforcement; ; w_f, effective width of FRP reinforcement; ε_{fu}*, ultimate rupture strain of FRP reinforcement declared by manufacturer.

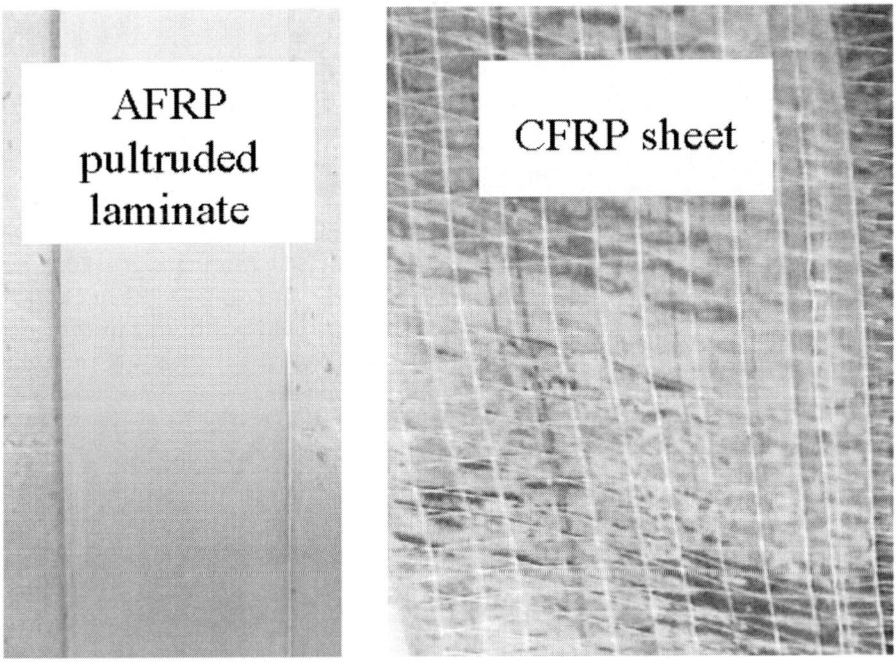

Figure 4. The appearances of longitudinal AFRP and transverse CFRP reinforcement.

2.3. Retrofitting Procedure

For enhancing the column flexural capacity, three specimens were retrofitted with longitudinal AFRP reinforcement, which was embedded within the concrete cover (after removal of poor concrete cover), and anchored to the existing footing. The retrofitting details and the application stages are shown in Figure 3 andFigure 5, respectively. As a first step, the concrete cover was removed for more effective utilization of AFRP reinforcement until the longitudinal steel bars were exposed (Figure 5a). This allowed AFRP reinforcement to be embedded within the cover thickness. This application may also cover a corrosion repair procedure, which is generally required in case of rehabilitation of old buildings, which were constructed with poor quality concrete. Then, a thin layer of SRM was applied to obtain a sound and smooth substrate over the internal steel reinforcing bars (Figure 5b). Afterwards, an epoxy-based primer was applied over the SRM to increase the adhesion between the AFRP reinforcement and SRM. Then, two pultruded AFRP laminates of 1.4 mm thickness and 42 mm width were placed symmetrically on each side of the columns in the strong direction, with an anchorage length of 300 mm (Figure 3 and Figure 5c), and the laminates were bonded to the substrate by using an epoxy adhesive. For the connection of the AFRP longitudinal reinforcement to the footing, conical holes were dug in the footing (Figure 3). The longitudinal AFRP reinforcement embedded in the columns was anchored directly to the footing in all retrofitted specimens. Different from the specimen LAM, for the specimen LAM-PB, the top 100 mm long part of the AFRP reinforcement in the conical hole was wrapped with a plastic insulating band in order to create a partially debonded anchor (Figure 3 and Figure 5d). Therefore, total bonded lengths of the AFRP reinforcement in the footing were 200 mm (rather than 300 mm) for the specimen LAM-PB. Additional AFRP anchorage reinforcement of 800 mm length was used at the column-footing interface in case of the specimen LAM-LAM (Figure 3 and Figure 5e). The 300 mm long part of the additional AFRP anchorage reinforcement was embedded into the conical hole. AFRP anchorage reinforcement and the main AFRP reinforcement were bonded to each other by using the same epoxy adhesive, and they were anchored to the footing (in the pre-dug conical hole) together by using epoxy grout. Same epoxy grout was used to fill the conical holes in the footings of specimens LAM and LAM-PB. After all, a layer of SRM was applied to bring the column cross-section to its original dimensions (Figure 5f). It should be noted that, buckling of NSM FRP bars was observed during the reversed cyclic tests carried out by Bournas and Triantafillou [47]. Therefore, after installation of longitudinal AFRP reinforcement and formation of concrete cover with SRM, as a final step, CFRP sheets were wrapped around the columns in transverse direction to enhance the deformability through confinement action and to avoid potential shear damages due to increased flexural strength as well as to prevent buckling of AFRP reinforcement in compression (Figure 5g). Additionally, CFRP confinement is believed to enhance the bond between the core concrete, SRM and AFRP reinforcement. Consequently, the external dimensions of the reference and retrofitted specimens were identical (200 mm × 300 mm). In other words, all retrofitting application was conducted within the thickness of the original concrete cover. Naturally, during practical applications, the safety of the footing and soil under the footing should be checked to confirm that they have also sufficient safety margin against increased internal forces.

Figure 5. (**a**) Removal of cover concrete, (**b**) application of a thin layer of cement based SRM to obtain a flat surface, (**c**) installation of the AFRP pultruded laminates, (**d**) laminate with isolated band (only for the specimen LAM-PB), (**e**) application of additional AFRP laminate anchorages (only for the specimen LAM-LAM), (**f**) application of last layer of structural repair mortar, and (**g**) wrapping with CFRP sheets in transverse direction.

2.4. Test Setup

The specimens were tested under constant axial load and reversed cyclic lateral loads in a quasi-static displacement-controlled manner. A displacement-based loading protocol was applied during testing of all specimens. Similar loading patterns were used by Matamoros and Sozen [65], Iacobucci *et al.* [13], Ilki *et al.* [20], Bournas and Triantafillou [66], and Goksu *et al.* [48] during testing of RC columns. Target lateral drift ratios were calculated as the ratio of the lateral displacement to the specimen height (at which actuator is connected) in both pushing and pulling directions (Figure 6). The axial load of 120 kN, which corresponded to 20% and 18% of the axial load capacity of the reference column without and with consideration of the capacity of the internal steel reinforcement, respectively, was applied to all specimens (Figure 7a). Other than a large number of displacement transducers, a number of strain gauges were also used on steel bars and AFRP reinforcement in longitudinal direction and on stirrups in transverse direction (Figure 7b).

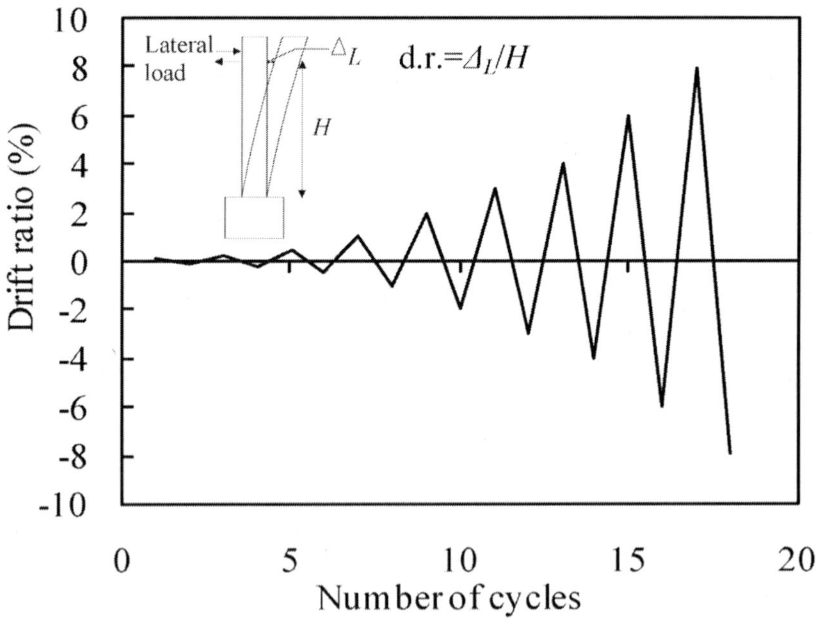

Figure 6. Loading protocol of the columns.

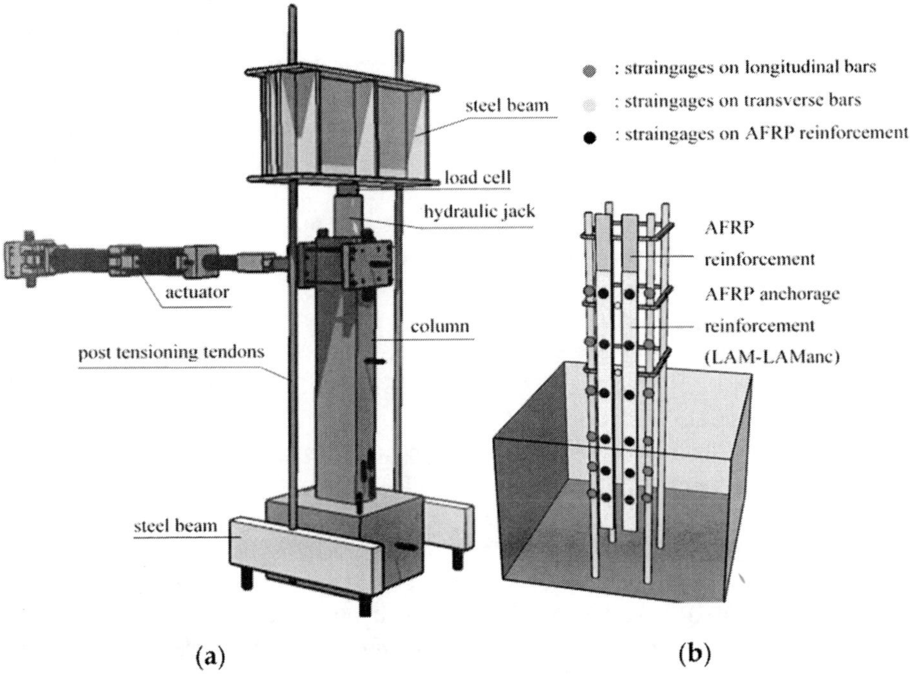

Figure 7. (**a**) The testing setup; and (**b**) Strain gauges on steel bars and AFRP reinforcement.

3. ANALYTICAL CONSIDERATIONS

3.1. Theoretical Flexural Strength and Displacement Capacity before and after Retrofit

The theoretical strength and local deformation characteristics are determined through moment–curvature relationships obtained by fiber analysis approach using the stress–strain relationships of materials representing monotonic increasing stresses. The monotonic moment–curvature relationships are obtained using XTRACT (2007) [67] software at the critical sections of each column for predicting the load–displacement relationships of the columns. Therefore, the theoretical load–displacement relationships approximately represent the envelopes of hysteretic curves. The effect of axial load is considered while obtaining the moment–curvature relationships. In the moment–curvature analysis, steel reinforcing bars are assumed to behave in an elasto–plastic manner with strain hardening. The stress–strain relationship of the steel reinforcing bar is defined according to actual uniaxial tension test results. Longitudinal AFRP reinforcement is taken into account as linear elastic material in tension since the buckling and debonding of AFRP reinforcement during cyclic loading is prevented by SRM and CFRP confinement. Nevertheless, the contribution of the AFRP reinforcement in compression is neglected. Since the retrofitted specimens, thereby the AFRP applications, were exposed to laboratory conditions for a very short time prior to testing, the design rupture strain of AFRP reinforcement (ε_{fu}) is assumed to be equal to ε_{fu}*, which is the manufacturer declared ultimate strain of FRP reinforcement *(*Equation (1)). In other words, the environmental reduction factor, C_E, is taken as 1.0. While ACI 440-2R-08 [5] recommends to reduce ε_{fu} by a factor of 0.70 for NSM FRP applications under monotonic flexural loading, ε_{fu} is reduced by a factor of 0.43 during the analysis presented here (the reason for this assumption is explained in the section titled "Strains of Steel and AFRP Reinforcement"). For obtaining the stress–strain relationship of CFRP confined concrete, the model proposed by Ilki *et al.* [68] is used. The stress–strain relationship of CFRP confined concrete is obtained for two different parts of the confined cross–section (for core concrete confined by transverse steel bars and external CFRP, and for SRM as concrete cover confined by only external CFRP). The stress–strain relationships of unconfined concrete, core concrete, SRM, internal steel and AFRP reinforcement used in calculation of moment capacities are presented inFigure 8. The marks in the figure correspond to the analytical values of the specimens during analysis. It should be noted that SRM causes an increase of 7% in theoretical moment capacity, while both SRM and AFRP cause an increase of 51% and 97% for the specimens LAM and LAM-LAM, respectively, in theoretical moment capacity in comparison with the reference column. The moment–curvature analysis is terminated when longitudinal FRP reinforcement ruptures (Figure 9).

$$\varepsilon_{fu} = C_E \cdot \varepsilon_{fu}^* \quad (1)$$

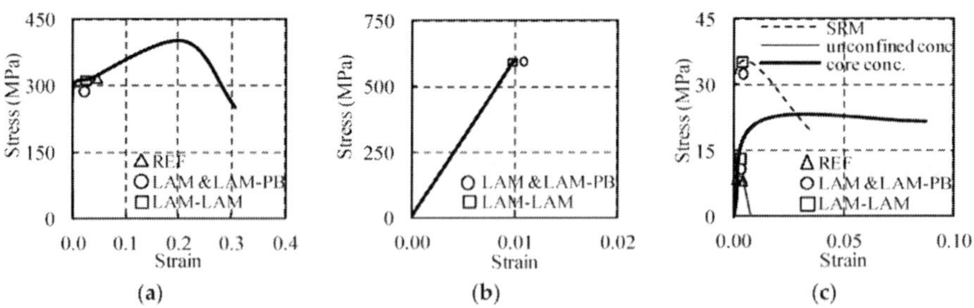

Figure 8. The theoretical stress–strain relationships of the (**a**) Internal steel reinforcement; (**b**) AFRP reinforcement; and (**c**) Concrete (conc., concrete).

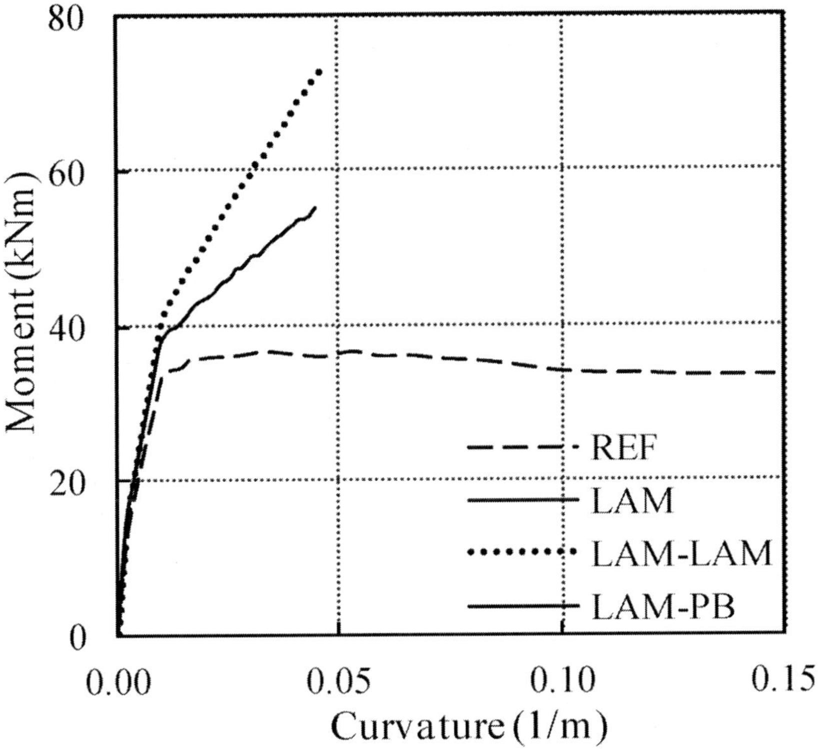

Figure 9. Theoretical moment–curvature relationships. (Note: The moment–curvature relationships for the specimens LAM and LAM–PB are the same).

After obtaining the moment–curvature relationships through fiber analysis approach with summarized assumptions, the total top displacements (δ_t) of the columns are determined considering the elastic and inelastic deformations. It should be noted that the shear deformations are neglected since they are very small with respect to contribution of flexure. As a first step, for calculating the yield displacements (δ_y) of the columns using Equation (2), the columns are discretized into 5 elements for considering effective flexural stiffness, rather than the gross cross-sectional stiffness, at different heights of the column as a function of bending moment when the longitudinal reinforcing bars yield at the critical maximum moment region (Figure 10). In Equation (2), χ_i is the elastic curvature (corresponding to respective bending moment on that discretized part), l_i is the length of the respective part, and h_i is the distance between the center of the respective part and the tip of the column, where the lateral load is applied. Sequentially, for obtaining the plastic component of the total top

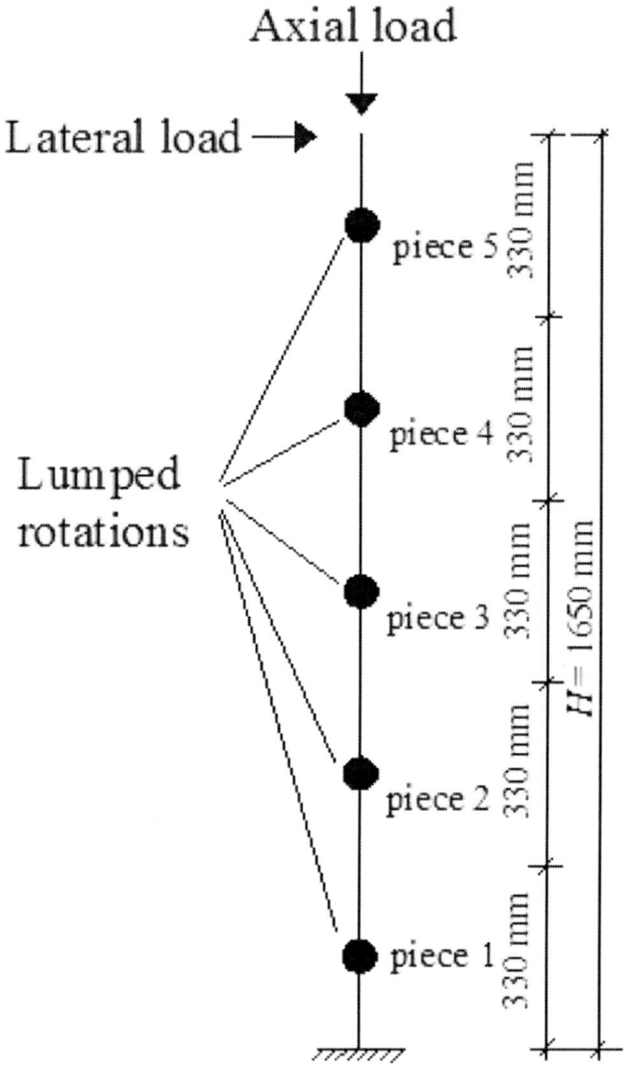

Figure 10. The locations of lumped rotations.

displacements (δ_p) of the columns, conventional plastic hinge approach is used (Equation (3)). In Equation (3), χ_p is the plastic curvature at the plastic hinge of the column, l_p is the plastic hinge length, H is the distance between the interface of the column and the footing and the point of application of the lateral load. The length of plastic hinge (l_p) of the column is assumed as $h/2$ (150 mm) according to TSDC [57], where h refers to the depth of the cross-section of the column. Finally, the total top displacements (δ_t) of the columns are determined using Equation (4). The theoretical lateral load–

displacement relationships are then obtained and compared with the experimental load–displacement relationships in the Section "Results and Discussions".

$$\delta_y = \sum_{i=1}^{5} \chi_i \cdot l_i \cdot h_i \tag{2}$$

$$\delta_p = \chi_p \cdot l_p \cdot (H - \frac{l_p}{2}) \tag{3}$$

$$\delta_t = \delta_y + \delta_p \tag{4}$$

3.2. Shear Strength before and after Retrofit

According to ACI 318M-14 [69], before retrofit, the shear strength, which includes contributions of concrete and transverse reinforcement, can be obtained using Equations (5) and (6). In these equations, V_s is the nominal shear strength provided by transverse reinforcement, A_v is the area of the transverse reinforcement, f_{yt} is the design yield strength of transverse reinforcement, d is the effective depth, s is the spacing of shear reinforcement, V_c is the nominal shear strength provided by concrete, N_u is the factored axial load normal to cross-section (to be taken as positive for compression), A_g is the gross area of the column section, f'_c is the design compressive strength of concrete, b_w is the width of the column,

$$V_s = \frac{A_v f_{yt} d}{s} \tag{5}$$

$$V_c = 0.166\left(1 + \frac{N_u}{14A_g}\right) \cdot \sqrt{f'_c}\, b_w d \tag{6}$$

According to ACI-440-2R-08 [5], the contribution of FRP reinforcement to shear strength can be obtained using Equations (7) and (8).

$$V_f = \frac{A_{fv} \cdot \varepsilon_{fe} \cdot E_f \cdot d_{fv}}{s_f} \tag{7}$$

$$A_{\mathrm{fv}} = 2n \cdot t_{\mathrm{f}} \cdot w_{\mathrm{f}} \tag{8}$$

where V_f is the FRP contribution to shear strength, A_{fv} is the area of FRP shear reinforcement, ε_{fe} is the effective strain of FRP sheet attained at failure, which was assumed to be 0.004 as recommended by ACI-440-2R-08 [5] and TSDC [57], E_f is the tensile modulus of elasticity of FRP shear reinforcement, d_{fv} is the effective depth of FRP shear reinforcement, s_f is the spacing of FRP reinforcement plies, n is the number of plies of FRP reinforcement, t_f is nominal thickness of one ply of FRP reinforcement, and w_f is the width of FRP reinforcement plies.

According to ACI-440-2R-08 [5], the total shear strength of the cross-section (V_n) resisted with FRP reinforcement, concrete and transverse reinforcement can be obtained using Equation (9).

$$V_{\mathrm{n}} = V_{\mathrm{c}} + V_{\mathrm{s}} + \psi_{\mathrm{f}} V_{\mathrm{f}} \tag{9}$$

where ψ_f is the FRP strength reduction factor and was assumed to be 0.95 as recommended by ACI-440-2R-08 [5] for shear fully wrapped sections.

Since the shear strengths of the columns (before and after retrofit) were remarkably higher than their flexural strengths, as expected, flexural effects dominated the behavior of the specimens (Table 3). This table includes the observed and predicted failure modes, which are consistent.

Table 3. Theoretical and experimental lateral load capacities and consecutive failure modes of the columns.

Specimens	REF	LAM and LAM-PB	LAM-LAM
Theoretical lateral load capacity (kN)	21.2	32.1	41.8
Experimental lateral load capacity (kN)	24.5	33.8 (for the specimen LAM) 34.6 (for the specimen LAM-PB)	45.1
Theoretical moment capacity (M_0) (kNm)	36.5	55.0	72.4
Theoretical failure mode	Y before C	Y, T	Y, then T
Observed failure mode	Y before C	Y, F	Y, FA
Shear strength (kN)	108.0 ($V_{c+}V_s$)	260.6 ($V_{c+}V_{s+}\psi_f \cdot V_f$)	

Y: Tensile reinforcement yields; C: Concrete crushes; T: AFRP reinforcement reaches its ultimate tensile capacity; F: AFRP reinforcement fractured; FA: Both AFRP reinforcement and anchorage fractured.

3.3. Anchorage Length for FRP Reinforcement in Footing

In order to utilize the full capacity of the FRP reinforcement, the failure of FRP reinforcement was desired to be due to rupture of the reinforcement rather than pullout, concrete splitting, concrete cone and spacing and edge cone failures, which have been listed as five primary failure modes in ACI-355-1R-91 [70]. Among these failures, the possibility of concrete cone, concrete splitting and edge cone failures was significantly reduced due to the constraining effect of steel reinforcing bars in the footing, where the conical anchor holes were dug in. Therefore, only possibilities of pullout failure and fracture of FRP reinforcement were taken into account by comparing the tensile strength of FRP reinforcement and the adhesion between concrete and epoxy grout (Equations (10) and (11)). In these equations, F_{bond} is the bond capacity, τ_b is the uniform bond strength along the anchorage length, l_{db} is the embedment length, while u is the perimeter, F_{frp} is the tensile capacity of the FRP reinforcement, ε_{fu} is the design rupture strain of FRP reinforcement, and A_{frp} is the cross-sectional area of the FRP reinforcement. For the selected embedment length (300 mm), the calculated value of F_{bond} (1012 kN) was far over F_{frp} (277 kN), which clearly eliminated the possibility of pullout failure. It should be noted that while calculating F_{bond}, τ_b was calculated to be 6.3 N/mm^2 based on ACI-355-1R-91 [70]. Moreover, test results of the retrofitted columns also confirmed that the embedment length of FRP reinforcement was sufficient. No debonding problem or damages resulting from concrete cone, concrete splitting, edge cone and pullout failures were observed during the tests.

$$F_{\text{bond}} = \tau_{\text{b}} \cdot l_{\text{db}} \cdot u \tag{10}$$

$$F_{\text{frp}} = E_{\text{f}} \cdot \varepsilon_{\text{fu}} \cdot A_{\text{frp}} \tag{11}$$

4. RESULTS AND DISCUSSIONS

The test results were outlined through hysteretic load–displacement curves, envelopes of these curves, strain distributions along longitudinal, transverse steel and longitudinal AFRP reinforcement, energy dissipation capacities, average section rotations in different gauge lengths and damage progression and distribution. Furthermore, experimental data was compared with the theoretical findings.

4.1. Lateral Load–Displacement Curves and Failure Modes

The hysteretic lateral load–displacement curves and their envelopes are presented in Figure 11 and Figure 12, respectively. The damage progression of the columns with increasing drifts is summarized in Table 4.

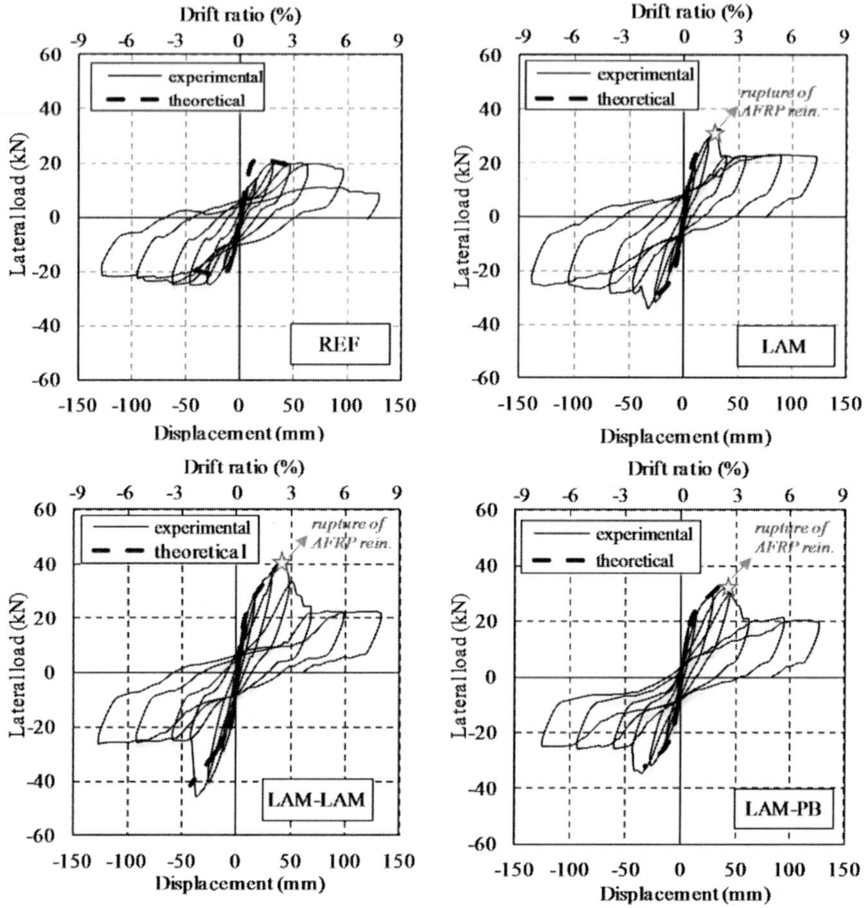

Figure 11. Experimental and theoretical lateral load–displacement curves for all specimens (rein., reinforcement).

Table 4. Damage progression of the columns (The numbers are drift ratios in %).

Damage mechanisms	Specimens			
	REF	LAM	LAM-LAM	LAM-PB
Yielding of longitudinal steel bars	2	3	3	3
Crushing of concrete cover	3	–	–	–
Spalling of concrete cover	−6	–	–	–
Fracture of AFRP reinforcement	–	3	3	3

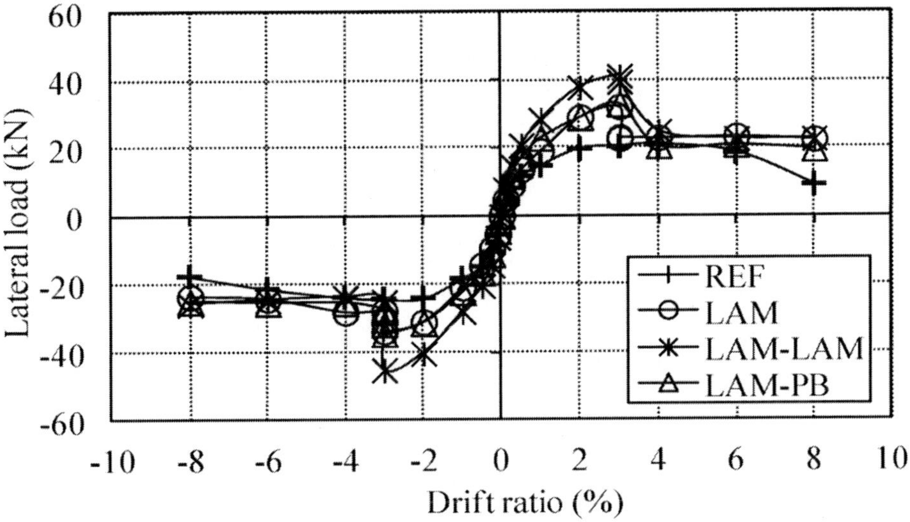

Figure 12. The envelopes of load–displacement relationships.

As seen in Figure 11 and Figure 12 and Table 3, the reference column (REF) reached its theoretical flexural capacity and exhibited a ductile behavior. The column LAM, retrofitted with AFRP reinforcement, experienced an enhancement in strength up to the drift ratio around 3%. At this drift, the enhancement in strength was around 38% with respect to the column REF. The sudden remarkable loss of strength upon exceeding the drift ratio around 3% was due to the fracture of AFRP reinforcement at the interface of the column and the footing (Figure 13a). As seen in Figure 11 and Figure 12, the column LAM-LAM exhibited a remarkably superior performance with respect to the reference and the other retrofitted columns by resisting lateral loads, approximately 1.9 and 1.3 times the loads resisted by the reference and other retrofitted columns, respectively. As seen in Figure 11, the column LAM-LAM also sustained its lateral load capacity until the drift ratio of 3%, around where the AFRP reinforcement and AFRP anchorage reinforcement fractured at the same section (Figure 13b). The higher strength of the column LAM-LAM is due to the contribution of additional AFRP anchorage reinforcement to the flexural capacity. The column LAM-PB behaved similar to the column LAM in terms of maximum lateral load and fracture of AFRP reinforcement around 3% drift ratio. At this drift ratio, the enhancement in strength was around 41% with respect to the column REF. It was observed that, while pulling, both AFRP reinforcements under tension were fractured upon exceeding the drift ratio of 3%. On the other hand, while pushing, only one AFRP reinforcement under tension fractured, whereas the other one did not fracture, instead, decomposed locally just above the isolated height (Figure 13c). Decomposition is the separation of the individual fibers after fracture of the epoxy matrix. Contribution of decomposed AFRP reinforcement to the behavior at larger drifts limited plastic residual deformations of the column LAM-PB. As seen in Figure 11, marginal residual displacements remained after unloading branches in pulling direction up to 8% drift ratio. Therefore, this type of anchorage system can be used to limit the residual displacement (and consequently residual damage). The retrofitted columns behaved similar in terms of strength degradation after the rupture of FRP reinforcement as the specimens tested by Ilki and Kumbasar [46], Bournas and Triantafillou [47], and Vrettos *et al.* [50], where the drift ratios at failure were approximately 2%, 4% and 2.5%, respectively. It is important to note that all significant damage was accumulated at the base of the columns, since the columns were wrapped with CFRP sheets in transverse direction along the full height. Consequently, the crack width reached several centimeters at the intersection of the column and the footing. This type of damage may be quite disadvantageous in case of earthquakes since the distribution of plastic deformations through the potential plastic hinge length is prevented due to presence of a rigid transverse CFRP jacket. The accumulation of a remarkable portion of plastic deformations only at the interface of the column and the footing may significantly reduce the overall drift capacity of the column. Since drift capacity is essentially important for a satisfactory seismic performance, this kind of reduction in drift capacity should be avoided. This problem can be overcome by application of transverse sheets in strips allowing distribution of damage (and plastic deformations) in the unconfined zones between transverse FRP strips.

Figure 13. Fracture of AFRP reinforcement during the test (**a**) the column LAM at 3% drift ratio, (**b**) the column LAM-LAM at 3% drift ratio, and (**c**) the column LAM-PB at 3% drift ratio.

Average experimental moment–curvature relationships obtained for critical sections of columns are presented in Figure 14. In this figure M (Second order effect is considered) is the experimental moment, M_0 is the theoretical moment capacity calculated per specimen, which is obtained by fiber analysis approach. The theoretical moment capacities (M_0) are presented in Table 3. For the calculation of experimental moment–curvature relationships, the average curvature values obtained along 0–20 mm, 20–150 mm, and 150–300 mm heights above the footing were taken into account. As seen in Figure 14, for the retrofitted columns, the average curvature values of the column sections measured along 0–20 mm height were in the order of 4.5×10^{-3} (1/mm), while the curvatures measured in 20–150 mm height above the footing were in the order of 4.5×10^{-4} (1/mm). Since the curvature values of the columns measured along 150–300 mm height were even smaller, these results are not presented in Figure 14. In agreement with these measurements, the damage was also accumulated particularly in 20 mm height of the column from top of the footing. During the autopsy after the tests, no concrete crushing and no cracking (other than the main crack at the interface of the column and the footing) were observed on the retrofitted columns. On the other hand, the deformations (as well as cracks) were distributed through the column height in case of the reference column.

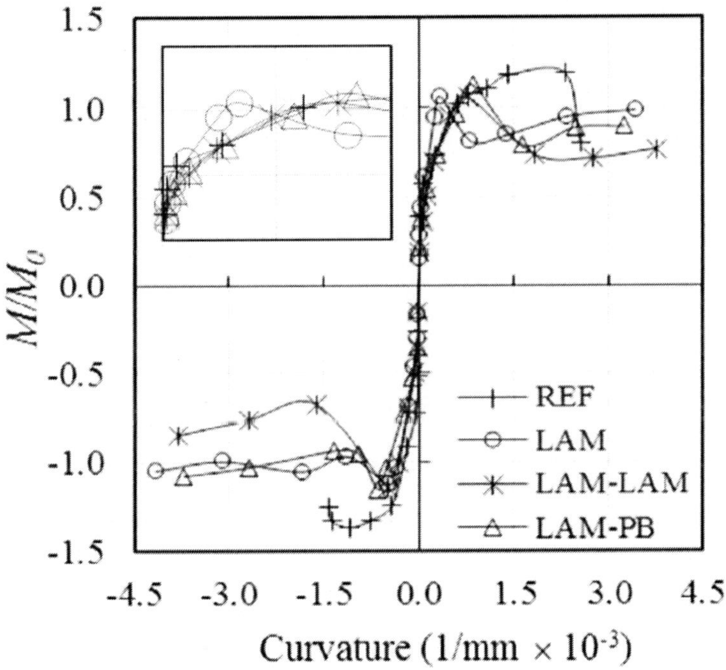

(a)

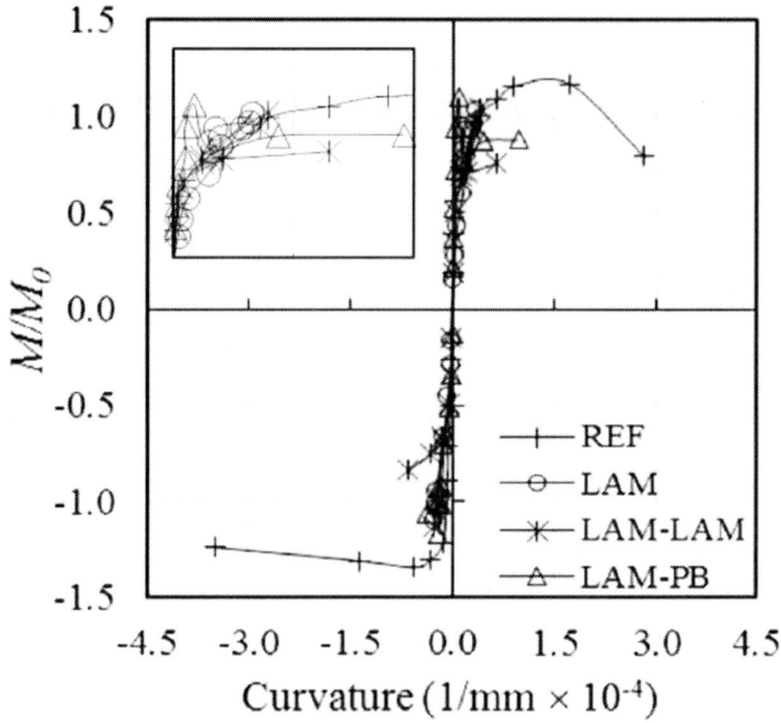

(b)

Figure 14. Moment–curvature relationships obtained in columns (**a**) 0–20 mm, and (**b**) 20–150 mm above the footing. (The 0.0–1.0 region of the figure is blown up at the left hand side of the figure).

4.2. Strains of Steel and AFRP Reinforcement

The strain values were obtained by making use of the strain gauges on longitudinal and transverse steel, and longitudinal AFRP reinforcement (Figure 15). The efficiency of AFRP reinforcement on flexural capacity can also be seen through the measured strains of longitudinal AFRP reinforcement. The maximum strains of longitudinal AFRP reinforcements in tension were measured to be in the range of 0.01 right before the rupture of the AFRP reinforcement. This value corresponded to approximately 43% of design rupture strain of FRP reinforcement. ACI 440-2R-08 [5] recommends to reduce ε_{fu} by using a factor of 0.70 for NSM application technique under monotonic flexural loading. This value is quite unconservative for the applied retrofitting technique in this study. It shall be much lower for cyclic loading based on the test results. In the study of Vrettos *et al.* [50], where the longitudinal CFRP sheets together with CFRP anchors were used for flexural retrofitting of RC columns, the measured strains were also far behind (approximately 38% of design rupture strain of FRP reinforcement) the strain reduction factor recommended by ACI 440-2R-08 [5]. It should be noted that, in the study of Vrettos *et al.*[50], the ultimate rupture strain of FRP reinforcement was not measured directly through pre-installed strain gauges but calculated through analysis of the cross-section at the column base by using conventional RC force equilibrium, strain compatibility, and material constitutive relationships based on the test results. As shown inFigure 15c,f, right after the rupture of the AFRP reinforcement, the strains decrease within the AFRP reinforcement, and corresponding stresses are transferred to the longitudinal reinforcing bars (the strains of steel bars increased remarkably along the measured height). The longitudinal steel bars of the column REF yielded at drift ratio of 2%, while that of retrofitted columns yielded at drift ratio around 3% right after the columns experienced strength degradation due to rupture of AFRP reinforcement. Since steel reinforcing bars in tension already reached their yield strain at around 3% drift ratio before rupture of AFRP reinforcement, the brittle failure of the column was avoided. It should be noted that drift ratios exceeding 3% may only be expected if substandard RC structures are subjected to severe earthquakes. Nevertheless, after AFRP reinforcements were fractured at around 3% drift ratio, the columns still resisted considerable lateral forces due to contribution of steel reinforcing bars until extremely large drift ratios (~6% to 8%) and kept their vertical load resisting capacities. According to strain gauge measurements, strains of longitudinal steel bars did not reach tensile strain corresponding to strain hardening. This observation is consistent with the nearly horizontal branch of the load–displacement relationships after fracture of AFRP reinforcement. On the other hand, none of the stirrups yielded (strains are in the range of 0.00004 and 0.0006) and no damage was observed on CFRP sheets in transverse direction due to relatively low level of applied axial load (20% axial load capacity of the column) and shear stresses.

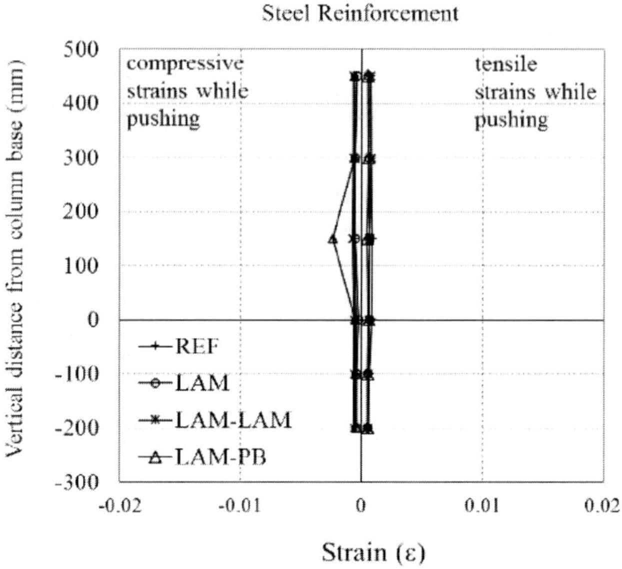
Steel Reinforcement
compressive strains while pushing
tensile strains while pushing
Vertical distance from column base (mm)
500
400
300
200
100
0
-100
-200
-300
REF
LAM
LAM-LAM
LAM-PB
-0.02
-0.01
0
0.01
0.02
Strain (ε)

(a)

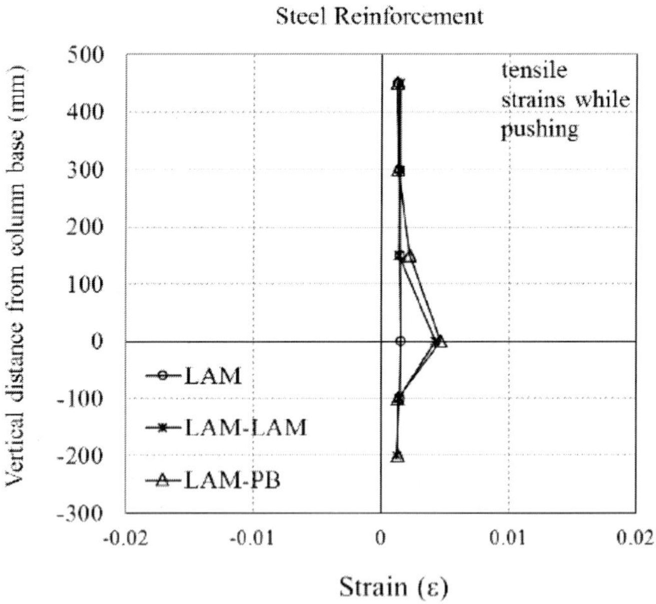
Steel Reinforcement
tensile strains while pushing
Vertical distance from column base (mm)
500
400
300
200
100
0
-100
-200
-300
LAM
LAM-LAM
LAM-PB
-0.02
-0.01
0
0.01
0.02
Strain (ε)

(b)

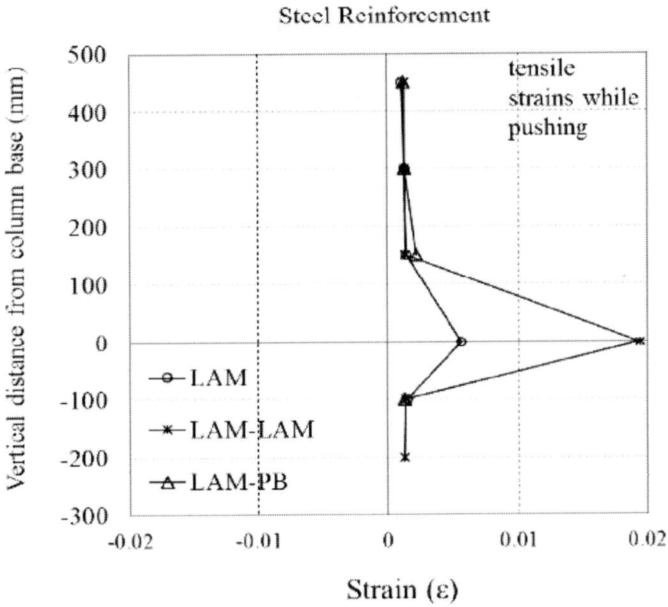
Steel Reinforcement
tensile strains while pushing
LAM
LAM-LAM
LAM-PB
Vertical distance from column base (mm)
500
400
300
200
100
0
-100
-200
-300
-0.02
-0.01
0
0.01
0.02
Strain (ε)

(c)

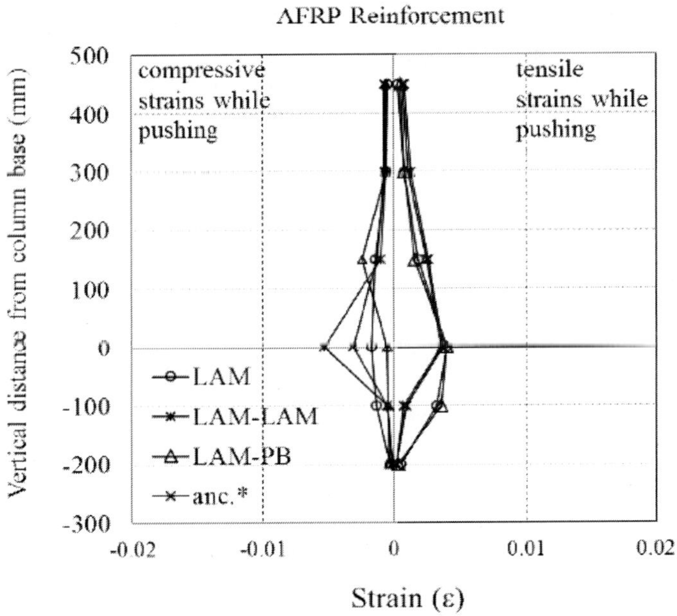
AFRP Reinforcement
compressive strains while pushing
tensile strains while pushing
LAM
LAM-LAM
LAM-PB
anc.*
Vertical distance from column base (mm)
500
400
300
200
100
0
-100
-200
-300
-0.02
-0.01
0
0.01
0.02
Strain (ε)

(d)

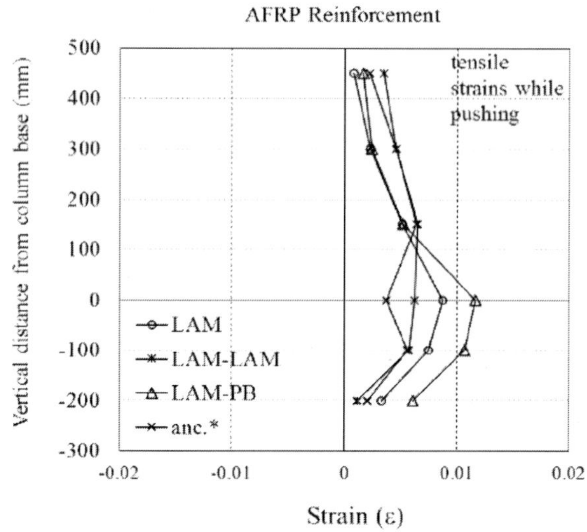

(e)

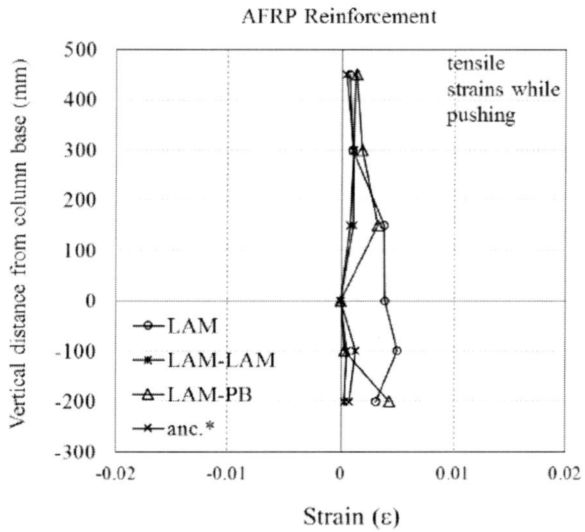

(f)

Figure 15. Strain distribution of longitudinal steel reinforcement (**a**) at ±0.01 drift ratio, (**b**) right before rupture of AFRP reinforcement (at around 3% drift ratio), (**c**) right after rupture of AFRP reinforcement (at around 3% drift ratio), and AFRP reinforcement (**d**) at ±0.01 drift ratio, (**e**) right before rupture of AFRP reinforcement (at around 3% drift ratio), (**f**) right after rupture of AFRP reinforcement (at around 3% drift ratio). * strain measurements obtained from additional AFRP anchorage reinforcements for the column LAM-LAM

4.3. Energy Dissipation Capacities

Energy dissipation capacities of the tested columns are calculated as the areas enclosed by the load–displacement hysteresis loops and presented in Figure 16. As seen in this figure, until the rupture of the AFRP reinforcement, the column LAM-LAM shows the highest energy dissipation capacity due to its higher lateral load capacity with respect to other columns. Moreover, the energy dissipation capacity of the column LAM is higher than the column LAM-PB. This is due to larger elastic deformations along the embedded unbonded length of AFRP reinforcement in case of column LAM-PB. After the rupture of the AFRP reinforcement, energy dissipation capacities of the columns LAM and LAM-LAM are approximately same, since the columns behave like the reference column in terms of sustained lateral load. As seen in Figure 16, the dissipated energy is significantly less for column LAM-PB due to elastic behavior of unbonded AFRP reinforcement. Nevertheless, with the gradual damaging of AFRP reinforcement, the contribution of AFRP reinforcement to the elastic behavior is diminished and plastic deformations increased causing more significant residual displacements with respect to previous loading cycles.

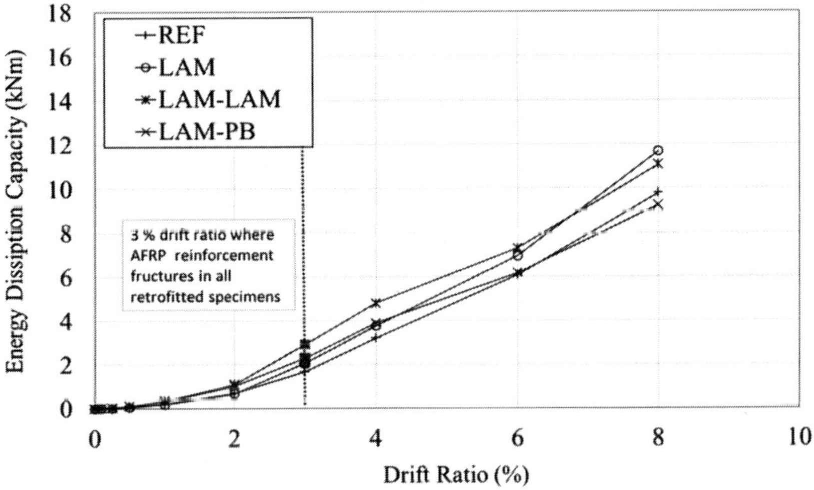

Figure 16. Energy dissipation capacities of the columns

5. CONCLUSIONS

In this study, the effects of using anchored and embedded AFRP laminate longitudinal reinforcement for the flexural retrofit of substandard RC columns were investigated. The tests were conducted under constant axial load and reversed cyclic lateral loads. Based on the experimental and theoretical studies, the following conclusions/observations can be listed:

- The reference column reached its theoretical flexural capacity, and exhibited a ductile behavior as foreseen during design.
- The retrofitted columns (LAM, LAM-LAM, LAM-PB) failed due to the rupture of AFRP reinforcement at around 3% drift ratio. At this drift ratio, the enhancement in strength was around 38%, 90% and 41% for the columns LAM, LAM-LAM, and LAM-PB, respectively, with respect to the reference column. It should be noted that drift ratios exceeding 3% may only be expected if substandard RC structures are subjected to severe earthquakes. Nevertheless, after AFRP reinforcements were fractured at around 3% drift ratio, the columns still resisted considerable lateral forces due to contribution of steel reinforcing bars until extremely large drift ratios (~6% to 8%). The column LAM-LAM exhibited a remarkably superior performance with respect to the reference and the other retrofitted columns due to additional AFRP anchorage reinforcement. When compared with the column LAM, the proposed anchorage type used for the column LAM-PB, limited the residual plastic deformations, while did not affect the lateral load capacity adversely. Therefore, this type of anchorage system can be utilized to limit the residual displacement and corresponding damage. It was observed that the predicted results through theoretical calculations satisfactorily captured the tested results by means of lateral load and deformation capacities, particularly, when 43% of the design rupture strain of FRP reinforcement was considered as the rupture strain of the FRP reinforcement during theoretical calculations.
- The AFRP reinforcement of the retrofitted columns ruptured at the interface of the column and the footing. The experimentally measured rupture strain of AFRP reinforcement was observed to be approximately 43% of design rupture strain of FRP reinforcement. Therefore, the strain reduction factor (70%) recommended by ACI 440-2R-08 [5] for NSM applications under monotonic loading conditions may lead to unconservative predictions of lateral load-deformation capacity for the proposed retrofitting technique in this study under cyclic loading conditions. Clearly, further investigations are also needed to accurately identify the theoretical seismic behavior of the columns retrofitted using NSM techniques under cyclic loading conditions.

It can be inferred from the results of this study that the proposed retrofitting method is promising for improving the seismic performance of substandard columns built with low quality concrete under reversed cyclic flexure and axial force. It should be noted that these conclusions are mainly based on the test results of this study, therefore, they are limited to the ranges of test parameters and details considered in this study.

ACKNOWLEDGMENTS

The financial support of the ITU Scientific Research Department (Scientific Research Project No:326024), BASF and ART-YOL Companies and the assistance of staff of Structural and Earthquake Engineering Laboratory and Building Materials Laboratory of Istanbul Technical University are acknowledged gratefully.

AUTHOR CONTRIBUTIONS

Alper Ilki coordinated the study. Ahmet Uzunhasanoglu worked on production of the specimens. All authors contributed to the execution of the tests. Engin C. Seyhan and Caglar Goksu contributed to the processing of the test data. The paper was written by Engin C. Seyhan, Caglar Goksu and Alper Ilki.

Notations

AFRP	Aramid fiber reinforced polymer
A_{frp}	Cross-section area of the FRP anchor reinforcement
A_{fv}	Area of FRP shear reinforcement
A_g	Gross area of the column section
A_u	Cross-sectional area of the transverse reinforcing bar
b_w	Web width of the column
C_E	Environmental reduction factor according to ACI 440-2R-08 2008
CFRP	Carbon fiber reinforced polymer
d	Distance from extreme compression fiber through centroid of tension steel reinforcement

d_{fv}	Effective depth of FRP shear reinforcement
E_f	Tensile elastic modulus of FRP reinforcement
f'_c	Design compressive strength of concrete
f_{max}	Tensile strength of longitudinal and transverse reinforcing bar
f_y	Yield stress of longitudinal reinforcing bar
f_{yt}	Design yield strength of transverse reinforcement
F_{bond}	Bond capacity of epoxy grout on AFRP reinforcement
F_{frp}	Tensile strength of FRP
FRP	Fiber reinforced polymer
GFRP	Glass fiber reinforced polymer
H	Distance between the interface of the column and the footing and the point of application of the lateral load
h_i	Distance between the center of the respective part and the tip of the column, where the lateral load is applied
l_{db}	Embedment length
l_i	Length of the respective part of the column
l_p	plastic hinge length
M	Experimental moment capacity
M_0	Theoretical moment capacity calculated by fiber analysis approach
n	Number of plies of FRP sheet

NSM	Near surface mounted
N_u	Factored axial load normal to cross-section (to be taken as positive for compression)
RC	Reinforced Concrete
s	Spacing of shear reinforcement measured in a direction parallel to the longitudinal reinforcement
s_f	Spacing of FRP reinforcement plies
t_f	Nominal thickness of one ply of FRP reinforcement
u	Perimeter of the FRP anchor
V_c	Nominal shear strength provided by concrete
V_s	Nominal shear strength provided by shear reinforcement
V_n	Nominal shear strength
$V_{r\text{-}frp}$	Nominal shear strength (with the contribution of FRP)
V_f	FRP contribution to shear strength
w_f	Effective width of FRP reinforcement
δ_t	Top displacement of the column
ϑ_p	Plastic rotation
ε_{c0}	Axial strain corresponding to unconfined concrete strength
ε_{cc}	Axial strain corresponding to confined concrete strength
ε_{fe}	Effective strain of FRP sheet attained at failure
ε_{fu}	Design rupture strain of FRP reinforcement

ε_{fu}^{*}	Ultimate rupture strain of FRP reinforcement declared by manufacturer
ρ_s	Existing volumetric ratio of transverse reinforcement of columns
ρ_{sm}	Required volumetric ratio of transverse reinforcement of columns
τ_b	Uniform bond strength along the anchorage length
χ_i	Curvature of the respective part of the column
χ_p	Curvature of the plastic part of the column
ψ_f	FRP strength reduction factor

REFERENCES

1. CEB-FIB. *Externally Bonded FRP Reinforcement for RC Structures*; Technical report for International Federation for Structural Concrete: Lausanne, Switzerland, 2001.
2. Pessiki, S.; Harries, K.A.; Kestner, J.T.; Sause, R.; Ricles, J.M. Axial behavior of reinforced concrete columns confined with FRP jackets. *J. Compos. Constr.* **2001**, *5*, 237–245.
3. Bakis, C.E.; Bank, L.C.; Brown, V.L.; Cosenza, E.; Davalos, J.F.; Lesko, J.J.; Machida, A.; Rizkalla, S.H.; Triantafillou, T.C. Fiber-reinforced polymer composites for construction-State-of-the-Art Review. *J. Compos. Constr.* **2002**, *6*, 73–87.
4. Lam, L.; Teng, J.G. Design-oriented stress-strain model for FRP-confined concrete. *Constr. Build. Mater.***2003**, *17*, 471–489.
5. ACI 440.2R-08. Guide for design and construction of externally bonded FRP systems for strengthening concrete structures. American Concrete Institute 440: Farmington Hills, MI, USA, 2008.
6. Bank, L.C. Progressive failure and ductility of FRP composites for construction: Review. *J. Compos. Constr.***2013**, *17*, 406–419.

7. Bennitz, A.; Schmidth, J.W.; Nilimaa, J.; Taljsten, B.; Goltermann, P.; Ravn, D.L. Reinforced concrete T-beams externally prestressed with unbonded carbon fiber-reinforced polymer tendons. *Struct. J.* **2012**, *109*, 521–530.

8. Lou, T.; Lopes, S.M.R.; Lopes, A. Numerical analysis of behaviour of concrete beams with external FRP tendons. *Constr. Build. Mater.* **2012**, *35*, 970–978.

9. Lou, T.; Lopes, S.M.R.; Lopes, A. External CFRP tendon members: Secondary reactions and moment redistribution. *Compos. Eng.* **2014**, *57*, 250–261.

10. Seible, F.; Priestley, M.J.; Hegemier, G.A.; Innamorato, D. Seismic retrofit of RC columns with continuous carbon fiber jackets. *J. Compos. Constr.* **1997**, *1*, 52–62.

11. Sheikh, S.; Yau, G. Seismic behavior of concrete columns confined with steel and fiber-reinforced polymers.*Struct. J.* **2002**, *99*, 72–80.

12. Antonopoulos, C.P.; Triantafillou, T.C. Experimental investigation of FRP-strengthened RC beam-column joints. *J. Compos. Constr.* **2003**, *7*, 39–49.

13. Iacobucci, R.D.; Sheikh, A.S.; Bayrak, O. Retrofit of square concrete columns with carbon fiber-reinforced polymer for seismic resistance. *Struct. J.* **2003**, *100*, 785–794.

14. Xiao, Y. Applications of FRP composites in concrete columns. *Adv. Struct. Eng.* **2004**, *7*, 335–343.

15. Prota, A.; Nanni, A.; Manfredi, G.; Cosenza, E. Selective upgrade of underdesigned reinforced concrete beam-column joints using carbon fiber polymers. *Struct. J.* **2004**, *101*, 699–707.

16. Bousias, S.N.; Triantafillou, T.C.; Fardis, M.N.; Spathis, L.; O'Regan, B.A. Fiber reinforced polymer retrofitting of rectangular reinforced concrete columns with or without corrosion. *Struct. J.* **2004**, *101*, 512–520.

17. Chang, S.; Li, Y.; Loh, C. Experimental study of seismic behaviors of as-built and carbon fiber reinforced plastics repaired reinforced concrete bridge columns. *J. Bridge Eng.* **2004**, *9*, 391–402.

18. Memon, M.S.; Sheikh, S.A. Seismic resistance of square concrete columns retrofitted with glass fiber-reinforced polymer. *Struct. J.* **2005**, *102*, 774–783.

19. Tsonos, A.G. Effectiveness of CFRP-jackets in post-earthquake and pre-earthquake retrofitting of beam-column subassemblages. *Struct. Eng. Mech.* **2007**, *27*, 393–408.

20. Ilki, A.; Demir, C.; Bedirhanoglu, I.; Kumbasar, N. Seismic retrofit of brittle and low-strength RC columns using fiber reinforced polymer and cementitious composites. *Adv. Struct. Eng.* **2009**, *12*, 325–347.

21. Garcia, R.; Hajirasouliha, I.; Pilakoutas, K. Seismic behaviour of deficient RC frames strengthened with CFRP composites. *Eng. Struct.* **2010**, *32*, 3075–3085.

22. Ilki, A.; Bedirhanoglu, I.; Kumbasar, N. Behavior of FRP retrofitted joints built with plain bars and low-strength concrete. *J. Compos. Constr.* **2011**, *15*, 312–326.

23. Wang, Z.; Wang, D.; Smith, S.T.; Lu, D. CFRP-confined square RC columns. I: Experimental investigation. *J. Compos. Constr.* **2011**, *16*, 150–160.

24. Jirawattanasomkul, T.; Zhang, D.W.; Ueda, T. Prediction of the post-peak behavior of reinforced concrete columns with and without FRP-jacketing. *Eng. Struct.* **2013**, *56*, 1511–1526.

25. Parvin, A.; Brighton, D. FRP composites strengthening of concrete columns under various loading conditions. *Polymers* **2014**, *6*, 1040–1056.

26. Demir, C.; Darılmaz, K.; Ilki, A. Cyclic stress–strain relationships of FRP confined concrete members.*Arabian J. Sci. Eng.* **2015**, *40*, 363–379.

27. Nanni, A.; Alkhrdaji, T.; Chen, G.; Barker, M.; Yang, X.; Mayo, R. Overview of testing to failure program of a highway bridge strengthened with FRP composites. In Proceedings of Fourth International Symposium on FRP for Reinforcement of Concrete Structures, Baltimore, MD, USA, November 1999; pp. 69–80.

28. De Lorenzis, L.; Nanni, A. Bond between near surface mounted fiber reinforced polymer rods and concrete in structural strengthening. *Struct. J.* **2002**, *99*, 123–132.

29. De Lorenzis, L.; Lundgren, K.; Rizzo, A. Anchorage length of near-surface mounted fiber reinforced polymer bars for concrete strengthening-experimental investigation and numerical modelling. *Struct. J.***2004**, *101*, 269–278.

30. Hassan, T.K.; Rizkalla, S.H. Bond mechanism of near-surface-mounted fiber-reinforced polymer bars for flexural strengthening of concrete structures. *Struct. J.* **2004**, *101*, 830–839.

31. El-Hacha, R.; Rizkalla, S.H. Near-surface-mounted fiber-reinforced polymer reinforcements for flexural strengthening of concrete structures. *Struct. J.* **2004**, *101*, 717–726.

32. Barros, J.A.O.; Dias, S.J.E. Near surface mounted CFRP laminates for shear strengthening of concrete beams. *Cem. Concr. Compos.* **2006**, *28*, 276–292.

33. Nordin, H.; Täljsten, B. Concrete beams strengthened with prestressed near surface mounted CFRP. *J. Compos. Constr.* **2006**, *10*, 60–68.

34. Liu, I.S.T.; Oehlers, D.J.; Seracino, R. Tests on the ductility of reinforced concrete beams retrofitted with FRP and steel near-surface mounted plates. *J. Compos. Constr.* **2006**, *10*, 106–114.

35. Teng, J.G.; de Lorenzis, L.; Wang, B.; Li, R.; Wong, T.N.; Lam, L. Debonding failures of RC beams strengthened with near surface mounted CFRP strips. *J. Compos. Constr.* **2006**, *10*, 92–105.

36. De Lorenzis, L.; Teng, J.G. Near-surface mounted FRP reinforcement: An emerging technique for strengthening structures. *Compos. Eng.* **2007**, *38*, 119–143.

37. Seracino, R.; Raizal Saifulnaz, M.R.; Oehlers, D.J. Generic debonding resistance of EB and NSM plate-to-concrete joints. *J. Compos. Constr.* **2007**, *11*, 62–70.

38. Castro, E.K.; Melo, G.S.; Nagato, Y. Flexural strengthening of RC "t" beams with near surface mounted (NSM) FRP reinforcements. In Proceedings of International Symposium on FRP Reinforcement of Concrete Structures (FRPRCS-8), Patras, Greece, 16–18 July 2007.

39. Chikh, N.; Foret, G.; Bousalem, B.; Firas, S. Analysis of reinforced concrete members strengthened with near-surface-mounted CFRP. In Proceedings of 11DBMC Conference on Durability of Building Materials and Components, Istanbul, Turkey, 11–14 May 2008.

40. El-Maaddawy, T.A.; El-Dieb, A.S. Near-surface-mounted composite system for repair and strengthening of reinforced concrete columns subjected to axial load and biaxial bending. *J. Compos. Constr.* **2011**, *15*, 602–614.

41. Sharaky, I.A.; Torres, L.; Comas, J.; Barris, C. Flexural response of reinforced concrete (RC) beams strengthened with near surface mounted (NSM) fibre reinforced polymer (FRP) bars. *Compos. Struct.* **2013**,*109*, 8–22.

42. Bilotta, A.; Ceroni, F.; Barros, J.; Costa, I.; Palmieri, A.; Szabó, Z.; Nigro, E.; Matthys, S.; Balazs, G.; Pecce, M. Bond of NSM FRP-

strengthened concrete: Round robin test initiative. *J. Compos. Constr.* **2015**.

43. Sena Cruz, J.M.; Barros, J.A.O.; Gettu, R.; Azevedo, A.F.M. Bond behavior of near surface mounted CFRP laminate strips under monotonic and cyclic loading. *J. Compos. Constr.* **2006**, *10*, 295–303.

44. Badawi, M.; Soudki, K. Fatigue behavior of RC beams strengthened with NSM CFRP rods. *J. Compos. Constr.* **2009**, *13*, 415–421.

45. Ceroni, F. Experimental performances of RC beams strengthened with FRP materials. *Constr. Build. Mater.***2010**, *24*, 1547–1559.

46. Ilki, A.; Kumbasar, N. The behavior of damaged reinforced concrete members strengthened by carbon fiber reinforced polymer composites. *Tech. J. Turk. Chamb. Civil Eng.* **2002**, *13*, 2597–2616, (In Turkish with English abstract).

47. Bournas, D.A.; Triantafillou, T.C. Flexural strengthening of reinforced concrete columns with near surface mounted FRP or stainless steel. *Struct. J.* **2009**, *106*, 495–505.

48. Goksu, C.; Polat, A.; Ilki, A. Attempt for seismic retrofit of existing sub-standard RC members under reversed cyclic flexural effects. *J. Compos. Constr.* **2012**, *16*, 286–299.

49. Fahmy, M.F.M.; Wu, Z. Retrofitting of existing RC square bridge columns using basalt FRP rebars. Available online: http://www.iifchq.org/proceedings/CICE_2012/02_FRP%20in%20Seismic%20Retrofitting/02_495_Fahmy,Wu_RETROFITTING%20OF%20EXISTING%20RC%20SQUARE.pdf (accessed on 27 November 2015).

50. Vrettos, I.; Kefela, E.; Triantafillou, T.C. Innovative flexural strengthening of reinforced concrete columns using carbon-fiber anchors. *Struct. J.* **2013**, *110*, 63–70.

51. Li, X.; Lv, H.L.; Zhang, G.C.; Sha, S.Y.; Zhou, S.C. Seismic retrofitting of rectangular reinforced concrete columns using fiber composites for enhanced flexural strength. *J. Reinf. Plast. Compos.* **2013**, *32*, 619–630.

52. Goksu, C.; Uzunhasanoglu, A.; Seyhan, E.C.; Ilki, A. Seismic retrofit of sub-standard RC columns with embedded aramid FRP reinforcement. In Proceedings of SMAR 2013 Second Conference on Smart Monitoring, Assessment and Rehabilitation of Civil Structures, Istanbul, Turkey, 9–11 September 2013.

53. Faustino, P.; Chastre, C. Flexural strengthening of columns with CFRP composites and stainless steel: Cyclic behavior. *J. Struct. Eng.* **2015**.

54. CNR-DT 200/2004. Guide for the design and construction of externally bonded FRP systems for strengthening existing structures. Available online: http://www.cnr.it/documenti/norme/IstruzioniCNR_DT200_2004_eng.pdf (accessed on 27 November 2015).

55. Akkaya, Y.; Demir, N. Durability issues in existing buildings and new infrastructure projects. In Proceedings of Joint Symposium on Concrete Engineering (JSCE), Istanbul, Turkey, 9 June 2010.

56. Eurocode 8-Part 3. Assessment and retrofitting of buildings. European Committee for Standardization: Brussels, Belgium, 2005.

57. TSDC. Regulations for buildings to be constructed in earthquake prone areas. Turkish Seismic Design Code: Ankara, Turkey, 2007.

58. Eurocode 2: Design of concrete structures. Part 1. General rules and rules for buildings. European Committee for Standardization: Brussels, Belgium, 2004.

59. TS EN 12390–3. Testing hardened concrete-Part 3: Compressive strength of test specimens. Turkish Standards Institute: Ankara, Turkey, 2003.

60. TS 708. Steel bars for concrete. Turkish Standards Institute: Ankara, Turkey, 1996.

61. Part 3: Structural and non-structural repair. In *Products and Systems for the Protection and Repair of Concrete Structures. Definitions, Requirements, Quality Control and Evaluation of Conformity*; BS EN 1504–3:2005; British Standard Institute: London, UK, 2006.

62. Part 4: Structural bonding. In *Products and Systems for the Protection and Repair of Concrete Structures. Definitions, Requirements, Quality Control and Evaluation of Conformity*; EN 1504–4:2004; British Standard Institute: London, UK, 2004.

63. TS-EN 196–1. Methods of testing cement-Part 1: Determination of strength. Turkish Standards Institute: Ankara, Turkey, 2009.

64. ASTM D3039. Standard test method tensile properties of polymer matrix composite materials. ASTM International: West Conshohocken, PA, USA, 2014.

65. Matamoros, A.B.; Sozen, M.A. Drift limits of high-strength concrete columns subjected to load reversals. *J. Struct. Eng.* **2003**, *129*, 297–313.

66. Bournas, D.A.; Triantafillou, T.C. Bond strength of lap spliced bars in concrete confined with composite jackets. *J. Compos. Constr.* **2011**, *15*, 156–167.

67. XTRACT 3.0.8. Cross-sectional structural analysis of components. TRC: Rancho Cordova, CA, USA, 2007.

68. Ilki, A.; Peker, O.; Karamuk, E.; Demir, C.; Kumbasar, N. FRP retrofit of low and medium strength circular and rectangular reinforced concrete columns. *J. Mater. Civil Eng.* **2008**, *20*, 169–188.

69. ACI-318M-14. Metric building code requirements for structural concrete and commentary. American Concrete Institute 318: Farmington Hills, MI, USA, 2014.

70. ACI-355–1R-91. State-of-the-art report on anchorage to concrete. American Concrete Institute 355: Farmington Hills, MI, USA, 1991.

Chapter 4

Concrete-Filled-Large Deformable FRP Tubular Columns under Axial Compressive Loading

Omar I. Abdelkarim and Mohamed A. ElGawady *

Department of Civil, Architectural, and Environmental Engineering, Missouri University of Science and Technology, Rolla, MO 65401, USA

ABSTRACT

The behavior of concrete-filled fiber tubes (CFFT) polymers under axial compressive loading was investigated. Unlike the traditional fiber reinforced polymers (FRP) such as carbon, glass, aramid, *etc.*, the FRP tubes in this study were designed using large rupture strains FRP which are made of recycled materials such as plastic bottles; hence, large rupture strain (LRS) FRP composites are environmentally friendly and can be used in the context of green construction. This study performed finite element (FE) analysis using LS-DYNA software to conduct an extensive parametric study on CFFT. The effects of the FRP confinement ratio, the unconfined concrete compressive strength (f'c), column size, and column aspect ratio on the behavior of the CFFT under axial compressive loading were investigated during this study. A comparison between the behavior of the CFFTs with LRS-FRP and those with traditional FRP (carbon and glass) with a high range of confinement ratios was conducted as well. A new hybrid FRP system combined with traditional and LRS-FRP is proposed. Generally, the CFFTs with LRS-FRP showed remarkable behavior under axial loading in strength and ultimate strain. Equations to estimate the concrete dilation parameter and dilation angle of the CFFTs with LRS-FRP tubes and hybrid FRP tubes are suggested.

Keywords: concrete-filled tube; large deformable FRP; large rupture strain; hybrid FRP; LS-DYNA

1. INTRODUCTION

Green buildings are environmentally sound buildings. The ideal green project preserves and restores the habitat that is vital for sustaining life by acting as a net producer and exporter of resources, materials, energy, and water rather than being a net consumer. The Environmental Protection Agency (EPA) suggests using recycled industrial goods such as demolition debris in construction projects for green buildings. Energy efficient building materials and appliances are promoted in the United States through energy rebate programs. However, using green materials in construction is usually costly. Recently, new fiber reinforced polymer (FRP) composites have been manufactured from recycled plastic bottles. They were introduced as alternatives to traditional FRPs such as glass, aramid, and carbon FRP. The new FRP composites are much cheaper than the traditional FRPs. These new FRP composites are made of polyethylene terephthalate (PET) and polyethylene naphthalate (PEN) fibers. The traditional FRP composites have linear elastic stress-strain relationships with a rupture failure strain ranged around 1.0% to 2.5%. However, the new FRP composites have bilinear stress-strain relationships with elastic modulus and tangent modulus. This bilinear stress-strain relation is because of the effect of amorphous phase motion and macromolecular chains sliding between LRS fibers and matrix [1]. However, the elastic modulus of the new FRP composites is, in general, lower than that of the traditional FRPs. They have much larger rupture strains, usually larger than 6.0%. Therefore, the new FRP composites were called large rupture strain FRPs (LRS-FRPs). PET polymers keep their mechanical strengths up to a temperature of 150–175 °C [2].

Use of the FRP in new structures has grown rapidly in the past two decades. The main purpose of using FRP is to enhance the strength and ductility of a structural member. Concrete-filled FRP tubes (CFFTs) have many benefits such as light weight-to-strength ratio, high confinement, and corrosion resistance. The FRP tube acts as a stay-in-place formwork, confines the concrete structural element, and increase its compressive strength. Several researchers investigated the behavior of CFFT columns using the traditional FRP tubes under different loadings [3,4,5,6,7,8,9,10]. Recently, some experimental works have been conducted to investigate the performance of the LRS-FRPs in jacketing concrete members to examine their behavior under different loading such as axial, flexural, shear loadings [1,11,12,13,14,15]. This research has shown that LRS-FRP jacketed concrete members had superior behavior compared to members retrofitted using conventional FRP. However, no studies were conducted to determine the benefits of combining both traditional and LRS-FRPs in a hybrid system.

The FRP confinement pressure (f_l) and the concrete dilation angle (ψ) are essential parameters in characterizing the performance of concrete under compression stress in the CFFTs. Confinement pressure is the lateral pressure from the FRP tube that confines the concrete core when the concrete material starts to expand. The confinement pressure and the confinement ratio can be calculated using Equations (1) and (2). The dilation angle is defined as the inclination of the failure surface towards the hydrostatic axis. Physically, the dilation angle is interpreted as a concrete internal friction. The dilation angle varies depending on the axial stress level and the FRP jacketing stiffness [16,17]. However, previous studies used the dilation angle to vary with the FRP jacketing stiffness and to be a constant value under varied axial load levels in the finite element analysis [18,19,20]. The finite element results of these studies agreed with the experimental results with reasonable accuracy. For unconfined concrete material, the dilation angle is usually taken between 36° to 40° with an average value of 38° [21,22,23].

$$(f_l) = \frac{2\, E_f\, \varepsilon_f\, t_f}{D} \tag{1}$$

$$Confinement\ ratio\ (CR) = \frac{f_l}{f_c'} \tag{2}$$

where Ef is the elastic modulus of the FRP tube in the confinement direction, εf is the ultimate tensile strain of the FRP in the confinement direction, tf is the FRP tube thickness, *D* is the column's diameter, and f'c the characterized concrete cylindrical strength at 28 days.

An extensive finite element (FE) study is presented to investigate the behavior of the CFFTs using LRS-FRP under axial compressive loading. LS-DYNA software [24] was used during this study. A high range of confinement ratios was investigated for the traditional FRP and LRS-FRP. New state-of-the-art CFFT columns using hybrid FRP tubes combined with traditional FRP and LRS-FRP are introduced. In addition, the effect of the concrete strength (f'c), column size, and column aspect ratio on the behavior of the CFFT were studied. This study introduces recommendations for using of the most effective FRP type in CFFT tubes. A new equation to estimate the dilation angle for the CFFT column with LRS-FRP tubes is suggested.

2. FINITE ELEMENT MODEL VALIDATION

FE modeling was used to analyze the behavior of CFFT with LRS-FRP under axial loading. The LS-DYNA 971 R3 software was used to design and validate the models against the experimental results that were gathered from 12 CFFT columns with LRS-FRP by Dai *et al.* [13]. Each column had a circular cross-section with an outer diameter of 150 mm and a height of 300 mm. These columns had a concrete compressive strength (f'c) that was between 32.5 and 39.2 MPa. Either PET-FRP or PEN-FRP was used to manufacture the FRP tubes (Table 1). These models were next used to conduct a parametric study investigating the differences between LRS-FRP, tradition FRP, and hybrid system of a combination of both by analyzing the effects of the confinement ratio, column's size, and the column's aspect ratio on the CFFT behavior under axial loading.

Table 1. Summary of cylinders variables (Modified from Dai *et al.* [13], copyright 2011 ASCE).

Cylinder Label	Diameter (mm) X Height (mm)	f'_c (MPa)	FRP Type	No. of Layers	Total Thickness (mm)
PEN-600-I	150 × 300	39.2	PEN-600	One	0.85
PEN-600-II				Two	1.70
PEN-600-III				Three	2.54
PET-600-I		32.5	PET-600	One	0.84
PET-600-II				Two	1.68
PET-600-III				Three	2.52
PET-900-I		39.2	PET-900	One	1.26
PET-900-II				Two	2.52
PET-900-III				Three	3.79

2.1. Geometry

The concrete cylinder and steel plates were modeled using solid elements (Figure 1). The outer FRP tube was simulated using shell elements. All solid elements were modeled with constant-stress and a one-point quadrature to reduce the computational time. Hourglass control was used to avoid spurious singular modes (*i.e.*, hourglass modes) for solid elements. The hourglass value for all models was taken as the default value of 0.10. Surface-to-surface contact elements were used to simulate the interface between the concrete cylinder and the outer FRP tube. Node-to-surface contact elements were used between the rigid plates and the cylinder. The coefficient of friction for all of the contact elements was taken as 0.6 [25].

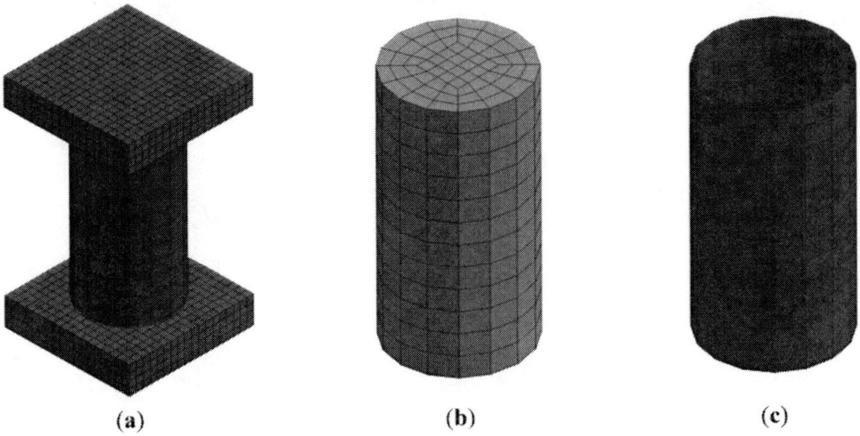

Figure 1. Finite element model components: (**a**) 3-D view; (**b**) concrete cylinder; and (**c**) FRP tube.

2.2. Concrete Material Model

Different material models are available in LS-DYNA to simulate concrete materials. Because the Karagozian and Case Concrete Damage Model Release 3 (K & C model) exhibited good agreement with the experimental results collected in previous studies, it was chosen for this study (e.g., [25]). The model was developed based on the theory of plasticity. The model has three shear failure surfaces: yield, maximum, and residual [26]. This material model has eighty parameters that can be either user defined or automatically generated. This study used the automatic generation option for the failure surface where f′c was the main input to the model. Another input to the model, the fractional dilation parameter (ω), considers any volumetric change in concrete. The fractional dilation parameter is related to the dilation angle by Equation (3). Youssf *et al.* [20] suggested an Equation (4) to calculate the dilation parameter to the CFFT with traditional FRP. Youssef *et al.*'s equation was modified to propose a new Equation (5) to calculate the dilation parameter of the CFFT with LRS-FRP based on the validation of the experimental results. In the case of a conventional concrete column without FRP confinement, the equation yields a dilation parameter of a constant value of 0.8, which is approximately equal to Tan 38°. This result agreed with the common value of the dilation angle of the concrete material without FRP confinement. The dilation parameter for the hybrid system of a combination of the LRS-FRP and the traditional FRP was equal to the summation of the two dilation parameters (Equation (6)).

$$\text{Dilation parameter } (\omega) = \tan\psi \tag{3}$$

$$\text{Dilation parameter in case of traditional FRP } (\omega_1) = -0.195\ln\frac{E_{f1}}{f_c'} + 0.6115 \tag{4}$$

$$\text{Dilation parameter in case of LRS-FRP } (\omega_2) = \frac{0.8-0.015\ E_1/f_c'}{1+0.075\ E_1/f_c'} \tag{5}$$

Dilation parameter in case of combination of traditional and

$$\text{LRS-FRP } (\omega_3) = (-0.195\ln\frac{E_{f1}}{f_c'} + 0.6115) + \frac{0.8-0.015\ E_1/f_c'}{1+0.075\ E_1/f_c'} \tag{6}$$

$$E_1 = \frac{2\ E_{f2}\ t_f}{D} \tag{7}$$

where E1 is the confinement modulus ratio, Ef1 is the elastic modulus of the traditional FRP, Ef2 is the tangent modulus of the LRS-FRP, tf is the thickness of the FRP, *D* is the column's diameter, and f'c is the characteristic cylindrical concrete strength at 28 days.

2.3. FRP Material Model

The material properties of PET-FRP and PEN-FRP composites have been studied by Dai *et al.* [13]. Such types of FRP have approximate bilinear stress-strain relationships that can be described in terms of two moduli of elasticity: the initial elastic modulus (E_{f1}) and the tangent modulus (E_{f2}). The material properties of PET-FRP and PEN-FRP are summarized in Table 2. The material properties of the glass and carbon FRP referenced in the manufacturer data sheet of Tyfo® SEH-51 and Tyfo® SCH-41 are summarized as well in Table 2. FRP composites were modeled as orthotropic materials using "108-ortho_elastic_plastic" material for LRS-FRP to simulate the bilinear behavior. Material model "002-orthotropic-elastic" was used for the traditional FRP to simulate the linear behavior. The "108-ortho_elastic_plastic" material model combines orthotropic, elastic, and plastic behaviors for shells only. This material is defined by the engineering constants: elastic modulus (E_{f1}), tangent modulus (E_{f2}), shear modulus (*G*), and Poisson's ratio (*PR*) in the two principle axes (*a* and *b*). Additionally, the fiber orientation is defined by a vector. However, the tangent modulus does not exist in the material model of "002-orthotropic-elastic". Failure criterion for FRP composites was defined using "000-add_erosion," to assign the ultimate strain of FRP in the "EFFEPS" card.

Table 2. Material properties of FRP composites (part of this reproduced after Dai *et al.* [13], copyright 2011 ASCE).

FRP Type	E_{f1} (GPa)	E_{f2} (GPa)	Tensile Strength (MPa)	Rupture Strain (%)
PET-FRP	17.9	8.3	750	8.7
PEN-FRP	27.0	12.0	760	6.3
Glass-FRP	26.1	–	575	2.2
Carbon-FRP	95.8	–	986	1.0

2.4. Loading and Boundary Conditions

Displacements and rotations in all directions were prevented at the bottom of the bottom plate. Displacements in *X* and *Y* directions were prevented for all of the nodes of the top plate. Monotonic downward (negative *Z*direction) displacement loading was applied on the top plate for axial compressive loading until failure occurred. Failure was defined as the rupture of the FRP or the crushing of the concrete cylinder.

2.5. Validation Results

Figure 2 illustrates the axial strain-axial stress relationships for all of the cylinders gathered from the FE and the experimental results. The axial strain of each cylinder was obtained by dividing the axial displacement of the loading plate by the cylinder's height of 300 mm. The axial stress of each cylinder was obtained by dividing the axial reactions at the bottom of the bottom plate by the cross-sectional area of the cylinder. All simulated columns behaved in a manner similar to the tested cylinders until failure. All of the cylinders failed by FRP rupture whether in the experimental or FE category (Figure 3). The FE's average error rates in predicting the ultimate axial stress and ultimate axial strain were 9% and 10%, respectively. The error was calculated as the absolute value of the difference between the experimental results and the FE results divided by the experimental results.

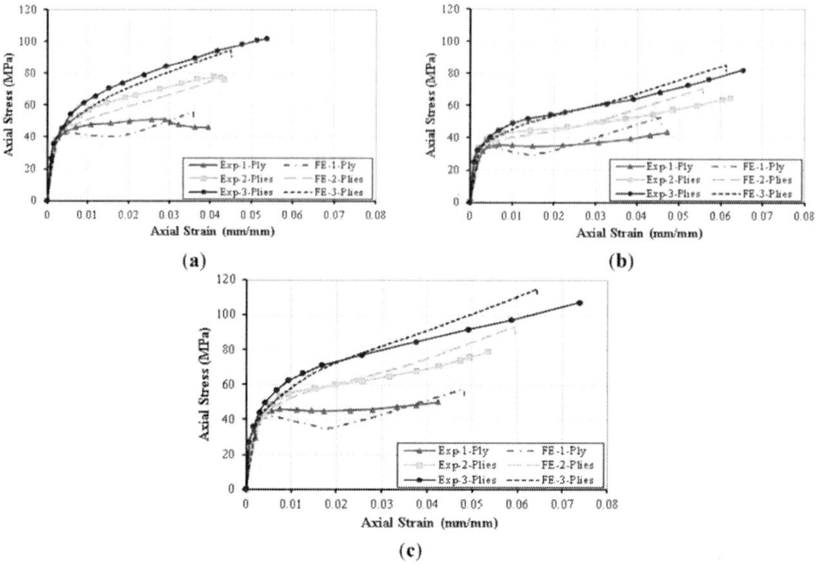

Figure 2. FE results *versus* experimental results by Dai *et al.* [13] of the CFFT with: (**a**) PEN-FRP-600; (**b**) PET-FRP-600; and (**c**) PET-FRP-900.

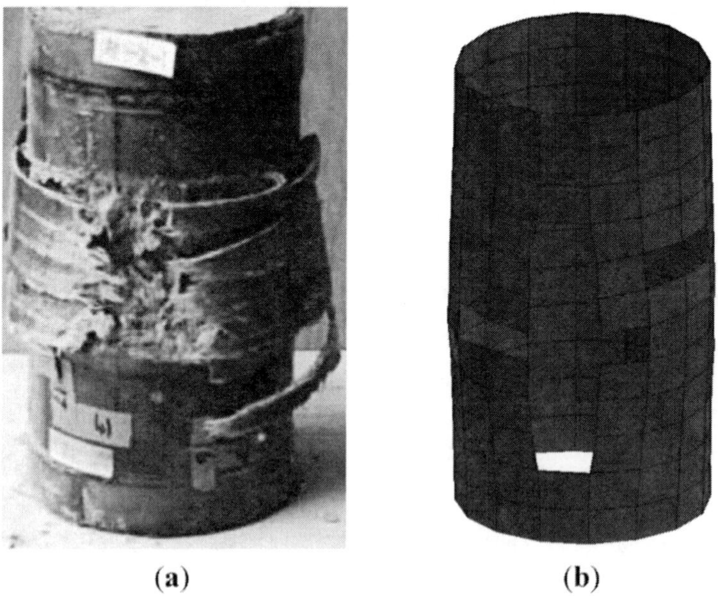

Figure 3. FRP rupture in FE analysis and experimental work by Dai *et al.* [13] (reprinted with permission from [13]. Copyright 2011 ASCE) of the PEN-600-I: (**a**) Experimental and (**b**) FE.

The FE predicted accurately the initial stiffness and stiffness degradation of all of the cylinders until the axial stress reached the unconfined concrete cylindrical strength (f′c). After this stress, the FE differentiated a little in values with the experimental results until failure. This difference in values was because the dilation angle was taken as a constant value in the FE. However, it would change with the axial stress level. However, the effect of the dilation did not significantly affect the overall behavior as the accuracy in predicting the ultimate strain was 91%, and the accuracy in predicting the ultimate stress was 90%.

3. PARAMETRIC STUDY

The LRS-FRP is a new composite that has only recently been investigated. Once the finite element model was validated, a comprehensive parametric study was conducted to numerically investigate the behavior of the CFFTs with LRS-FRP. The behavior of the CFFT using different FRP types, including traditional FRP and LRS-FRP, was investigated. A new hybrid system of FRP composites was investigated by combining traditional FRP with LRS-FRP to confine the concrete. The influence of fiber stacking sequences was investigated by placing the PET-FRP layers in the inner surface of the FRP tube and placing the traditional FRP in the outer tube for some columns and vice versa for others. In addition, the effects of the confinement ratio, the unconfined concrete nominal compressive strength (f′c), the column's size, and the column's aspect ratio were investigated.

All of the investigated columns had a diameter of 150 mm, a height of 300 mm, and an aspect ratio of 2 except columns C44 to C48. Four different column sizes with aspect ratios of 2 were investigated during this study. The diameters × heights ranged from 150 mm × 300 mm to 1500 mm × 3000 mm. Three different column aspect ratios ranging from 2 to 10 were investigated. Seven different confinement ratios ranging from 0.3 to 1.2 were investigated for PET-FRP, PEN-FRP, Glass-FRP, and Carbon-FRP. Five concrete cylindrical compressive strengths (f′c) ranging from 27.6 MPa to 82.8 MPa were examined. Each parameter was studied independently, resulting in an analysis of 49 columns. Table 3 summarizes the investigated columns' variables.

Table 3. Parametric study columns' variables.

Parameter	Col. Label	Diameter (mm)	Height (mm)	Aspect Ratio	LRS-FRP Thickness (mm)	Traditional FRP Thickness (mm)	f'_c (MPa)	Conf. Ratio	Dilation Parameter
PET-FRP	C0	150	300	2	3.4	0	27.6	1.20	0.29
	C1	150	300	2	2.97	0	27.6	1.05	0.33
	C2	150	300	2	2.55	0	27.6	0.90	0.37
	C3	150	300	2	2.13	0	27.6	0.75	0.41
	C4	150	300	2	1.7	0	27.6	0.60	0.46
	C5	150	300	2	1.28	0	27.6	0.45	0.52
	C6	150	300	2	0.85	0	27.6	0.30	0.60
PEN-FRP	C7	150	300	2	3.1	0	27.6	1.20	0.23
	C8	150	300	2	2.7	0	27.6	1.05	0.26
	C9	150	300	2	2.32	0	27.6	0.90	0.30
	C10	150	300	2	1.92	0	27.6	0.75	0.34
	C11	150	300	2	1.54	0	27.6	0.60	0.40
	C12	150	300	2	1.15	0	27.6	0.45	0.47
	C13	150	300	2	0.78	0	27.6	0.30	0.55
GLASS-FRP	C14	150	300	2	0	4.33	27.6	1.20	−0.17
	C15	150	300	2	0	3.8	27.6	1.05	−0.14
	C16	150	300	2	0	3.25	27.6	0.90	−0.11
	C17	150	300	2	0	2.7	27.6	0.75	−0.08
	C18	150	300	2	0	2.15	27.6	0.60	−0.03
	C19	150	300	2	0	1.62	27.6	0.45	0.02
	C20	150	300	2	0	1.07	27.6	0.30	0.10
CARBON-FRP	C21	150	300	2	0	2.6	27.6	1.20	−0.32
	C22	150	300	2	0	2.27	27.6	1.05	−0.30
	C23	150	300	2	0	1.94	27.6	0.90	−0.27
	C24	150	300	2	0	1.62	27.6	0.75	−0.23
	C25	150	300	2	0	1.3	27.6	0.60	−0.19
	C26	150	300	2	0	0.98	27.6	0.45	−0.13
	C27	150	300	2	0	0.65	27.6	0.30	−0.05
PET/Glass (inside/outside)	C28	150	300	2	2.55	1.07	27.6	1.20	0.47
	C29	150	300	2	1.7	2.15	27.6	1.20	0.43
	C30	150	300	2	0.85	3.25	27.6	1.20	0.48
Glass/PET (inside/outside)	C31	150	300	2	2.55	1.07	27.6	1.20	0.47
	C32	150	300	2	1.7	2.15	27.6	1.20	0.43
	C33	150	300	2	0.85	3.25	27.6	1.20	0.48
PET/Carbon (inside/outside)	C34	150	300	2	2.55	0.65	27.6	1.20	0.31
	C35	150	300	2	1.7	1.3	27.6	1.20	0.27
	C36	150	300	2	0.85	1.95	27.6	1.20	0.33
Carbon/PET (inside/outside)	C37	150	300	2	2.55	0.65	27.6	1.20	0.31
	C38	150	300	2	1.7	1.3	27.6	1.20	0.27
	C39	150	300	2	0.85	1.95	27.6	1.20	0.33
Concrete Strength (f'_c)	C40	150	300	2	5.1	0	41.4	1.20	0.29
	C41	150	300	2	6.8	0	55.2	1.20	0.29
	C42	150	300	2	8.5	0	69	1.20	0.29
	C43	150	300	2	10.2	0	82.8	1.20	0.29
Column size	C44	200	400	2	4.55	0	27.6	1.20	0.29
	C45	300	600	2	6.8	0	27.6	1.20	0.29
	C46	1500	3000	2	34	0	27.6	1.20	0.29
Column aspect ratio	C47	300	1500	5	6.8	0	27.6	1.20	0.29
	C48	300	3000	10	6.8	0	27.6	1.20	0.29

3.1. LRS-FRP versus Traditional FRP

The CFFTs with LRS-FRP and with traditional FRP were investigated with different confinement ratios ranging from 0.3 to 1.2. Figure 4 illustrates the typical axial strain-normalized strength behavior of the CFFTs with LRS-FRP and with traditional FRP. The normalized strength was calculated as the axial stress divided by the f′c. All of the columns failed by FRP rupture. However, the CFFTs with traditional FRP behaved, as expected, with bilinear strain-stress relationships. The CFFTs with LRS-FRP behaved with trilinear behavior. This behavior of CFFTs with LRS-FRP was because of the effect of the bilinear behavior on the LRS-FRP instead of the linear behavior in the traditional FRP. All of the columns had the same initial stiffness. The reason was that the effect of the FRP confinement did not appear until the axial stress reached to almost the f′c when the concrete volume change started to become positive (expansion; reference). The CFFTs with traditional FRP continued with the secant modulus until failure occurred. The CFFTs with LRS-FRP showed a stiffness degradation after axial strain of approximately 0.016 and 0.013 for PEN-FRP and PET-FRP, respectively. The CFFTs with LRS-FRP showed higher ultimate strain and lower secant stiffness than those with traditional FRP. As expected, the CFFTs with carbon FRP tubes showed higher secant stiffness and lower ultimate strain. The CFFTs with PET-FRP showed higher ultimate strain and lower secant stiffness. The CFFTs with LRS-FRP showed a higher strength than those with traditional FRP. The reason was the high hoop rupture strain of the LRS-FRP reached 8.7 times that of the carbon FRP and 2.9 times that of the glass FRP. The axial strength of the CFFTs with PEN-FRP and PET-FRP was almost the same. However, the axial strength of the CFFT with PET-FRP was approximately 1.25 times that of the CFFT with PEN-FRP.

Figure 5 illustrates the relation between the confinement ratio and the normalized strength and between the confinement ratio and the ultimate axial strain for all of the FRP composites. This figure illustrates the efficiency of the different types of FRP in normalized strength and ultimate axial strain at the same confinement ratio. It is very clear in the figure that the CFFTs with LRS-FRP were more efficient than those with traditional FRP. This clearly indicated the great effect high rupture strain had on the confinement.

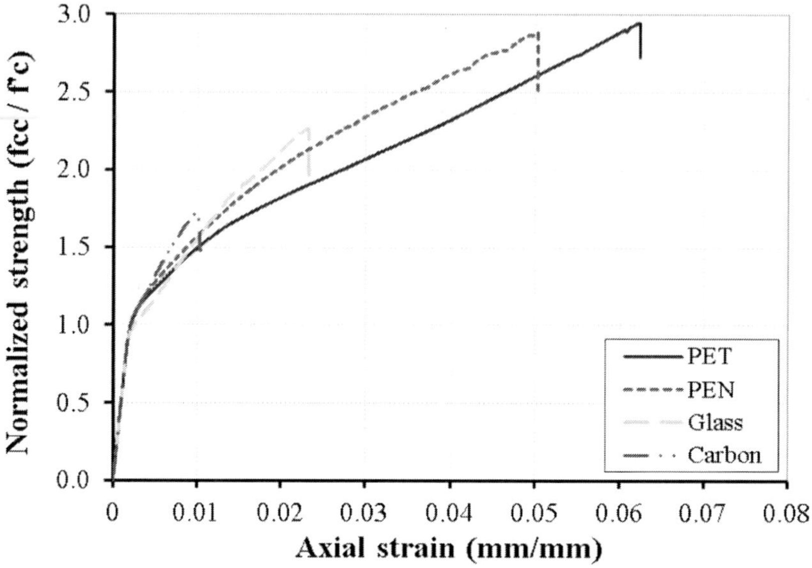

Figure 4. Axial strain-normalized strength (fcc/f'c) relationship for PET, PEN, Glass, and Carbon FRP with same confinement ratio of 0.9.

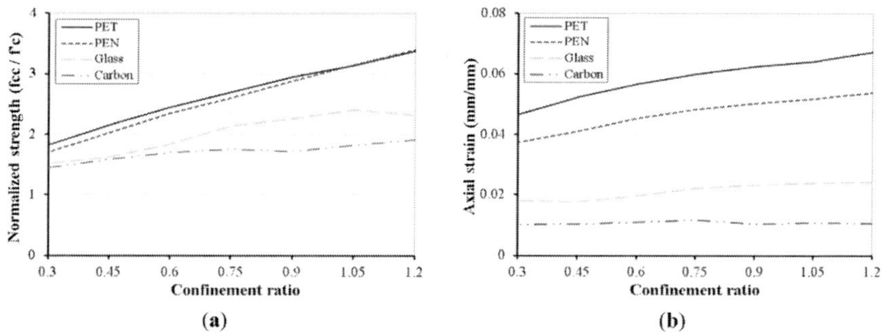

Figure 5. Efficiency of the traditional *versus* LRS-FRP composites in: (**a**) normalized axial strength; (**b**) axial strain.

Figure 6 illustrates the axial strain-normalized strength for the CFFTs with traditional FRP, LRS-FRP, and hybrid FRP with the same confinement ratio of 1.2. Fiber stacking sequences were investigated by placing PET-FRP in the inner surface of the FRP tube and placing glass or carbon FRPs in the outer surface and vice versa.Figure 6a illustrates the PET, glass, and hybrid PET/glass where the PET was in the inner surface. In general, placing the LRS-FRP in the inner surface and the traditional FRP in the outer surface revealed a better performance than placing the traditional FRP in the inner surface. The reason for this behavior was that the rupture strain of the traditional FRP is much lower than that of the LRS-FRP. Therefore, the traditional FRP ruptured earlier than the LRS-FRP. Hence, when the traditional FRP was placed in the inner surface, the LRS-FRP was controlled by the traditional FRP rupture strain, and it ruptured directly after the rupture of the traditional FRP. However, when the traditional FRP was placed in the outer surface, the LRS-FRP was controlled by it, and it continued until ruptured with high hoop strains. Therefore, the hybrid FRP of PET/traditional FRP reached higher hoop strains than the traditional FRP. However, the hybrid FRP ruptured at a lower strain than that of the LRS-FRP alone because of the synergistic effect from hybridization.

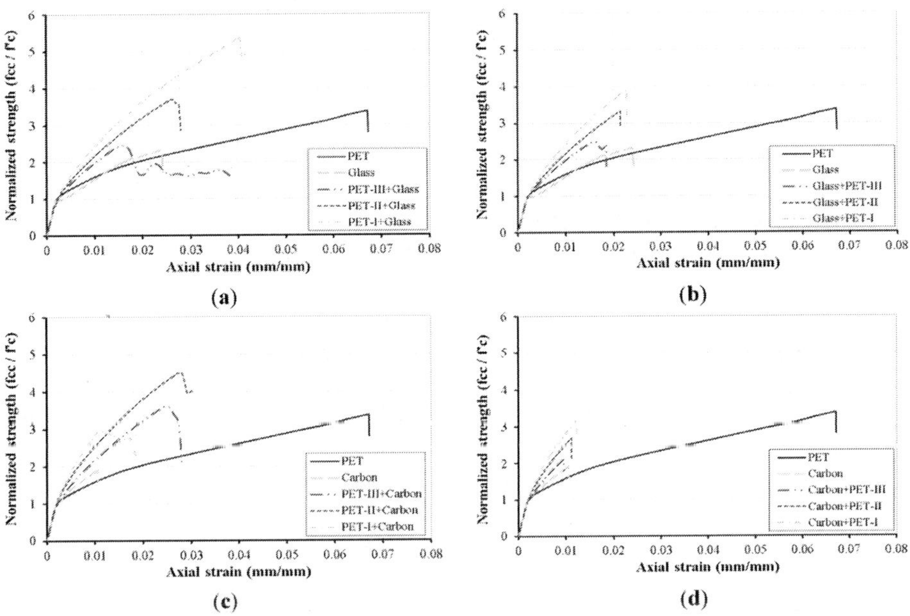

Figure 6. Axial strain-normalized strength relationship of the traditional, LRS, and hybrid FRP: (**a**) PET/Glass-FRP; (**b**) Glass/PET; (**c**) PET/Carbon-FRP; and (**d**) Carbon/PET-FRP. Note: PET/Glass-FRP is the stacking sequence of inside/outside FRP and vice versa.

In general, the hybrid of PET/glass performed better than PET/carbon. The reason for that was the large difference in rupture strains between the PET and carbon. In general, the axial strain-normalized strength relation of the CFFTs with hybrid FRP was nonlinear instead of bilinear in the case of LRS-FRP alone. The relation was linear in the case of traditional FRP alone. The strength and ultimate axial strain of the CFFTs with hybrid FRP increased when the traditional FRP was increased. This indicated that using few layers of LRS-FRP with the traditional FRP would improve the CFFT's performance a lot. However, the difference in the confinement ratio contribution of the LRS-FRP has to be considerable in order to avoid sudden failure as in the case of (PET-I + Carbon) in Figure 5c. When the carbon FRP reached its ultimate strain (1% only), it failed and one layer of PET-FRP was not enough to continue to confine the concrete core, which led to rupture of the PET layer as well.

3.2. Unconfined Concrete Compressive Strength (f'c)

Five columns were studied with different concrete unconfined compressive strengths (f'c) ranging from 27.6 MPa to 82.8 MPa. Figure 7 illustrates the axial strain-normalized strength relation of the CFFTs with different f'c. In general, changing the f'c did not affect the normalized strength or the ultimate axial strain because the columns had the same FRP confinement ratios. However, when the concrete core was high-strength (f'c ≥ 55.2 MPa), the strength and ultimate axial strain were inversely proportional with the f'c. The lateral concrete expansion is dependent on the concrete mechanical properties. Therefore, the lateral expansion of high-strength concrete is significantly higher than that of the normal strength concrete, which reduces the effect of FRP confinement. The ultimate axial strain and the normalized strength decreased by 14.6% and 9.0%, respectively when the f'c of the high-strength concrete increased by 25% (from 55.2 MPa to 69.0 MPa). The ultimate axial strain and the normalized strength decreased by 21.1% and 24.9%, respectively when the f'c of the high-strength concrete increased by 50% (from 55.2 MPa to 82.8 MPa).

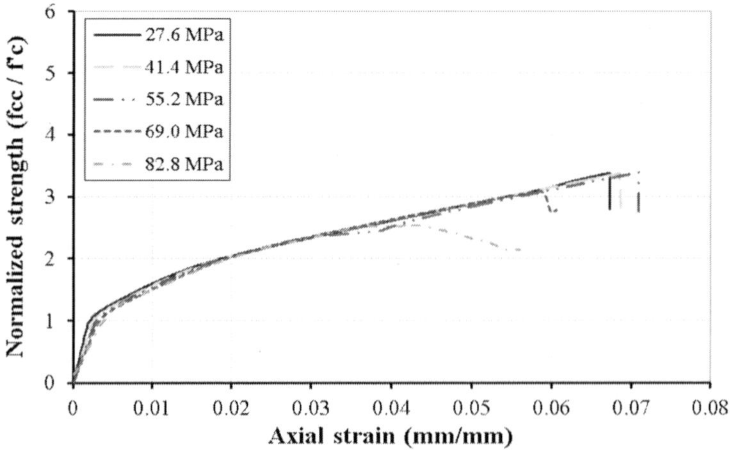

Figure 7. Axial strain-normalized strength relationship of CFFT with LRS-FRP with different concrete compressive strength (f'c).

3.3. Column Size

Four columns with sizes ranging from 150 mm × 300 mm to 1500 mm × 3000 mm were studied. Figure 8illustrates the axial strain-normalized strength relation of the CFFTs with different column sizes. In general, the strength was reduced when the column size was increased as the FRP confinement could not affect the whole cross-section. Figure 9 shows the axial stress distribution of all of the columns in the mid and top cross-sections. It is very clear that the FRP confinement affected a zone along the outer perimeter in the cross-section, and this zone decreased when the column diameter increased. However, the behavior of the columns with dimensions of 150 mm × 300 mm and 200 mm × 400 mm was almost the same in axial strain-normalized strength as the behavior in cross-section. This behavior was because both dimensions were considerably low for a confinement ratio of 1.2.

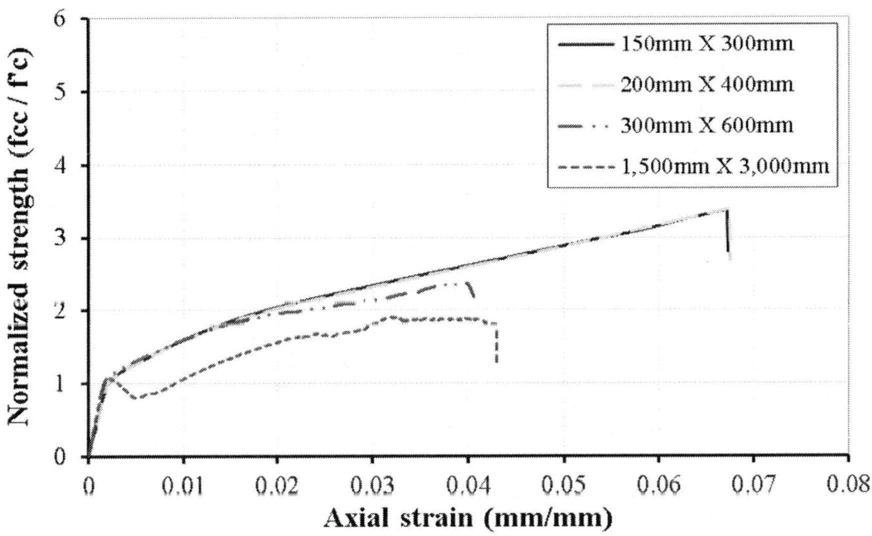

Figure 8. Axial strain-normalized strength relationship of CFFT with LRS-FRP with different column sizes.

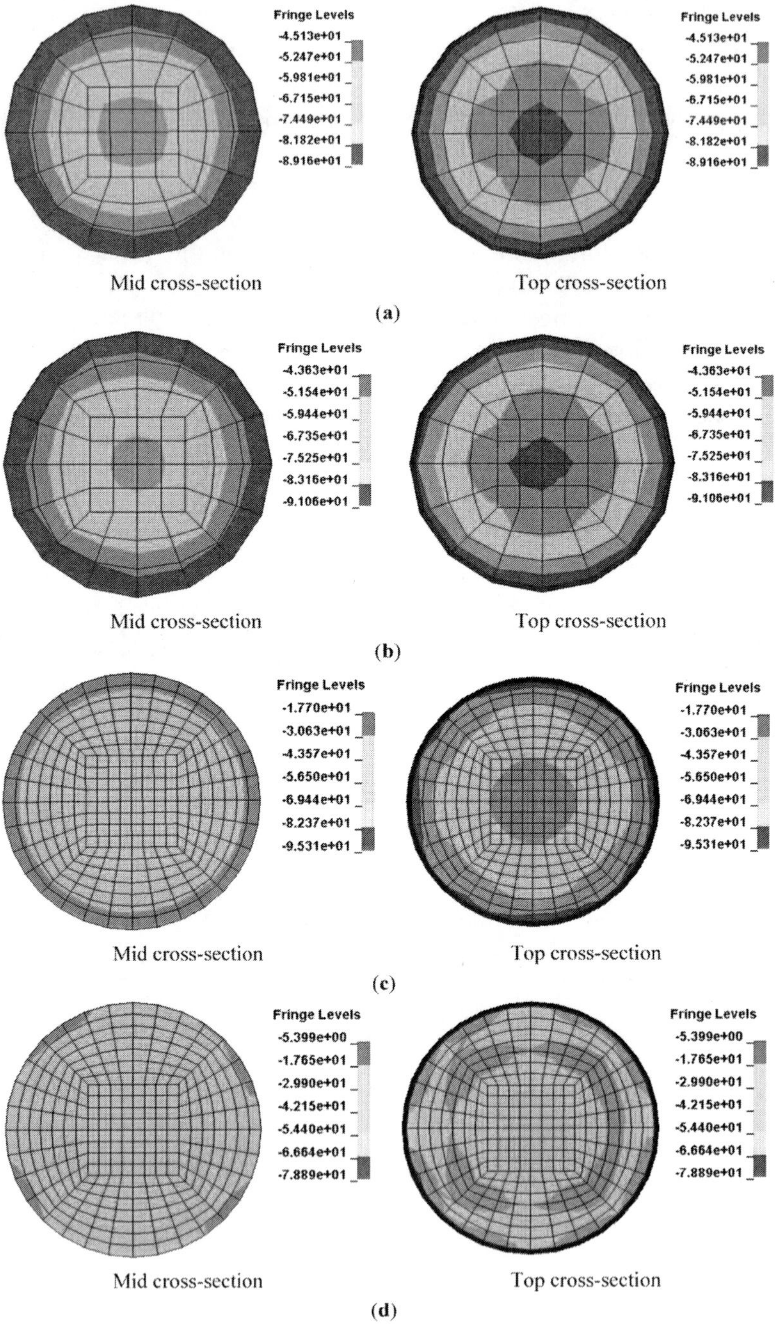

Figure 9. Axial stress in cross-section of the columns of sizes: (**a**) 150 mm × 300 mm; (**b**) 200 mm × 100 mm; (**c**) 300 mm × 600 mm; and (**d**) 1500 mm × 3000 mm. Note: Fringe levels shows the axial stresses legend in MPa.

3.4. Column Aspect Ratio

Three columns with different column aspect (height-to-diameter) ratios ranging from 2 to 10 were studied.Figure 10 illustrates the axial strain-normalized strength relation of the CFFTs with different aspect ratios. The ultimate axial strain and axial strength decreased when the column's aspect ratio increased. The column with an aspect ratio of 2 failed by FRP rupture. However, the columns with aspect ratios of 5 and 10 failed by compression failure. Figure 11 illustrates the column's deformation for different aspect ratios. Figure 11a illustrates the global buckling that occurred in the column with an aspect ratio of 10, leading to compression failure. Figure 11b illustrates the deformation of the column, with an aspect ratio of 5, that bulged in the top and bottom thirds leading to compression failure. Figure 11c,d illustrate the common failure of the confined short column of FRP rupture at the middle part. The ultimate axial strain decreased by 26% and the axial strength of the CFFT with LRS-FRP decreased by 48% when the aspect ratio increased from 2 to 5. The ultimate axial strain decreased by 63% and the axial strength of the CFFT with LRS-FRP decreased by 58% when the aspect ratio increased from 2 to 10.

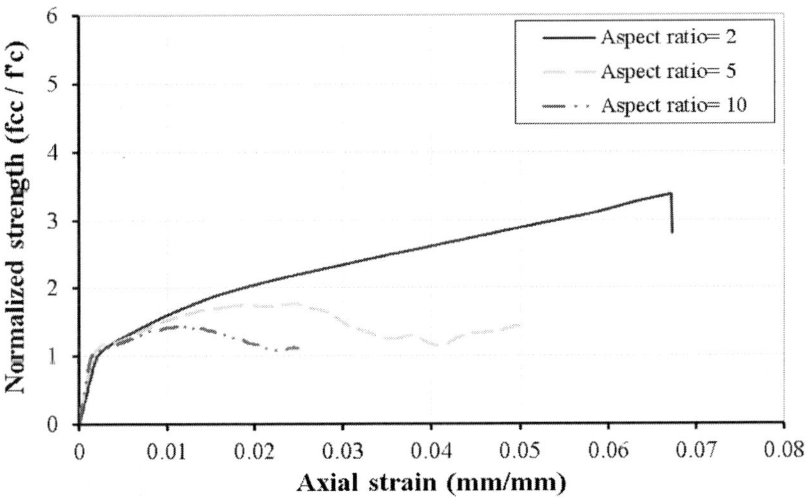

Figure 10. Axial strain-normalized strength relationship of CFFT with LRS-FRP with different column aspect ratios.

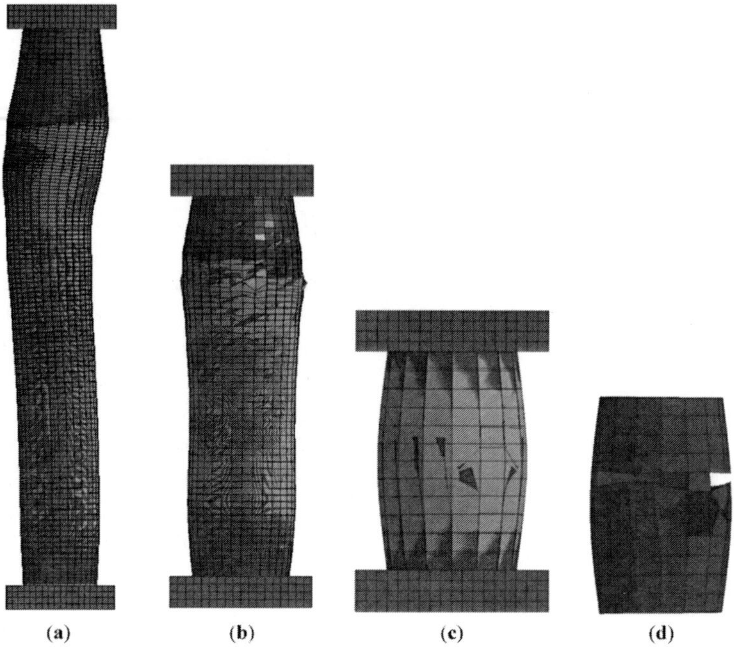

Figure 11. Column deformation with different aspect ratios: (**a**) aspect ratio of 10; (**b**) aspect ratio of 5; (**c**) aspect ratio of 2; and (**d**) FRP rupture of column with aspect ratio of 2.

4. FINDINGS AND CONCLUSIONS

The behavior of the concrete-filled fiber tubes (CFFT) with new high deformable fiber reinforced polymers under axial compressive loading was investigated. Unlike the traditional fiber reinforced polymers (FRP) like carbon, glass, aramid, *etc.*, the new FRP composites have a large rupture strain and are made with cheap materials. The large rupture strain (LRS) FRP composites are made with polyethylene naphthalate (PEN) and polyethylene terephthalate (PET) fibers. The PEN and PET fibers can be used in green buildings. They are environmentally friendly as they are made from recycled materials (e.g., bottles). They have a high ultimate strain (>5.0%), however their elastic modulus is low. This study used finite element (FE) analysis using LS-DYNA software to conduct an extensive parametric study to investigate the behavior of the CFFTs with the LRS-FRP under axial compressive loading. Forty-nine columns were investigated to determine important factors may affect the behavior of the CFFTs under axial compressive loading. A high range of FRP confinement ratios was investigated. In addition, the effects of the unconfined concrete compressive strength (f′c), column size, and column aspect ratio on the behavior of the

CFFT were studied. A comparison between the behavior of the CFFTs with LRS-FRP and the traditional FRP (carbon and glass) with a high range of confinement ratios was conducted as well. This paper introduced a new state-of-the-art hybrid FRP to be used for the CFFT columns by investigating different combinations of traditional FRP with LRS-FRP. Generally, the CFFTs with LRS-FRP showed a remarkable behavior under axial loading in strength and ultimate strain. The LRS-FRP composites were more efficient than the traditional FRP composites in strength and ultimate strain. The behavior of the hybrid FRP with a stacking sequence of LRS/glass (inner/outer of the tube) showed much better behavior in strength than the traditional FRP or the LRS-FRP. However, this hybrid FRP showed a higher ultimate axial strain than the traditional FRP. The LRS alone was better in the ultimate axial strain. A new equation to estimate the concrete dilation parameter and dilation angle of the CFFT columns with LRS-FRP tubes or hybrid FRP tubes was suggested.

In conclusion LRS-FRP is a promising family of new material; however, more research is still required to characterize their fire resistance and durability. Their behavior with different matrices and their bonding with concrete members should be investigated as well.

ACKNOWLEDGMENTS

This research was supported by the National University Transportation Center (NUTC) at Missouri University of Science and Technology (Missouri S & T). However, any opinions, findings, conclusions, and recommendations presented in this paper are those of the authors and do not necessarily reflect the views of the sponsors.

AUTHOR CONTRIBUTIONS

Omar Abdelkarim carried out the finite element modeling presented in this manuscript. Mohamed ElGawady was the initiator of the research idea and supervisor of all of the technical work reported in this manuscript.

REFERENCES

1. Lechat, C.; Bunsell, A.R.; Davies, P. Tensile and creep behaviour of polyethylene terephthalate and polyethylene naphthalate fibres. *J. Mater. Sci.* **2011**, *46*, 528–533.

2. Venkatachalam, S.; Nayak, S.G.; Labde, J.V.; Gharal, P.R.; Rao, K.; Kelkar, A.K. *Degradation and Recyclability of Poly (Ethylene*

Terephthalate); Available online: http ://www. intechopen.com / books/ polyester/ degradation-and-recyclability-of-poly-ethylene-terephthalate- (accessed on 1 August 2015).

3. Qasrawi, Y.; Heffernan, P.J.; Fam, A. Performance of concrete-filled FRP tubes under field close-in blast loading. *J. Compos. Constr.* **2015**, *19*.

4. Moon, J.; Lehman, D.E.; Roeder, C.W.; Lee, H.-E. Strength of circular concrete-filled tubes with and without internal reinforcement under combined loading. *J. Struct. Eng.* **2012**, *139*.

5. ElGawady, M.A.; Sha'lan, A. Seismic behavior of self-centering precast segmental bridge bents. *J. Bridge Eng.* **2010**, *16*, 328–339.

6. Sadeghian, P.; Fam, A. Bond-slip analytical formulation toward optimal embedment of concrete-filled circular FRP tubes into concrete footings. *J. Eng. Mech.* **2010**, *136*, 524–533.

7. ElGawady, M.; Booker, A.J.; Dawood, H.M. Seismic behavior of posttensioned concrete-filled fiber tubes. *J. Compos. Constr.* **2010**, *14*, 616–628.

8. Shao, Y.; Mirmiran, A. Experimental investigation of cyclic behavior of concrete-filled fiber reinforced polymer tubes. *J. Compos. Constr.* **2005**, *9*, 263–273.

9. Fam, A.; Flisak, B.; Rizkalla, S. Experimental and analytical modeling of concrete-filled FRP tubes subjected to combined bending and axial loads. *ACI Struct. J.* **2003**, *100*, 499–509.

10. Zhang, W.; Shahrooz, B.M. *Analytical and Experimental Studies into Behavior of Concrete-Filled Tubular Columns*; Cincinnati Infrastructure Institute, University of Cincinnati: Cincinnati, OH, USA, 1997.

11. Ispir, M. Monotonic and cyclic compression tests on concrete confined with pet-FRP. *J. Compos. Constr.***2015**, *19*.

12. Bai, Y.-L.; Dai, J.-G.; Teng, J. Cyclic compressive behavior of concrete confined with large rupture strain FRP composites. *J. Compos. Constr.* **2014**, *18*.

13. Dai, J.-G.; Bai, Y.-L.; Teng, J. Behavior and modeling of concrete confined with FRP composites of large deformability. *J. Compos. Constr.* **2011**, *15*, 963–973.

14. Anggawidjaja, D.; Ueda, T.; Dai, J.; Nakai, H. Deformation capacity of rc piers wrapped by new fiber-reinforced polymer with large fracture strain. *Cem. Concr. Compos.* **2006**, *28*, 914–927.

15. Jirawattanasomkul, T.; Dai, J.-G.; Zhang, D.; Senda, M.; Ueda, T. Experimental study on shear behavior of reinforced-concrete members fully wrapped with large rupture-strain FRP composites. *J. Compos. Constr.* **2013**, *18*.

16. Mirmiran, A.; Zagers, K.; Yuan, W. Nonlinear finite element modeling of concrete confined by fiber composites. *Finite Elem. Anal. Des.* **2000**, *35*, 79–96.

17. Yu, T.; Teng, J.; Wong, Y.; Dong, S. Finite element modeling of confined concrete-i: Drucker-prager type plasticity model. *Eng. Struct.* **2010**, *32*, 665–679.

18. Rousakis, T.C.; Karabinis, A.I.; Kiousis, P.D.; Tepfers, R. Analytical modelling of plastic behaviour of uniformly FRP confined concrete members. *Compos. Part B Eng.* **2008**, *39*, 1104–1113.

19. Karabinis, A.I.; Rousakis, T.C.; Manolitsi, G.E. 3D finite-element analysis of substandard RC columns strengthened by fiber-reinforced polymer sheets. *J. Compos. Constr.* **2008**, *12*, 531–540.

20. Youssf, O.; ElGawady, M.A.; Mills, J.E.; Ma, X. Finite element modelling and dilation of FRP-confined concrete columns. *Eng. Struct.* **2014**, *79*, 70–85.

21. Jankowiak, T.; Lodygowski, T. Identification of parameters of concrete damage plasticity constitutive model. *Found. Civ. Environ. Eng.* **2005**, *6*, 53–69.

22. Kmiecik, P.; Kamiński, M. Modelling of reinforced concrete structures and composite structures with concrete strength degradation taken into consideration. *Arch. Civ. Mech. Eng.* **2011**, *11*, 623–636.

23. Rezazadeh, M.; Costa, I.; Barros, J. Influence of prestress level on NSM CFRP laminates for the flexural strengthening of rc beams. *Compos. Struct.* **2014**, *116*, 489–500.

24. Hallquist, J.O. *LS-DYNA Theory Manual*; Livermore Software Technology Corporation: Livermore, CA, USA, 2006; Volume 3.

25. Abdelkarim, O.I.; ElGawady, M.A. Analytical and finite-element modeling of FRP-concrete-steel double-skin tubular columns. *J. Bridge Eng.* **2015**, *20*.

26. Malvar, L.J.; Crawford, J.E.; Wesevich, J.W.; Simons, D. A plasticity concrete material model for DYNA3D.*Int. J. Impact Eng.* **1997**, *19*, 847–873.

Chapter 5

Unsymmetrical Fibre-Reinforced Plastics for the Production of Curved Textile Reinforced Concrete Elements

Henrik L. Funke, Sandra Gelbrich, Andreas Ehrlich, Lars Ulke-Winter, Lothar Kroll

Institute of Lightweight Structures, Technische Universität Chemnitz, Chemnitz, Germany

ABSTRACT

A new constructive and technological approach was developed for the efficient production of large- dimensioned, curved freeform formworks, which allow the manufacturing of single and double- curved textile reinforced concrete elements. The approach is based on a flexible, multi-layered formwork system, which consists of glass-fibre reinforced plastic (GFRP). Using the unusual structural behavior caused by anisotropy, these GFRP formwork elements permit a specific adjustment of defined curvature. The system design of the developed GFRP formwork and the concrete-lightweight-elements with stabilized spacer fabric was examined exhaustively. Prototypical curved freeform surfaces with different curvature radii were designed, numerically computed and produced. Furthermore, the fabric's contour accuracy of the fabric was verified, and its integration was adjusted to loads.

Keywords: Anisotropic Formworks, Textile-Reinforced Concrete, Fibre-Reinforced Plastics, Curved Concrete

1. INTRODUCTION

Research in the fields of innovative concrete structures with high potential for lightweight design, and of textile reinforcement for special applications has been object of intensive scientific and application-oriented efforts for a couple of years [1] -[3] . There is a lack of appropriate formwork systems to implement light shell structures of this kind, with the need for flexibly moldable, reusable systems being particularly tense [4]. Textile reinforced concrete offers a high range of variation and thus facilitates a flexible adjustment of the form and textile reinforcement to the relevant conditions (cf. e.g. [5]) and the citations listed there). Another advantage in comparison to ordinary reinforced concrete is that corrosion can be largely excluded. In this way, filigree constructions of minimal thickness can be realized [6] -[8].

A crucial technological objective of textile reinforced concrete elements is the development of complex solid preform-structures. These are produced by processing flat structures through appropriate cutting [9]. The soft- elastic behavior of the 3D-textiles can be influenced to a large extent by modifying parameters such as stiffness, alignment and concentration of pile threads. In this way, it can be adjusted to the defined curvature. Although selectively deformable textile 3D-structures, for instance spacer fabric, for the reinforcement of concrete light- weight elements exist, the corresponding formwork elements are only in an early stage of development. These elements are essential for the realization of concrete shell structures that can be curved in any way.

Currently, the only possible shapes of shell structures are that of domes, hyperbolic paraboloids and conoids. Their production furthermore entails a considerable amount of material and high costs [4] [10] [11]. Also the mathematical description of the predetermined complex freeform surfaces and the anisotropic material characteristic is difficult [12] [13]. In the field of computer-aided visualization, different methods have been developed for the shape optimization of surfaces and their static construction calculation and design. However, these technologies are until now not applied in formwork production [14]. Among the common formwork techniques are conventionally segmented steel and wood systems, pneumatically supported and modeled formworks and combinations of them [15] [16].

Close examinations of the production of double-curved concrete-lightweight-elements based on flexible formwork systems made from GFRP are yet to come. The focus of this research work is on the numerical calculation and experimental verification of flexible, anisotropic GFRP formworks and the production of prototypical double-curved concrete-lightweight-elements with integrated stabilizing spacer fabric.

2. MATERIALS AND METHODS

2.1. Components for Glass-Fibre Reinforced Plastic

The unidirectional (UD) reinforced fabric UT-E500 by GURIT Holding AG was used for the production of the GFRP formwork. UT-E500 consists of aluminoborosilicate ("E glass") and has an area density of 500 g/sqm. For the thermosetting resin matrix, the epoxy resin Epilox® T19-27 by LEUNA-Harze GmbH was used. By means of manual laminating, UD single layers were made from the UD fabric and the epoxy resin. They had a fibre volume content of 30%. Within this UD single layer, the independent parameters: longitudinal (E1) and transverse (E2) moduli of elasticity, Poisson's ratio (ν_{12}), shear modulus (G_{12}) and the coefficients of linear thermal expansion α_1 and α_2 between +20°C and +120°C were determined experimentally. Table 1 shows the results of these tests.

2.2. Adjustment and Determination of Defined Curvatures

The determined basic parameters of the unidirectional single layer (Table 1) were used for the analytical and numerical calculation of multi-layered formworks, which showed both a balanced symmetrical and an asymmetrical construction. The GFRP formwork with a plate size of 50 × 50 cm^2 was made of 11 UD single layers with total thickness of 3.3 mm (Figure 1).

After the production of the GFRP bond, the thermosetting matrix was cured in a heating cabinet over a period of 6 hours at a constant temperature of 120°C. After curing, the GFRP bond were cooled to room temperature (20°C). Due to this difference in temperature of −100 K, residual stress caused a small curvature of the anisotropic layer structure. Afterwards, high curvatures were caused by external preloading (up to a material load of R = 0.95 after CUNTZE criterion), utilizing the coupling effects that result from the GFRP bond (Figure 2).

The calculations of the coupling effects caused by anisotropy were conducted analytically with the Classical Laminate Theory (CLT) and the First Order Shear Deformation Theory (FSDT). For the experimental verification of the anisotropic coupling effects calculated beforehand, selected GFRP bonds segments were produced and tested in the institute's own structural test bench with the ABD-testing device (Figure 3).

Table 1. Parameters of the unidirectional single layer.

E_1 GPa	E_2 GPa	ν_{12} -	G_{12} GPa	$\alpha_{1(20/120)}$ $10^{-6}\cdot K^{-1}$	$\alpha_{2(20/120)}$ $10^{-6}\cdot K^{-1}$
23.7	6.4	0.3	1.6	7	130

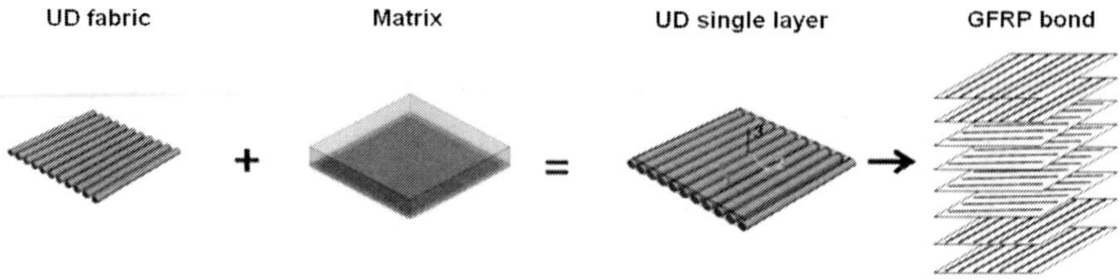

Figure 1. Process chain of anisotropic GFRP layered bond structures.

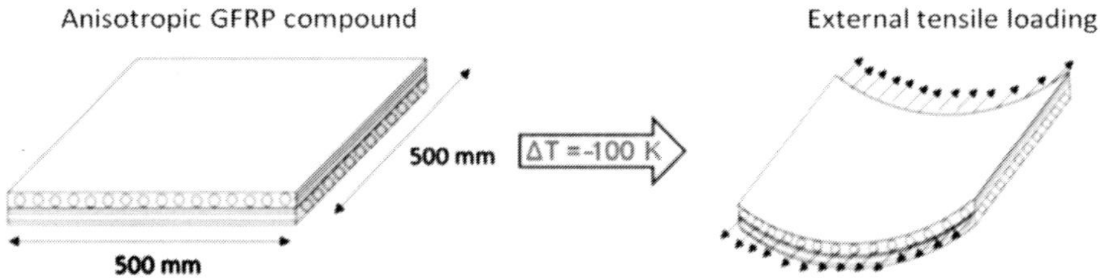

Figure 2. Temperature and tensile load of anisotropic GFRP layered bond structures for the adjustment of curvature states.

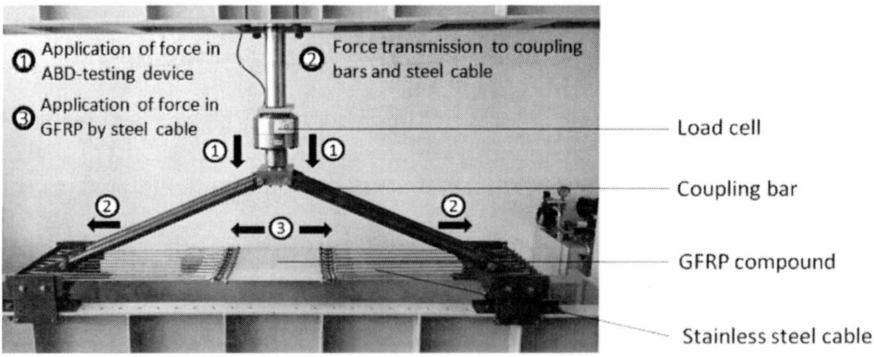

Figure 3. ABD-testing device.

The results were used during the further procedure for the determination of the functional relations of curvatures, process parameters and geometrical parameters. Based on that, the analysis and identification of single and double curved basic forms was conducted. Their combination resulted in a maximum of defined freeform surfaces.

The experimental analysis of the curves was conducted with the aid of the optical forming-analysis-systems ARGUS and ARAMIS by GOM (Figure 4).

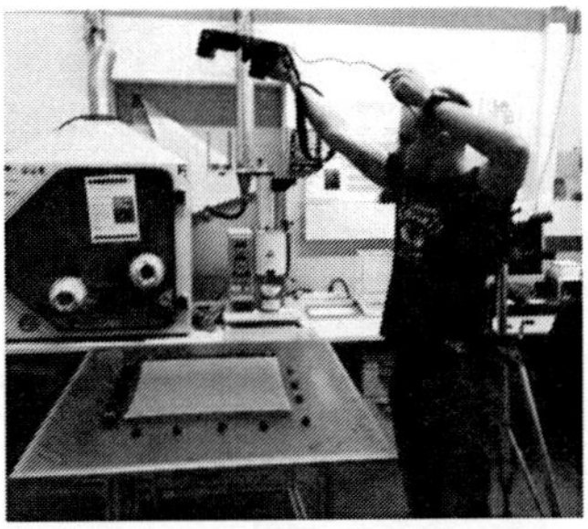

Figure 4. Curved GFRP layered bond structure plate with coded reference mark (picture on the left) and test set-up of the optical forming-analysis-system ARGUS (picture on the right).

Figure 5. 3D-fabrics "SITGrid" by V. Fraas Solutions in Textile GmbH.

Table 2. Composition of the fine grained concrete mix.

Component	Content in kg/m^3	Mass fraction in %
White cement CEM I 52.5 R	500	21.39
Amorphous aluminosilicate	150	6.42
Dolomite sand 0.5/1.0 (×50 = 0.71 mm)	1270	54.32
Dolomite filler (×50 = 70 μm)	150	6.42
Water	240	10.27
AR-glass fibres (12 mm, integral)	18	0.77
superplasticizers	10	0.43

2.3. Polymer-Bound Stabilization of 3D Spacer Fabrics for the Integration into Concrete

For the production of the textile reinforced concrete elements, the textile 3D-fabric "SITgrid" by V. Fraas Solutions in Textile GmbH was used. It is made from alkali-resistant glass (AR glass), with 2400 tex in warp and weft (Figure 5), which had a tensile strength of 978 MPa. The fabric was placed on the formwork elements to show and stabilize the curvatures accurately. Thermosetting and thermoplastic resin systems were used for the stabilization. The resin systems were applied by spraying, rolling or by using of capillary effect of the pile thread on the pre-curved 3D-fabric. By means of the Fourier transform infrared spectroscopy with attenuated total reflection (ATR-FTIR), the impact of the pile thread on capillary suction with the epoxy resin Indufloor-IB1240 of the company SCHOMBURG GmbH could be studied. For this purpose the ATR-FTIR testing apparatus ALPHA-P by BRUKER DALTONIK GmbH was used.

2.4. Textile Reinforced Concrete

The development of the fine grained concrete was focused on the workability of the fresh concrete as well as the durability and good bonding between concrete matrix and textile reinforcement. Table 2 shows the qualitative and quantitative composition of the fine grained concrete mix. Apart from white Portland cement type 52.5 R, the fine concrete contained an amorphous aluminosilicate as puzzolanic binder (Table 2). Dolomite sand with a grain size of 0.5 to 1.0 mm was used as aggregate and dolomite powder with an average grain size of 70 μm was used as filler. The alkali-resistant (AR) short glass fibres (16 M.-% ZrO2) were 12 mm long and had a length mass of 45 tex. The high-performance superplasticizers had a polycarboxylate ether (PCE) contentof 30 M.-%. The water-cement-ratio was 0.37.

2.5. Production Tests GFRP Formworks/Concrete-Lightweight-Elements

Selected representative freeform surfaces, double-curved with different radii of curvature, were produced in the production tests for the system structure GFRP formworks/concrete-lightweight-elements. The adjustment of the curves is carried out via the new flexible GFRP formwork elements. The basic proceeding when conducting the production tests comprised the production of the flexible formwork with reference curvature or preload curvature. After the production of the formwork, the spacer (Figure 6—left) of the reinforcement was positioned and the spacer fabric was installed, fixed and stabilized (Figure 6—middle). After positioning and stabilizing the textile 3D reinforcement structure from Figure 5, the fine concrete was laminated (Figure 6—right), using the adapted concrete composition displayed in Table 2.

Figure 6. Production test of the curved textile reinforced concrete elements: textile spacer (picture on the left), spacer fabric with textile spacer (picture on the middle), application of fine grained concrete (picture on the right).

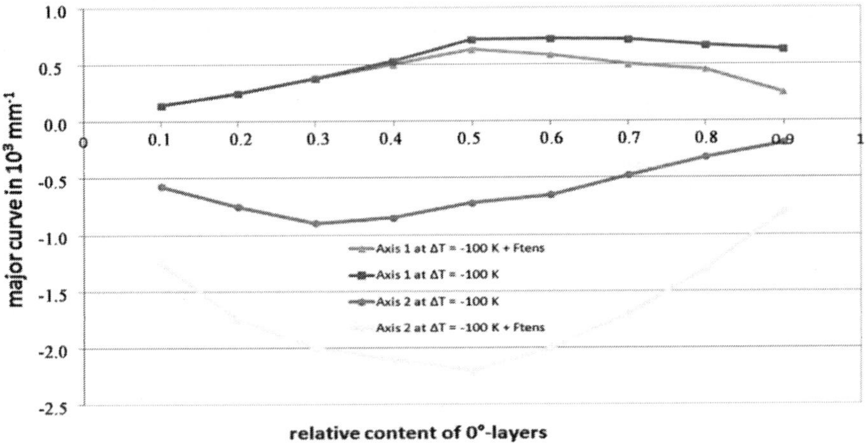

Figure 7. Theoretical major curve of an asymmetrical layer structure ($90_n/0_m$), depending on the portion of 0°-layer, with pure temperature load ($\Delta T = -100$ K) and superimposed temperature and tensile load.

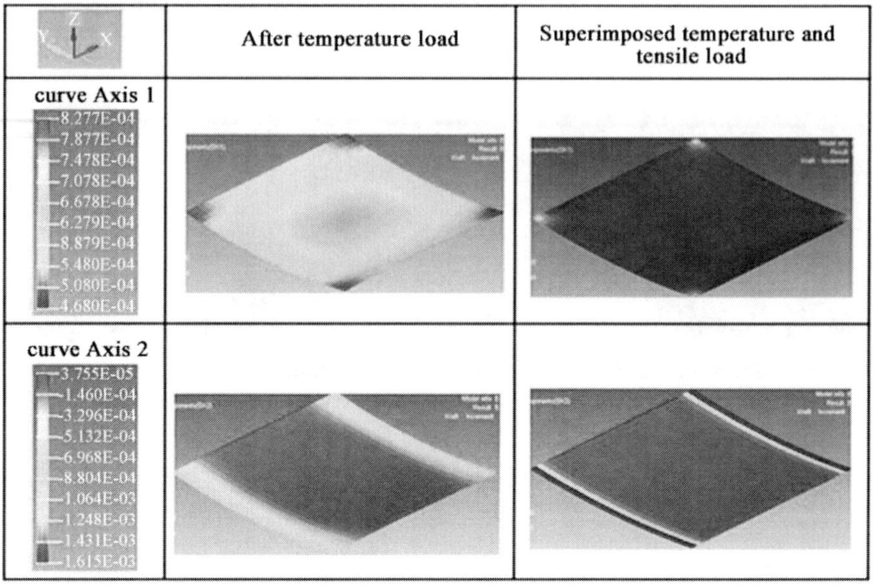

Figure 8. Theoretical major curve of the asymmetrical layer structure 905/06 depending on the portion of 0°-layer, with pure temperature load (ΔT = −100 K) and superimposed temperature and tensile load.

3. RESULTS AND DISCUSSION

3.1. Theoretical Verification of Major Curves

The theoretical major curves (for calculation approach cf. [12]) of an asymmetrical layer structure with 0°- and 90°-layer content ($90_n/0_m$) are exemplarily depicted for two loading cases in Figure 8, dependent on the 0°-layer content. The curves increased with increasing layer content accompanied by increasing anisotropy, both in the 1- and in the 2-axis (Figure 7).

The highest anisotropy was found with a relative 0°-layer content of about 50 percent. This caused the largest curvature around both axes (Figure 7 and Figure 8). A further increase of the 0°-layer content caused a decrease of curvature, because the anisotropy of the GFRP layered bond structure decreased.

Due to external preload forces, the tensile load caused an increase of curvature around the 2-axis. At the same time, curvature around the 1-axis decreased, due to traction.

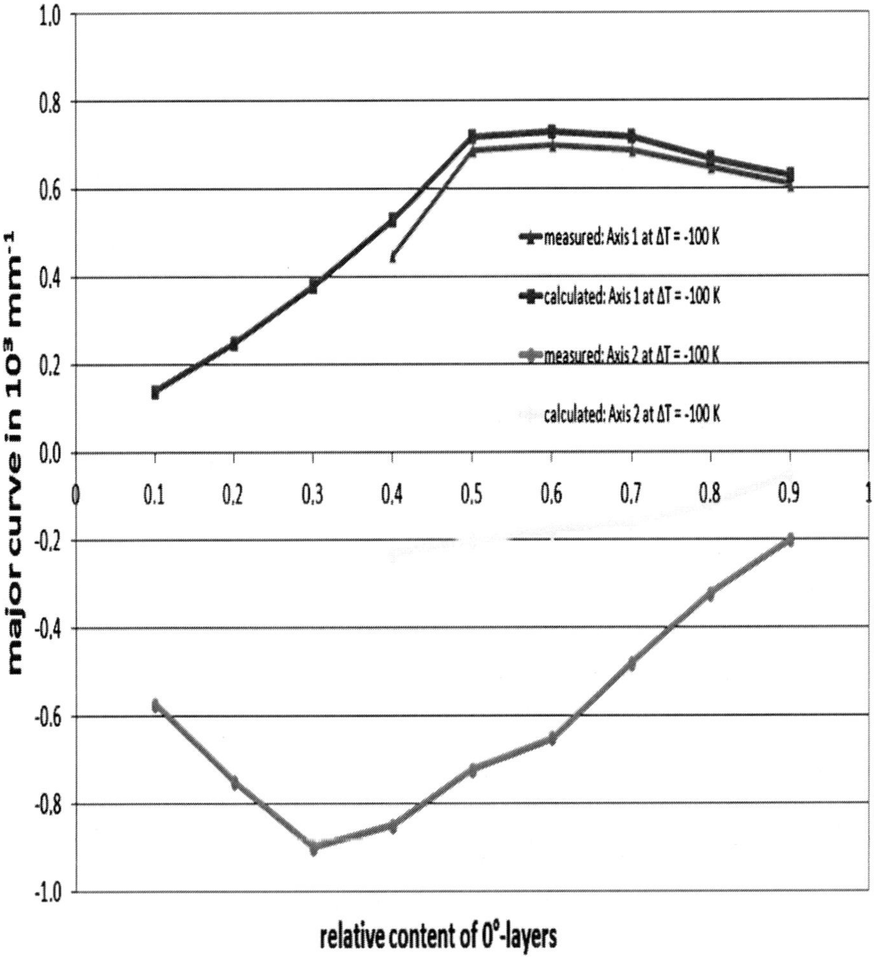

Figure 9. Comparison between calculated and experimentally verified major curvatures at $\Delta T = 100$ K.

3.2. Experimental Verification of Major Curves

Figure 9 displays the theoretical and experimentally verified major curves of the asymmetrical layer structure ($90_n/0_m$), depending on 0˚-layer content with a pure temperature load of ($\Delta T = -100$ K). The major curves around the 1-axis that were determined experimentally agreed qualitatively and approximately also quantitatively with the major curve that was calculated previously (Figure 9 andFigure 10). The minor quantitative differences between the calculated and the experimentally verified major curves around the 1-axis were caused by physico-chemical reactions of the thermosetting matrices. In the calculations, these matrices could be included only insufficiently. Apart from chemical shrinkage, they included residual stress caused by swelling after increased water absorption in or among the molecular chains of the thermosetting matrices [17] .

In contrast to that, there are greater differences between the calculated and experimentally verified major curves around the 2-axis (Figure 9 and Figure 10). In reality however, only one of the major curves prevails, due to stability problems (problems with the transmission/distribution of forces). In this case, this is the curvature around the 1-axis.

3.3. Polymer-Bound Stabilization of 3D-Fabric

Technique and technology from textile manufacturing had an influence on shape through coating with a thermosetting resin system. Mechanical characteristics were adapted to the variety of shapes and drapery. This resulted in an exact contour adapting (Figure 11—left). Reverse deformation was less than 5 percent one day after the stabilization (Figure 11—right). Thus, an exact depiction of the curvature state could be ensured.

Further production studies aimed on an additional stabilization with the cold-setting epoxy resin Indufloor-IB1240 by using of the capillary effect of the pile threads. However, it was not possible to prove an increase in stiffness of the 3D-fabric, due to the insufficient capillary effect of the spacer threads (Figure 12). That can be recognized on the sample "pile thread without epoxy resin contact", which had no common quantitative characteristic peaks with the sample "epoxy resin". As opposed to this, the sample "pile thread with epoxy resin contact" had a high quantitative and qualitative peak alignment with the sample "epoxy resin" (Figure 12).

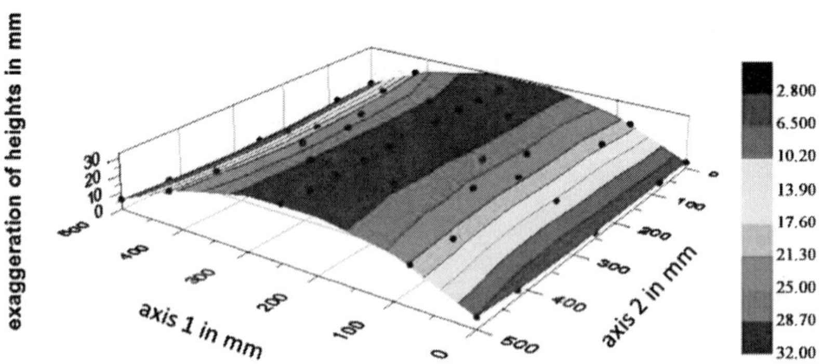

Figure 10. Experimentell curvature of the asymmetrical layer structure 905/06 after temperature load (ΔT = −100 K).

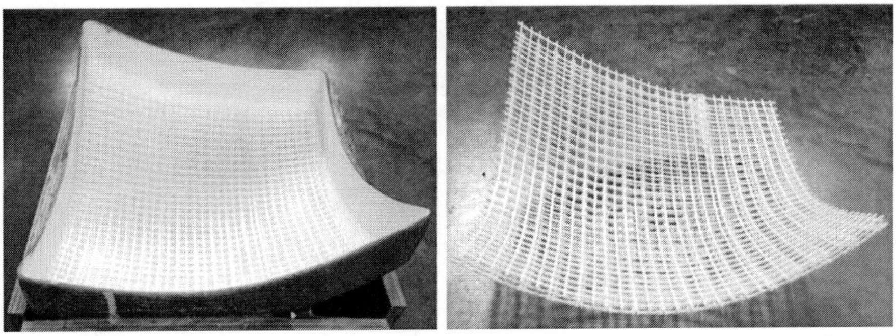

Figure 11. Polymer-bound stabilization of 3D-fabrics with a thermosetting resin system: flexible 3D-fabric in curved GFRP formwork (picture on the left), stabilized 3D-fabric (picture on the right).

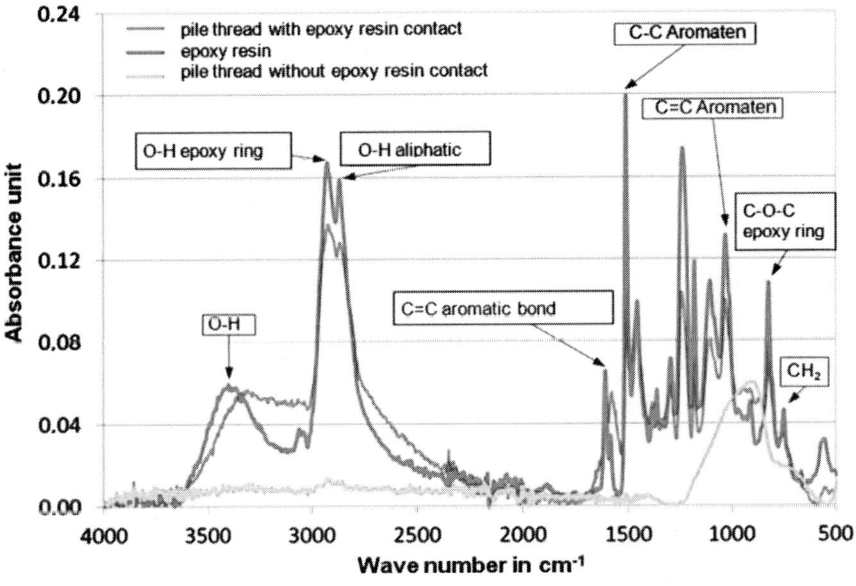

Figure 12. Fourier transform infrared spectroscopy with attenuated total reflection for the validation of polymer-bound stabilization of 3D-fabric.

Figure 13. Production tests of concrete lightweight elements.

Table 3. Characteristics of fresh and hardened fine grained concrete after 28 days.

Characteristic	Fresh concrete	Hardened concrete
geometric bulk density	2.32 g/cm³	2.24 g/cm³
air content	2.5 Vol.-%	-
linear shrinkage	0.71 mm/m	
compressive strength	-	109.3 MPa
3-point bending tensile strength	-	14.74 MPa
CDF-Test	-	M 28 = 1172 g/m² Ru, 28 = 100%

3.4. Properties and Production Tests of Concrete Lightweight Elements

Table 3 shows the characteristics of fresh concrete and hardened concrete after 28 days. In the fresh concrete, an air content of 2.5% by volume and a geometrical density of 2.32 g/cm^3 were found by the use of an air entrainment meter. The total shrinkage deformation, measured with a shrinkage drain, was 0.71 mm/m. The compressive strength and 3-point bending tensile strength were 109.3 MPa respectively 14.74 MPa. For the textile reinforced concrete, a 3-point bending tensile strength of 41.51 MPa was measured.

Of special importance during the production tests was to ensure the evenness of the concrete layer thickness, consistently good surface quality, sufficient stability of the GFRP formwork and to avoid critical cracks both in the concrete and in the formwork system. Also essential were good sheeting qualities (Figure 13).

4. CONCLUSION

The aim of this research project was to develop a flexible, multi-layered formwork system made from glass-fibre reinforced plastic, which allows for a specific adjustment of defined curvature states, utilizing the structural behavior influenced by anisotropy. The adjustment of the coupling effects, which are induced by anisotropy, was calculated in advance analytically by means of the extended laminate theory and numerically by means of the Finite Element Method. A good correspondence of the respective results for the representative shell structures was proved. An experimental verification of these intrinsic coupling phenomena has been conducted with specifically produced textile reinforced concrete-lightweight-elements. Based on the yielded results, ideal layer constructions for the key curvature states and their variation range could be set. Beyond the efficient production of curved concrete-lightweight-elements, GFRP formworks employ excellent concrete qualities on highest classes of face concrete. This contributes to the generation of new forms of architecture and buildings. The intensively conducted numerical, technological and experimental tests show that combinations of concrete and stabilizing spacer fabrics permit the implementation of single and double curved, multi-axially loaded surface structures. Furthermore, the flexible GFRP formwork design allows not only a location-independent implementation of freeform surfaces following the principle "form follows form" but also results in thin-walled and thus extremely light concrete shell structures.

ACKNOWLEDGEMENTS

This work was supported by the Priority Program SPP 1542 of the German Research Foundation (DFG). The authors would like to acknowledge with gratitude the foundation's financial support.

REFERENCES

1. Curbach, M. and Scheerer, S. (2011) Concrete Light—Possibilities and Visions. In: Šrůma, V., Ed., Proceedings of the fib Symposium Prague 2011: Concrete Engineering for Excellence and Efficiency, Prague, 8-10 June 2011, 29-44.

2. Brameshuber, W. (2006) Textile Reinforced Concrete. RILEM Report 36. State-of-the-Art Report of RILEM Technical Committee, TC 201-TRC.

3. Funke, H., Gelbrich, S., Ehrlich, A. and Kroll, L. (2014) Rheological and Mechanical Development of a Fiber-Reinforced Concrete for an Application in Civil Engineering. SOJ Materials Science & Engineering, 2, 1-4.

4. Hofstadler, C. (2008) Schalarbeiten. Springer-Verlag, Berlin, Heidelberg.

5. Curbach, M. and Jesse, F. (2009) Verstärken mit Textilbeton. In: Bergmeister, K., Fingerloos, F. and Wörner, J.-D., Eds., Betonkalender 2010, Teil 1, Abschn. VII. Ernst & Sohn, Berlin.

6. Funke, H., Gelbrich, S., Ehrlich, A. and Kroll, L. (2014) A Fibre-Reinforced Architectural Concrete for the Newly Designed Façade of the Poseidon Building in Frankfurt am Main. Journal of Materials Science Research, 3, 33-39.

7. Greiner, S. (2007) Zum Tragverhalten von Schalen aus ultrahochfestem Faserfeinkornbeton (UHFFB). Z. Beton- und Stahlbetonbau, 100, 77-80.

8. Funke, H., Gelbrich, S. and Ehrlich, A. (2013) Development of a New Hybrid Material of Textile Reinforced Concrete and Glass Fibre Reinforced Plastic. In: Hufenbach, W.A. and Gude, M., Eds., Procedia Materials Science, Materials Science Engineering, Symposium B6—Hybrid Structures, 2, 103-110.

9. Curbach, M., Ortlepp, S., Brückner, A., Kratz, M., Offermann, P. and Engler, T. (2003) Entwicklung Einer Großformatigen, Dünnwandigen, Textilbewehrten Fassadenplatte. Z. Beton- und Stahlbetonbau, 98, 345-350.

10. Herzog, T. and Moro, J.L. (1992) Gespräch mit Felix Candela. In: Arcus 18: Zum Werk von Felix Candela—Die Kunst der leichten Schalen, Verlagsgesellschaft Rudolph Müller, Köln, 10-22.

11. Preisinger, C., Harrer, J., Ressl, C. and Kollegger, J. (2005) Stahlbetonschalen ohne Schalung—Ein neuer Weg im Schalenbau. Beton- und Stahlbetonbau, 100, 31-38.

12. Kroll, L. (2005) Berechnung und technische Nutzung von anisotropiebedingten Werkstoff—Und Struktureffekten für multifunktionale Leichtbauanwendungen. Habilitationsschrift, TU Dresden, Dresden.

13. Kaufmann, J. (2014) Beitrag zu anisotropiebedingten Koppeleffekten bei rotationssymmetrischen mehrschichtigen Faserverbundbauteilen. Promotionsschrift, TU Chemnitz, Chemnitz.

14. Dallinger, S., Pardatscher, H. and Kollegger, J. (2009) Zweifach gekrümmte Schalen aus Betonfertigteilen. Z. Forschung & Entwicklung für Zement und Beton, 5, 32-33.

15. Zapf, W. (1990) Schalung für großformatige gekrümmte Stahlbetonfertigteile. German Patent No. DE 3841579 A1.

16. Henri, V. (1987) Verfahren und Vorrichtung zum Formen von gebogenen Sektionen aus Beton. European Patent No. EP 0 238 168 A1.

17. Schürmann, H. (2007) Konstruieren mit Faser-Kunststoff-Verbunden, Springer Verlag, Berlin, Heidelberg, New York.

Chapter 6

RC Columns Strengthened with Novel CFRP Systems: An Experimental Study

Annalisa Napoli and Roberto Realfonzo *

Department of Civil Engineering, University of Salerno, Via Giovanni Paolo II 132, Fisciano (SA) 84084, Italy

ABSTRACT

This paper presents an experimental study undertaken to investigate the seismic behavior of full scale square (300 mm × 300 mm) reinforced concrete (RC) columns strengthened with novel systems employing carbon fiber-reinforced polymers (CFRP) wraps. Experimental tests were carried out by subjecting specimens to a constant axial load and a cyclically reversed horizontal force applied in displacement control. Results have allowed for investigating the influence of the used strengthening systems on the specimens' performance in terms of flexural strength and ductility as well as on the exhibited failure modes. The effectiveness of the studied techniques is also evaluated by comparing the performance of tested specimens with that of companion columns strengthened with alternative CFRP systems investigated in a previous experimental campaign.

Keywords: RC columns; FRP composites; external strengthening; experimental tests; strength; ductility

1. INTRODUCTION

Recent earthquakes have frequently highlighted the vulnerabilities of existing reinforced concrete (RC) structures to seismic deformation and shear demands. It is known that the most building heritage dating back to the 1970s was designed in order to withstand only gravity loads or according to outdated seismic rules. In particular, these "under-designed" structures are often characterized by an unsatisfactory weak column-strong beam behavior that, under a seismic event, most likely yields to the formation of local hinges in the columns and to a consequent low available global ductility.

In order to improve the strength and mainly the rotational ductility of RC columns, external confinement systems employing fiber-reinforced polymers (FRP) are now frequently proposed as a viable alternative to traditional strengthening measures, mainly consisting of steel or concrete jacketing.

The growing interest in the FRP materials is confirmed by both the great variety of experimental investigations performed by researchers and the number of analytical models published in the literature to predict the compressive strength, the corresponding ultimate axial strain and the stress-strain constitutive law of the FRP confined concrete. National and International design guidelines are also available [1,2,3].

Most experimental tests were performed in uniaxial compression on small scale concrete specimens variably confined by using carbon (C) or glass (G) FRP layers applied via wet-layup; the efficacy of the reinforcement on different cross-sections was examined with emphasis on the circular shape for which an updated database of experimental results can be found in the literature [4,5].

A relatively limited number of experimental studies were carried out on full scale FRP confined circular or rectangular RC columns subjected to axial load and cyclic flexure [6,7,8,9], and few studies take into account the presence of smooth steel rebars as longitudinal reinforcement of specimens [10,11,12,13].

In these studies, it has been widely proven that the FRP confinement significantly improves the ductility of columns; conversely, in presence of cyclic bending, it is not sufficient by itself to provide a comparable increase of strength which is often desired by structural engineers; the reduced performance in terms of strength gain is mainly experienced when the axial load level in the column is low and even more evident in the case of non-circular cross sections.

To address this issue, in a previous experimental investigation the authors have examined the effectiveness of a strengthening layout characterized by the combined use of FRP confinement and longitudinal L-shape steel profiles connected to the concrete stub by means of a steel anchorage system [12,13]. It was demonstrated that such a technique is capable of significantly increasing the flexural strength but a reduction of the available ductility with respect to columns confined with FRP only is observed due to the premature failure of the profile's anchorage system.

In order to overcome this drawback and the issue related to the not ease of installation of the studied technique (in particular of the profile's connection systems), the authors have experimentally investigated the feasibility of two alternative strengthening solutions for full scale RC columns, and the results are discussed in this paper. The first solution consists of placing four steel threaded rods inside concrete grooves which were adhesively bonded into the foundation for a length of about 250 mm; then, the strengthening is completed by external CFRP confinement. The second solution, instead, differs from the first one in a way that additional L-shape steel profiles have been used at the column corners before applying the CFRP confinement.

Columns included in the experimental campaign were reinforced by using deformed steel rebars and manufactured with medium/low concrete compressive strength; the reinforcement details (*i.e.*, lap splice lengths, hoop spaces, *etc.*) were arranged following design rules used in the past and without keeping into account any seismic details.

Tests were carried out at the Laboratory of Materials and Structural Testing of the University of Salerno by subjecting specimens to a constant axial load and a cyclically reversed horizontal force applied in displacement control.

The experimental results have allowed for investigating the influence of the two abovementioned strengthening systems on the specimen performances in terms of flexural strength and ductility as well as on the exhibited failure modes. Finally, the analysis of the nonlinear cyclic behavior of strengthened members has allowed to draw useful information about the stiffness degradation and the energy dissipation phenomena. The performance of tested specimens was also compared with that of FRP strengthened columns examined in the previous experimental campaign [12], and advantages and drawbacks of the explored techniques were also highlighted.

2. EXPERIMENTAL SECTION

The experimental program included 5 full scale RC columns subjected to cyclic flexure under constant axial load. Of these, four members were strengthened by CFRP systems whereas the remaining one was used as control (unstrengthened) specimen; the latter, after being damaged, was also repaired, strengthened with FRP and re-tested. Details about RC specimens are reported in the following.

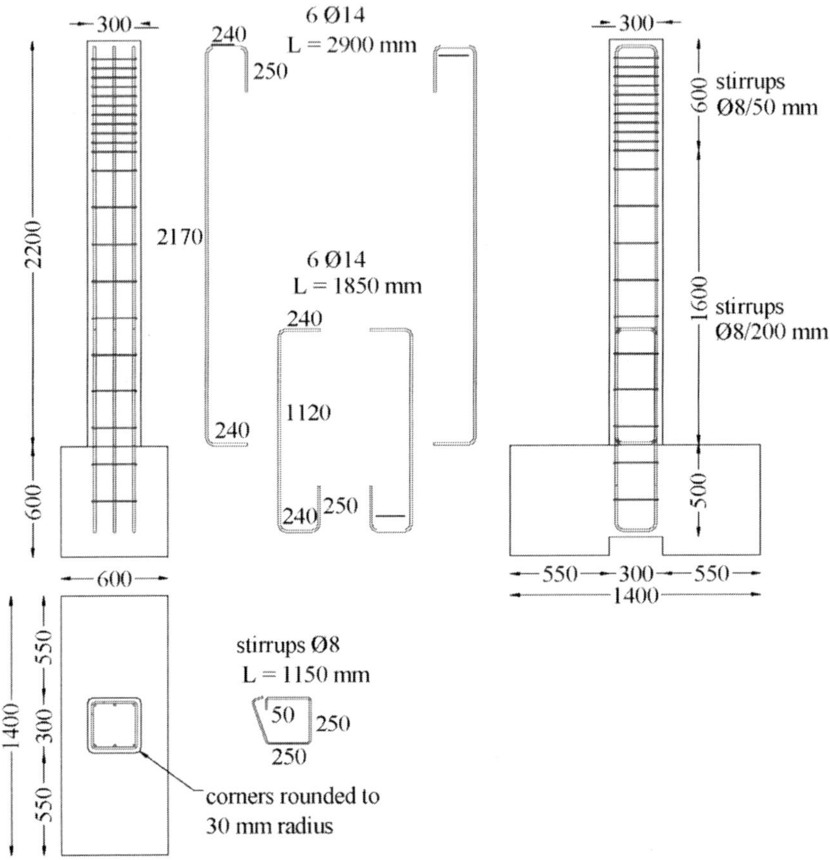

Figure 1. Details of RC specimens.

2.1. Specimens

Test specimens have a square 300 × 300 mm^2 cross section, a length of 2200 mm and a concrete stub of 1400 × 600 × 600 mm^3. They were designed to simulate structural components belonging to gravity load designed existing buildings. For this reason, the concrete mix was chosen to obtain a mean value of the compression strength, f_{cm} ≈ 16–19 MPa, which is frequently found in RC buildings built in the 1960–1970s; the actual value of the concrete strength per column was estimated by testing in compression a set of three 150 mm edge cubic samples, cast along with the column and cured under the same environmental conditions.

In addition, the steel reinforcement details were arranged without any particular care for seismic considerations. Figure 1 illustrates the configurations of the longitudinal and transverse deformed steel reinforcement adopted for the specimens which are the same already considered in the previous experimental campaign [12]. As shown, the longitudinal reinforcement, made of steel grade "B450C" [14], consisted of six 14 mm diameter rebars having a lap splice length at the column-base joint equal to about 40 bar diameters (≈600 mm), as frequently used in Italy; a clear concrete cover of 27 mm was considered at the both sides of the cross section. Tensile tests were performed on small samples to characterize the yield (f_{sy}) and the ultimate strength (f_{su}) of the steel longitudinal rebars and the corresponding strains (ε_{sy} and ε_{su}); from these tests the following average values were found: f_{sy} = 490 MPa and ε_{sy} = 0.23%; f_{su} = 619 MPa and ε_{su} = 11.78%.

The transverse reinforcement, made of the same steel grade as for longitudinal rebars, consisted of 8 mm diameter steel stirrups; 200 mm spaced and closed with 90° hooks at both ends.

2.2. Strengthening Layouts

Figure 2 shows the two strengthening layouts investigated in the experimental campaign which were designed in collaboration with Italian company "Interbau Srl" [15].

The system "type A" (Figure 2a) entailed the use of four (2 + 2) M16 threaded steel rods (tensile strength ≈ 635 MPa), which were placed with mortar inside grooves cut into the concrete. These rods integrating the pre-existing longitudinal steel reinforcement (3 + 3Φ14), crossed the column for a height of about 1450 mm and were anchored with epoxy resin into the concrete foundation for a length of about 250 mm. Then, the column strengthening was completed by a passive confinement obtained with unidirectional CFRP layers. In particular, starting from the column base, the first portion of the member (of about 450–500 mm) was continuously wrapped by using one or two CFRP plies, while the remaining part was strengthened employing a single layer of 150 mm spaced strips, each having a width of 100 mm; the overlap length between layers in the transverse direction was about 300 mm. It is highlighted that the layout of the CFRP confinement was the same already used in the previous experimental campaign [12], and the length of the region with continuous wrapping was based on previous theoretical estimates performed on the development of the plastic hinge zone. In addition, the choice of using two CFRP plies for the continuous wrapping was always based on considerations drawn by experimental tests previously performed by the authors; indeed, in those tests it was shown that the employment of only CFRP confinement with two layers was sufficient by itself to provide an appreciable increase of the flexural strength of the columns. In the experimental campaign discussed herein, instead, it was decided to also investigate the change in the load-carrying capacity of the column when only one CFRP layer was employed as continuous wrapping, since the system "type A" included the use additional steel rods.

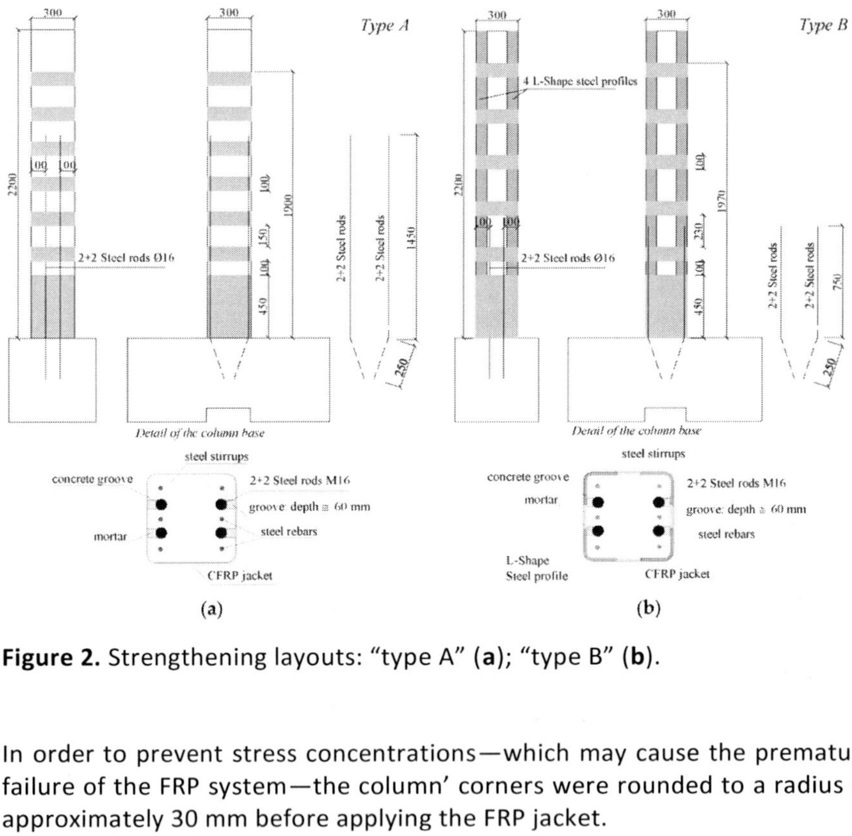

Figure 2. Strengthening layouts: "type A" (**a**); "type B" (**b**).

In order to prevent stress concentrations—which may cause the premature failure of the FRP system—the column' corners were rounded to a radius of approximately 30 mm before applying the FRP jacket.

As provided by supplier, the employed CFRP sheets had an elastic modulus of 390 GPa, a tensile strength of 3000 MPa and an ultimate strain of 0.80%; the thickness of the single layer was ≈0.22 mm. The thixotropic, two part adhesive used to bond the external CFRP reinforcement, instead, was characterized by an elastic modulus equal to 9600 MPa and 11,200 MPa in compression and tension, respectively, and required a 7-days curing time to achieve optimal performances; additional information on the mechanical properties of the used adhesive can be found in the technical sheets provided by the manufacturer [16].

The system "type B" (Figure 2b), instead, also entailed the use of longitudinal cold bent steel profiles along the member corners before applying the external wrapping made of two CFRP layers; they were four 80 mm × 80 mm × 6 mm equal leg angles made of structural steel S235. The angles were glued to the concrete substrate by an epoxy adhesive and were properly manufactured in order to have a rounded corner with a radius of about 30 mm. After placing the angles and before manufacturing the FRP jacket, a 6 mm layer of epoxy mortar was applied on each column side, in order to fill the gap between the profiles (see the detail in Figure 2b).

Figure 3 and Figure 4 show some details about the strengthening procedures used for the considered layouts "type A" and "type B", respectively. In particular, Figure 3a illustrates the shape of the threaded rods used for internally reinforcing the RC columns, pre-bent at one end in order to be inserted into the concrete foundation;Figure 3b,c show the status of the column after: (a) cutting the outer portion of the concrete surface; and (b) filling the grooves and covering the threaded rod with mortar. Finally, Figure 3d depicts the column at the end of the strengthening procedure "type A".

Figure 3. Strengthening procedure: layout "type A".

Figure 4. Strengthening procedure: layout "type B".

Regarding the strengthening procedure "type B", Figure 4a,b illustrate the placing of the steel rods inside the concrete groves, the filling with mortar and the arrangement of the steel angles at the columns corners; Figure 4c,d show, respectively, the subsequent operations dealing with the application of a 6 mm mortar layer on the column and the CFRP external confinement. Finally, Figure 4e depicts the column at the completion of the strengthening procedure "type B".

2.3. Test Set-Up

The test set-up, as shown in Figure 5, is very similar to that already used in the previous experimental campaigns; therefore, specific details, omitted for the sake of brevity, can be found in [12].

Tests were carried out by subjecting the specimens to combined axial and lateral loads. The specimens were restrained to the lab's floor by means of a steel system which consisted of: (a) two transverse beams placed on the RC foundation and fixed to the floor by means of four high strength thread rods that were properly pre-tensioned in order to avoid any stub rotation; and (b) two stiff steel plates fixed onto the ground and placed orthogonally to the load direction at the bottom of the stub in order to prevent any horizontal movement.

The axial load (N) was applied before the horizontal one by pre-tensioning a pair of 40 mm diameter high strength steel rods with a 2000 kN MOOG hydraulic actuator: this actuator, placed at the top of the column, kept the axial load constant during each test. A value of the normalized compression load "v" equal to 0.40 was considered for five tests, whereas for the remaining one v = 0.14 was adopted; this parameter is given by $N/(A_g f_{cm})$, where A_g is the column cross-section area.

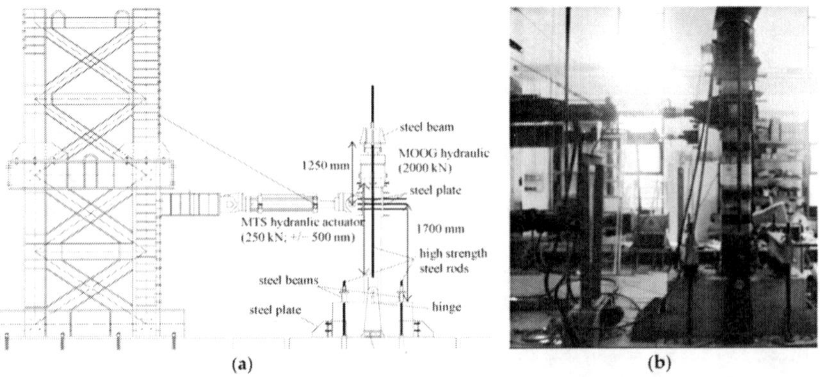

Figure 5. Test set-up.

The horizontal action, instead, was cyclically applied in displacement control by using a 250 kN MTS hydraulic actuator, mounted at 1700 mm from the column base and fixed to a reaction steel frame. An increment of the imposed horizontal displacement every three cycles was considered in order to evaluate the strength and stiffness degradation at repeated lateral load reversals. In particular, after initial cycles performed at 1, 2, 4, and 6 mm, the displacement amplitude was given as fraction of the estimated top yield displacement of the column, Δ_y($\approx$20 mm for columns tested under ν = 0.40 and $\approx$17 mm for the column tested under ν = 0.14). Two different displacement rates were considered during the tests: 0.1 mm/s before the achievement of Δ_y and 1 mm/s after Δ_y.

Tests were stopped well beyond a predetermined "conventional collapse" corresponding to the 15% strength degradation evaluated on the monotonic envelope of the load-displacement curves. Loads, strains, displacements and crack widths were measured during tests as already done in the previous experimental campaign [12]. In particular: (a) horizontal and vertical strains have been monitored using several surface strain gauges placed of the CFRP wrap at about 100 mm from the stub face; (b) longitudinal strains of the steel angles were also recorded by positioning strain gauges on each profile; (c) one potentiometer was used to measure lateral displacements at 1700 mm from the column base (*i.e.*, where the lateral load is applied); and (d) vertical displacements and crack widths at the column-stub interface were monitored by using four LVDTs placed at about 100 mm from the base. In particular, two LVDTs for each side were positioned and one of them had the pin located at 30 mm from the column base; in this way, the difference between the readings of two LVDTs allowed estimating the crack opening at the base and the slip of the outer rebars.

3. RESULTS

Table 1 summarizes the main data and results of the six tests performed in the latest experimental campaign.

Each test is identified by a label providing the following information: the column number (from C1 to C5) with the addition of "N" (*i.e.*, "new") to distinguish these tests from those of the previous experimental campaign [12]; the strengthening system (type ''A'' or ''B'' in Figure 2a,b, respectively). It is noted that the test C1Nr-A refers to the control specimen C1N that, after being tested, was repaired and FRP retrofitted according to the layout "type A". The repairing procedure consisted of: (a) removing the damaged concrete surface in the first portion of the column ($\approx$200 mm), where a slight phenomenon of rebars' buckling was observed at the end of the test C1N;

(b) cleaning and filling the missing concrete with proper fiber reinforced mortar, and (c) subsequently applying the FRP system "type A".

In addition to information about the FRP strengthening layout and the number of FRP layers employed for the continuous wrapping of the column, Table 1 reports: the actual value of f_{cm} estimated per each column by testing in compression a set of three 150 mm edge cubic samples; the normalized value of ν and the corresponding applied axial load N; the peak lateral strengths in the two directions of loading (F^{+}_{max} and F^{-}_{max}), the corresponding displacements (Δ^{+} and Δ^{-}) and the maximum displacements of the column ($\Delta^{+}_{85\%}$ and $\Delta^{-}_{85\%}$) measured at the conventional collapse.

Table 1. Test results.

Strengthening System	Test	# FRP Layer	f_{cm} (MPa)	ν -	N (kN)	F^{+}_{max} (kN)	F^{-}_{max} (kN)	Δ^{+} (mm)	Δ^{-} (mm)	$\Delta^{+}_{85\%}$ (mm)	$\Delta^{-}_{85\%}$ (mm)
unstrengthened	C1N	-	19.4	0.40	700	68.7	69.3	30.0	29.9	61.0	54.0
Type A	C1Nr-A	2	19.4	0.40	700	96.8	100.1	49.9	50.0	109.2	102.9
Type A	C4N-A	1	15.3	0.40	550	90.3	87.3	40.0	40.0	77.6	80.2
Type A	C3N-A	2	19.0	0.40	680	97.8	106.2	50.0	50.0	118.9	86.4
Type B	C2N-B	2	17.9	0.40	645	107.9	113.2	29.0	40.0	70.3	104.4
Type B	C5N-B	2	17.4	0.14	220	78.7	82.2	33.5	34.0	78.7	65.8
unstrengthened	C9-D	-	31.8	0.13	365	71.08	66.32	42.6	42.2	-	-
Type A1	C15-D-A1	2	22.0	0.14	278	110.7	110.8	50.9	50.9	92.4	92.8
Type A1	C24-D-A1	2	15.0	0.40	540	119.8	112.4	49.9	49.9	134.3	103.5

With the aim to better compare the effectiveness of the considered strengthening techniques, Table 1 also reports data and results of three tests performed in the previous experimental campaign, namely C9-D, C15-D-A1 and C24-D-A1 [12], for which the maximum displacements reported herein were computed at 85% conventional collapse (not available for the test C9-D). These specimens had the same geometry and steel reinforcement configurations of those used for the new test members. Except for the specimen C9-D, that was unstrengthened and used herein as control specimen for the only test performed under ν = 0.14 (*i.e.*, test C5N-B), the other members were strengthened using a system "type A1" (see Figure 6a,b). As mentioned earlier, such a system is characterized by the combined use of CFRP confinement and steel profiles placed at the column's corners and connected to the foundation by using an L-shape steel system. The layout of FRP confinement and the arrangement of steel profiles are the same of those in Figure 2b, whereas details on the anchorage system of such profiles, illustrated in Figure 6b, can be found in [12].

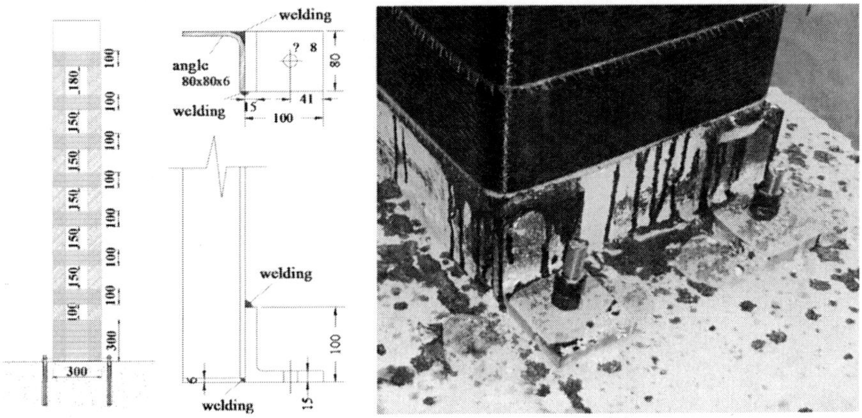

Figure 6. Strengthening systems "type A1" (**a**); details of steel profile's anchorage system (**b**).

It is worth noting that, as it can be observed in Table 1, the compressive strengths of concretes used for manufacturing the specimens C15-D-A1 and C24-D-A1 are rather comparable to those of the "new" test specimens; conversely, the specimen C9-D is characterized by an f_{cm} value significantly greater than the other concrete strengths thus affecting a direct comparison with the test C5N-B. In addition, it is highlighted that a value of ν slightly lower than 0.14 (ν = 0.13) was used for the test C9-D, due to technical issues of the testing machine.

The following sections better detail the discussion of the test results.

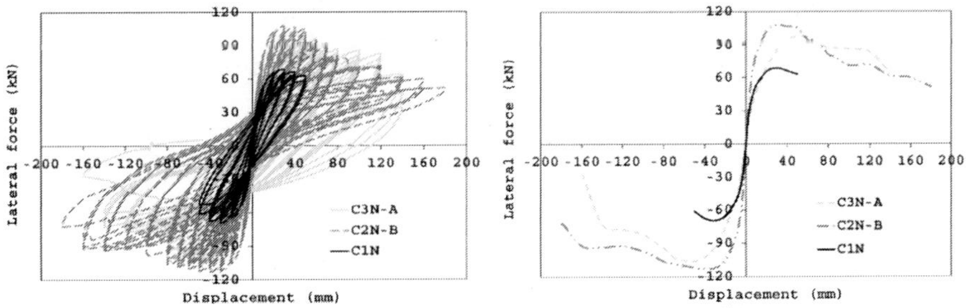

Figure 7. *F*–Δ hysteresis curves (**a**) and envelopes (**b**): columns C1N, C2N-B and C3N-A.

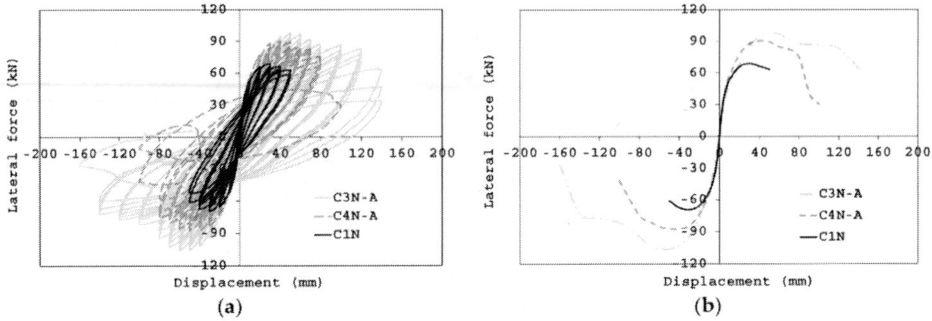

Figure 8. *F*–Δ hysteresis curves (**a**) and envelopes (**b**): columns C1N, C3N-A and C4N-A.

3.1. Cyclic Curves

Figure 7 and Figure 8 show the lateral force (*F*)–displacement (Δ) hysteretic curves and the corresponding *F*–Δ envelopes for some tested specimens. In particular, in Figure 7a the cyclic response of the control specimen (C1N) is compared with the performance of two members strengthened with the FRP systems "type B" and "type A" (samples C2N-B and C3N-A, respectively).

As expected, the reference column experienced a rapid degrading behavior in the post-peak phase and achieved the collapse under low values of the imposed displacement. Disregarding the type of strengthening system, the specimens C2N-B and C3N-A have exhibited significant increases of strength over the control specimen (about 60% and 48%, respectively). The improvement of the column's confinement through the use of steel profiles avoided the rupture of the FRP wrapping and delayed the crushing of concrete; indeed, after achieving the conventional collapse, the sample C2N-B showed a less degrading behavior with respect to the sample C3N-A and was capable of still undergoing significant lateral displacements. It is highlighted that the failure of two additional steel rods (on the same column's side), due to a slight slippage combined to a fatigue phenomenon, was observed for the specimen C2N-B at imposed displacements of 90 and 120 mm (corresponding to lateral loads in the post-peak phase equal to about 78 and 70 kN, respectively.

Figure 8a,b shows the dependence of the cyclic response on the number of FRP layers employed for the continuous wrapping at column base; the *F*–Δ comparisons clearly highlight the beneficial role of using a double layer of CFRP layers in improving the deformation capacity of specimens. In particular, it is observed that the lateral displacement exhibited by the specimen C3N-A at the conventional collapse is about 30% over that measured for the counterpart confined with a single FRP layer; an increase of flexural strength equal to about 15% is also computed.

Figure 9a shows the *F*–Δ monotonic envelopes of test specimens subjected to ν = 0.40 which are also compared with the curve of the column C24-D-A1 [12]. Furthermore, in order to by-pass the dependence of the response on the concrete strength value, in Figure 9b the experimental results have been represented in terms of normalized flexural strength–drift ratio (μ–δ) curves. μ and δ are given by:

$$\mu = \frac{F \cdot L_S}{B \cdot H^2 \cdot f_{cm}} = \frac{M}{B \cdot H^2 \cdot f_{cm}}; \Delta = \frac{\Delta}{L_S} \tag{1}$$

where *B* and *H* are the width and depth of the column cross section, respectively; *F* is the horizontal force applied by the actuator and Δ the corresponding displacement; L_s (=1700 mm) is the column's shear span.

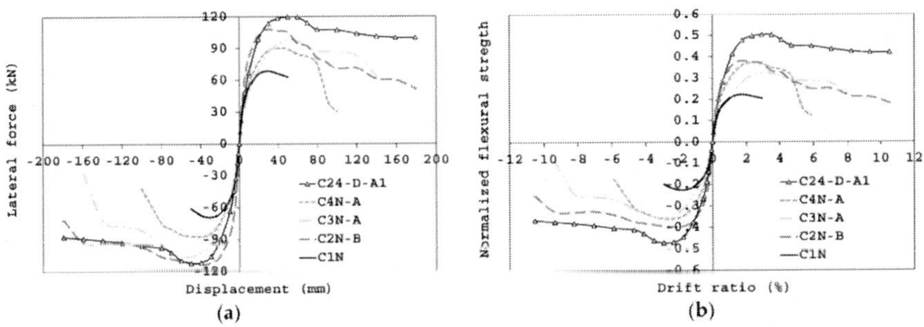

Figure 9. *F*–Δ envelopes (**a**) and μ–δ envelopes (**b**) under ν = 0.40.

From the experimental *F*–Δ comparisons it is noted that the strengthening system "type A1" provides increases of flexural strength and ductility slightly greater than those obtained through the layout "type B"; conversely, as mentioned earlier, this beneficial effect is offset by the need of time consuming and labor-intensive operations to provide a proper installation of the anchor system of steel profiles; on site, such installation is even more difficult when the anchoring of column's steel profiles is made into a beam member rather than a concrete foundation. In addition, the strength increase progressively reduces with the occurrence of the slipping of the steel connector linking the profile's anchor system to the concrete foundation, as shown in Figure 10 [12]; therefore, the performance provided by the use of a system "type A1" strongly depends on the proper design of the steel profile's anchoring to the concrete beam/foundation.

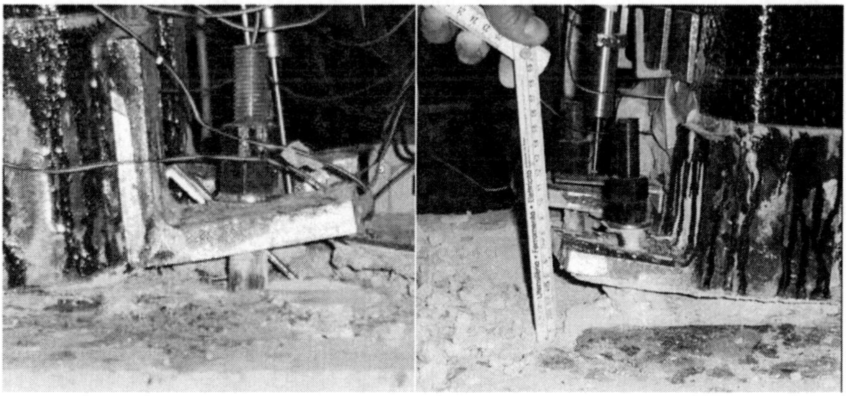

Figure 10. Pull-out of the steel connector from the concrete foundation.

Figure 11 allows for investigating the influence of the axial load level ν on the experimental response of specimens strengthened with the system "type B". For this purpose, Figure 11a shows the comparison between the *F*–Δ cyclic curve of column C2N-B and that of the specimen C5N-B, subjected to ν equal to 0.40 and 0.14, respectively; Figure 11b, instead, depicts the corresponding *F*–Δ envelopes, with the addition of the curve related to the reference specimen C1N, available for ν = 0.40.

As expected, the presence of a higher axial load approximately produces a 37% strength increase, thus confirming the beneficial effect of the CFRP confinement for significant values of the acting axial load.

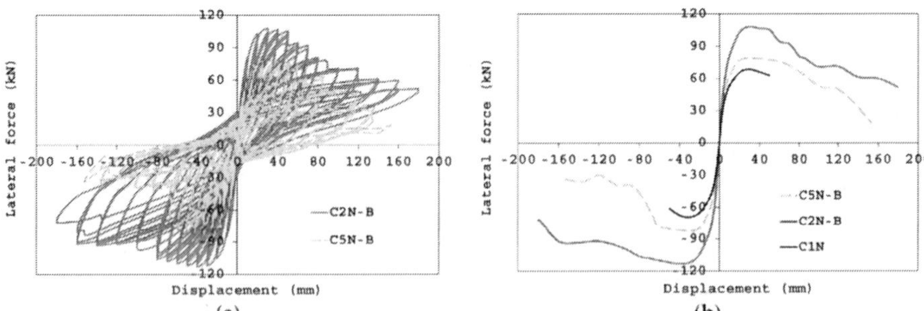

Figure 11. *F*–Δ hysteresis curves (**a**) and envelopes (**b**): columns C2N-B (ν = 0.40) and C5N-B (ν = 0.14).

Furthermore, to better examine the performance of the specimen C5N-B subjected to a low value of the axial load, in Figure 12a,b the cyclic response and the corresponding envelope curve are compared with those of the control (unstrengthened) member C9-D and the strengthened column C15-D-A1, tested in the previous experimental campaign [12]. It is worth highlighting that, for a better comparison of the experimental behaviors, both graphs have been represented in normalized terms μ–δ.

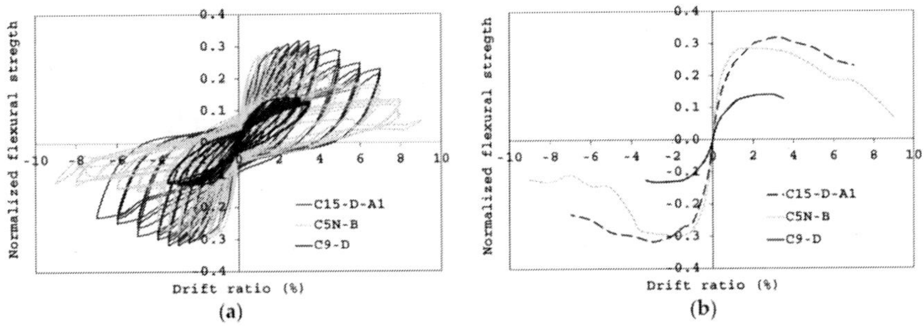

Figure 12. Normalized cyclic (**a**) and envelope (**b**) *F*–δ curves for specimens subjected to ν = 0.14.

The plots allow for verifying the beneficial effects of the studied strengthening technique even in presence of low axial load. On average, the flexural strength of the sample C5N-B is about twice that characterizing the control member and only 10% lower than the strength shown by the counterpart C15-D-A1. However, the specimen C5N-B exhibited an abrupt strength degradation in the post-peak phase caused by the rupture of the additional rods arranged in the column, first of which occurred at a displacement of about 70 mm.

Finally, in Figure 13, the performance of the unstrengthened specimen C1N is compared with that of the repaired one C1Nr-A. The plots allow for verifying the effectiveness of the system "type A" as a repairing technique of damaged columns; as shown, indeed, the application of this system is not only capable of restoring the performance of the original specimen, but also of improving the experimental response in terms of both flexural strength and deformation capacity. By looking at the plot in Figure 13b, it is observed that, except for the reduced initial stiffness (as shown in the *F*–Δ enlargement), the member C1Nr-A exhibited an increase of strength and ductility over the counterpart C1N of about 43% and 84%, respectively.

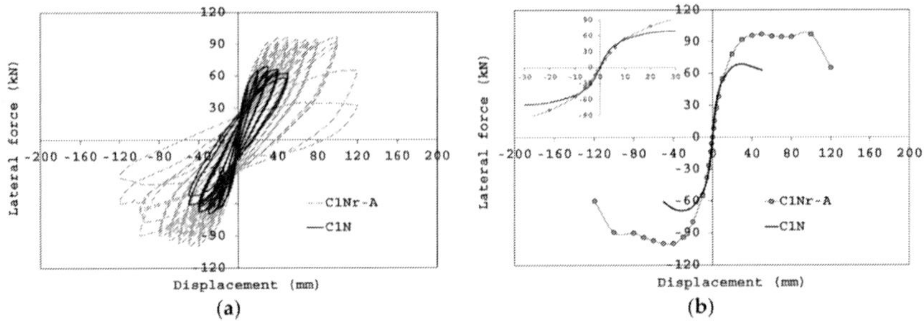

Figure 13. *F*–Δ hysteresis curves (**a**) and envelopes (**b**): columns C1N and C1Nr-A.

3.2. Failure Mode

Figure 14, Figure 15 and Figure 16 show the crack pattern and failure mode exhibited by test specimens. In particular, as already observed in the previous experimental campaign [12], the unstrengthened member C1N experienced cracking phenomena and significant damages concentrated in the first 500 mm from the column base (Figure 14). A flexural crack first developed at the column-foundation interface at a displacement value of 20 mm; the width of this crack did not significantly increase during the test since the specimen rapidly achieved the collapse. Other flexural cracks evenly distributed along the member at a distance of about 200 mm, which approximately corresponds to the stirrups spacing (Figure 14a). Lacking adequate confinement, the specimen was rapidly involved by the development of vertical cracks due to the incipient buckling of steel rebars (Figure 14b). Relevant signs of damage and concrete spalling occurred in the first 200 mm from the column base at an imposed displacement of 40 mm (Figure 14c); the test was stopped soon to avoid severe damage of the member since it had to be repaired and re-tested.

Figure 14. Crack pattern: control specimen C1N.

Disregarding the type of strengthening system, the presence of a FRP jacket has contributed to: inhibit the crack opening; delay the buckling of longitudinal rebars. Figure 15a,b illustrate the damage exhibited by strengthened specimens C4N-A, C3N-A and C1Nr-A. As expected, the use of a single FRP layer for the external confinement (test C4N-A) strongly penalized the performance of the column since a first rupture of the sheet prematurely occurred in the first 100 mm from the base (see Figure 15a taken at a displacement of 50 mm). After the FRP system failure, the column rapidly achieved the collapse with the buckling of steel rebars and of the additional rods, stirrups opening and crushing of concrete (Figure 15b). The cyclic behavior of the strengthened specimen significantly improved by doubling the number of FRP layers (test C3N-A); in this case, the FRP rupture occurred in the first 50 mm from the column base only at a displacement value of 120 mm. Figure 15c shows the status of damage at the column base at the end of the test: fracture and buckling of some rebars and rods and stirrups opening were observed.

(a) (b) (c)

Figure 15. Damage of specimens strengthened with the system “type A”.

The presence of longitudinal steel profiles (system “type B”) has further inhibited the damage and crack propagation. Independently on the applied axial load v (0.14 or 0.40) the arrangement of steel profiles avoided the FRP rupture at the base even for high values of the imposed displacement, as shown in Figure 16a. Approaching the end of the test, the FRP exhibited only a bulging in the first 50 mm from the base, whereas inside the wrap the concrete column was severely damaged (Figure 16b). Specimens C5N-B and C2N-B have both experienced the rupture of additional steel rods at column-foundation interface.

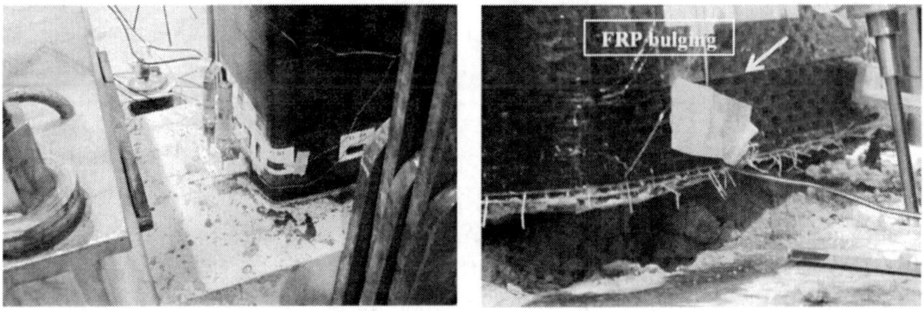

(a) (b)

Figure 16. Damage of specimens strengthened with the system "type B".

3.3. Stiffness Degradation and Energy Dissipation

Based on the experimental results, it was possible to evaluate the mean value of stiffness for the *i*-th cycle by using the following ratio [17]:

$$k = \frac{\left|F^{+}_{\max,i}\right| + \left|F^{-}_{\max,i}\right|}{\left|s^{+}_{\max,i}\right| + \left|s^{-}_{\max,i}\right|} \tag{2}$$

where F+max,i and F−max,i denote the maximum values of the lateral force at the *i*-th cycle in the push and pull directions, and s+max,i and s−max,i are the corresponding displacements.

By normalizing the stiffness of each displacement cycle *k* with respect to that of the first cycle k_I, it was possible to obtain a measure of the stiffness degradation.

Figure 17a depicts the relationship between the normalized stiffness k/k_I and the drift ratio for all the specimens subjected to the higher axial load (ν = 0.40). As expected, lacking the external confinement, the unstrengthened column C1N experienced a more rapid strength degradation with respect to the other specimens. Conversely, it seems that the column C3N-A, although not provided with longitudinal steel profiles, exhibited a strength degradation lower than the counterparts C2N-B and C24-D-A1. This result can be motivated by observing Figure 17b showing that the specimen C3N-A was characterized by an initial stiffness already reduced and significantly lower than that of the control specimen C1N; this evidence might have affected the corresponding strength degradation making it lower than the effective one. However, from Figure 17b it can be observed that also the column C4N-A showed an initial stiffness lower than the unstrengthened specimen. For both columns C3N-A and C4N-A, this reduced initial stiffness can be partly motivated in the installation procedure of the system "type A" which, by requiring the cutting of concrete grooves for introducing the additional steel rods, inevitably induces a pre-damage of the specimens; probably, this pre-damage is not relevant in the case of the specimen C2N-B where the presence of steel angles in some way does not affect the initial stiffness.

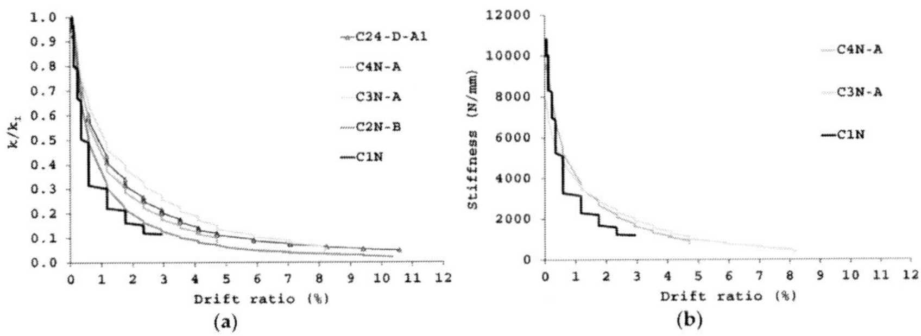

Figure 17. Stiffness degradation (**a**) and stiffness-drift ratio relationship (**b**) specimens subjected to ν = 0.40.

In Figure 18 the stiffness degradation of columns strengthened with the system: "type B" is compared with that exhibited by the counterparts previously strengthened with the alternative system employing steel profiles at the columns corners (*i.e.*, systems "type A1"). As shown, regardless of the axial load level, the specimens C15-D-A1 (ν = 0.14) and C24-D-A1 (ν = 0.40) exhibited a lower stiffness degradation with respect to the corresponding members C5N-B (ν = 0.14) and C2N-B (ν = 0.40); this result implies that the introduction of internal steel rods in addition to steel profiles and CFRP confinement was less effective than the use of a more complex system like the type A1 one.

Finally, Figure 19 depicts the relationships between the cumulative dissipated energy (*E*) and the imposed drift ratio. This energy parameter was calculated from the area under the lateral load-displacement response enclosed within one complete cycle up to the achievement of the conventional collapse for each specimen. In particular, the plot in Figure 19a refers to all the specimens subjected to the higher axial load (ν = 0.40), whereas that in Figure 19b allows for a comparison between systems "type B" and "type A1".

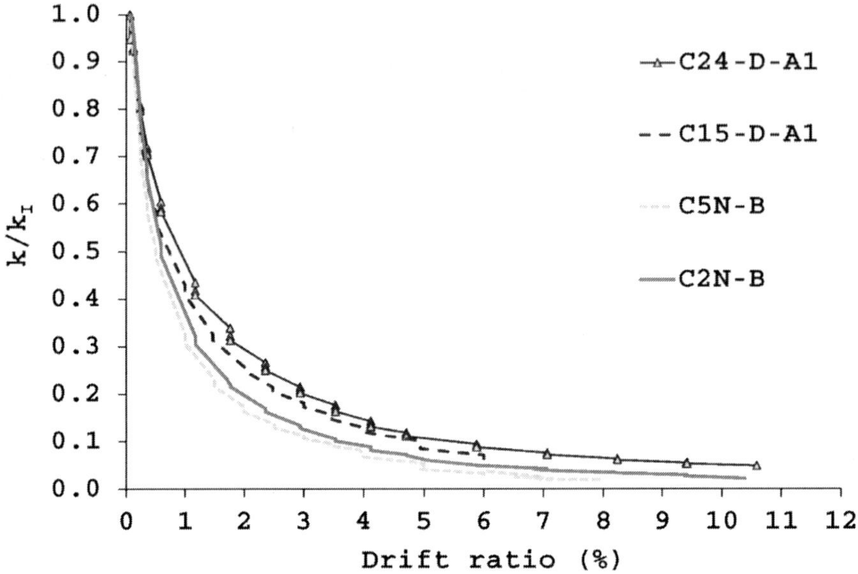

Figure 18. Stiffness degradation: specimens strengthened with the systems "type B" and "type A1".

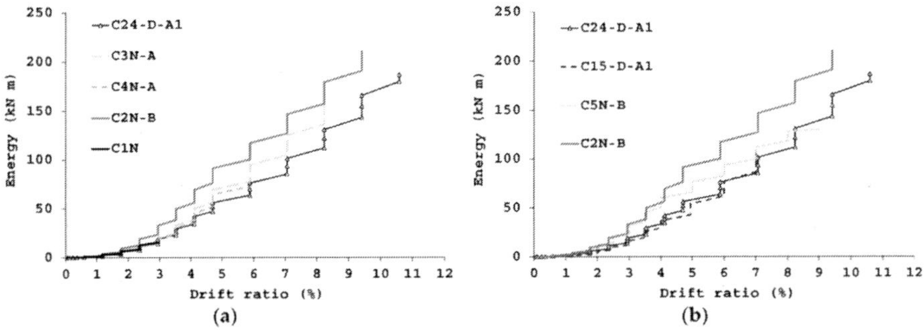

Figure 19. Energy dissipation: specimens subjected to ν = 0.40 (**a**) and comparison between systems "type B" and "type A1" (**b**).

As already highlighted in [12], from Figure 19a it is observed that the CFRP system produces a significant increase of the cumulative dissipated energy; however, in comparison with the unconfined member C1N, there is no remarkable improvement of the dissipated energy at a given displacement cycle (except for the specimen C2N-B). In the case of the strengthening system "type A", it is noted that doubling the number of the FRP layers (from one to two) produced only a small increase of the energy performance (compare test C4N-A and C3N-A). Furthermore, disregarding the presence of steel profiles at column corners, the plot in Figure 19a shows that, in terms of dissipated energy, the CFRP strengthening systems "type A" and "type B" are more effective than the previous solution "type A1", whose width of the hysteresis cycles is strongly affected by the performance of steel profile's anchoring to the concrete beam/foundation. This experimental evidence is also confirmed by observing the plot in Figure 19b where, regardless of the applied axial load level, the energy dissipated by columns strengthened with the system "type B" is significantly greater than that computed for the counterparts C15-D-A1 and C24-D-A1.

3.4. Strain Values

As already mentioned, measurements of the hoop FRP strain were recorded during tests by eight strain gauges (labelled H1, H2 and H8) arranged on the FRP jacket at a distance of about 100 mm from the column base. In addition, in the case of the strengthening system "type B", two couples of strain gauges (labelled V1–V2, and V3–V4) were placed on the respective steel L-shape profiles (at about 600 mm from the column base) to monitor the strain distribution during tests.

Figure 20 shows for the strengthened members the envelopes of the FRP transverse strains (ε_h) plotted under both positive (push) and negative (pull) directions of the imposed displacement Δ; such plots allow for obtaining a qualitative trend of the strains achieved by the FRP jacket. In particular, the strain diagrams refer to the maximum values recorded at the first of the three cycles performed at each level of the imposed displacement, up to the end of the test (when the measures were available); negative and positive values of ε_h indicate compressive and tensile strains of the FRP jacket, respectively.

Under Δ = 0, *i.e.*, at the origin of the ε–Δ axes, the plots report the small values of the FRP strains exhibited after the application of pure axial load on the columns.

Observing the plots in Figure 20, it is highlighted that, disregarding the type of strengthening system, the maximum tensile strains never attained the ultimate FRP strain value indicated by the supplier ($\varepsilon_{f,u}$ = 8‰); in particular, except for measures provided some strain gauges, the higher strains were below the 2‰–2.4‰ threshold (which is only 1/4–1/3 of $\varepsilon_{f,u}$).

By comparing the plots in Figure 20a,b it is observed that, at a given level of the imposed displacement, the tensile strains recorded in the case of the test C4N-A were generally higher than those measured by doubling the number of layers for the continuous FRP jacket (test C3N-A), with maximum values recorded by the strain gauge H2.

The FRP strains typically increase in the case of the strengthening system "type B" (see test C2N-B in Figure 20c), where the values recorded by strain gauges H6 and H8 are well over the 2‰ threshold. Conversely, reduced values were, on average, recorded for the specimen C5N-B (Figure 20d) where a reduced efficacy of the FRP jacket is obtained under lower axial load levels (ν = 0.14).

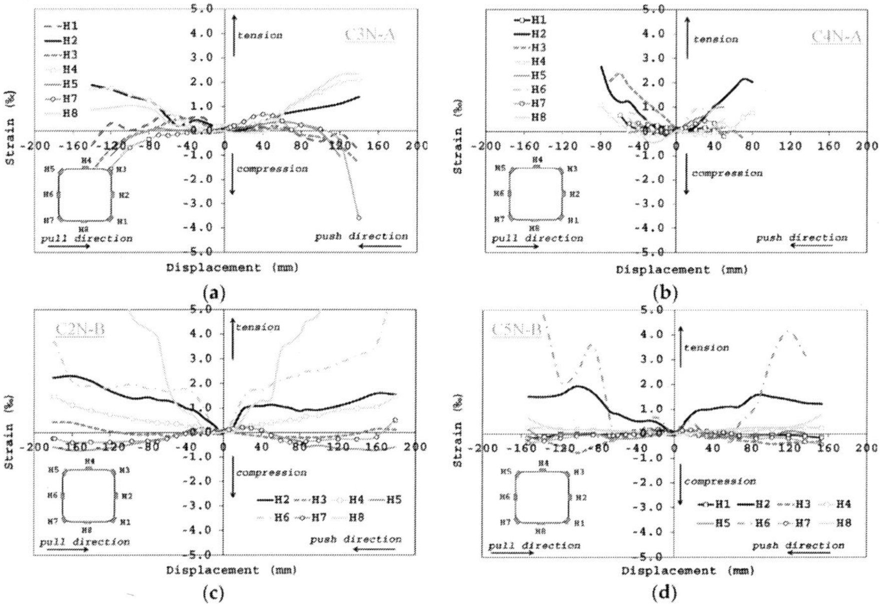

Figure 20. FRP strains: tests C3N-A (**a**); C4N-A (**b**); C2N-B (**c**); and C5N-B (**d**).

Finally, by focusing on the strengthening system "type B", Figure 21 shows the envelopes of the longitudinal strains (ε_v) recorded by the couples of strain gauges V1, V2 and V3, V4 placed on the L-shape steel profiles (the measures were not always available); negative and positive values of ε_v indicate compressive and tensile strains in the profiles, respectively. A rather symmetric response of the strains in the push and pull directions is observed from the plots, with maximum values recorded under the higher axial load (test C2N-B in Figure 21a); however, it is highlighted that such strains were always below 0.9‰, thus implying that the use of steel profiles mainly contributes to improving the external confinement of the columns rather than significantly increasing their flexural strength.

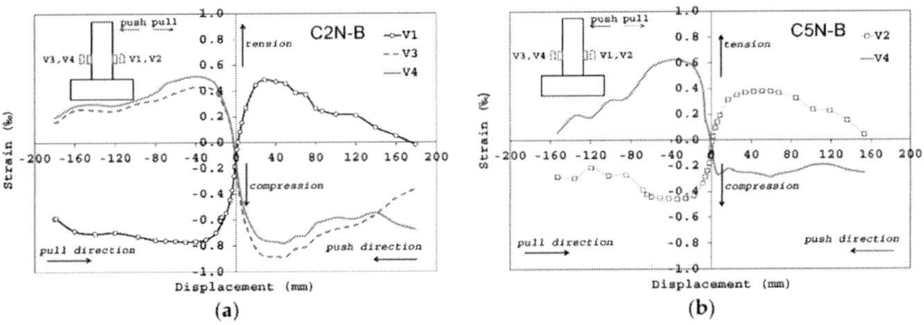

Figure 21. Strains on the L-shape steel profiles: tests C2N-B (**a**) and C5N-B (**b**).

4. CONCLUSIONS

In this paper, results from cyclic tests on RC columns strengthened with novel CFRP systems have been presented and compared with those of companion specimens tested in a previous experimental campaign. In particular, two strengthening solutions were investigated herein, namely "type A" and "type B". The first solution ("type A") consisted of placing four steel threaded rods inside concrete grooves which were adhesively bonded into the foundation for a length of about 250 mm; then, the strengthening was completed by external CFRP confinement. The second solution ("type B"), instead, differed from the first one for the additional use of L-shape steel profiles at the column corners before applying the CFRP confinement.

From the performed tests it has been highlighted that in presence of high axial load (v = 0.40), the strengthened specimens have exhibited increases of strength ranging from 48% to 60% with respect to the control sample. In the case of samples strengthened by the system "type A", a delayed FRP failure has been noticed by doubling the number of FRP layers, and 30% greater displacements at collapse were measured. The addition of steel profiles at column corners (system "type B") avoided the failure of the FRP wrap itself and delayed the crushing of the concrete; indeed, beyond the conventional collapse, the specimens were still capable of undergoing significant lateral displacements.

Furthermore, it has been shown that the columns strengthened with the system "type B" have provided increases of flexural strength and ductility slightly lower than those obtained through the layout "type A1", examined in the previous experimental campaign. However, it is worth noting that this beneficial effect is offset by the need of time consuming and labor-intensive operations to provide a proper installation of the anchor system of steel profiles; on site, such installation is even more difficult when the anchoring of column's steel profiles is made into a beam member rather than a concrete foundation. Finally, even though a greater stiffness degradation was computed, in terms of dissipated energy, the CFRP strengthening systems "type A" and "type B" were more effective than the previous solution "type A1", whose width of the hysteresis cycles is strongly affected by the performance of steel profile's anchoring to the concrete beam/foundation.

ACKNOWLEDGEMENTS

The Authors gratefully acknowledge Interbau Srl (Milan, Italy) for the technical and financial support to the experimental program.

AUTHOR CONTRIBUTIONS

Annalisa Napoli and Roberto Realfonzo have evenly contributed to all the steps needed to produce this article, from conceiving, designing and performing tests, to analyzing the experimental results and writing the paper.

REFERENCES

1. American Concrete Institute (ACI) Committee 440. *Guide for the Design and Construction of Externally Bonded FRP Systems for Strengthening Concrete Structures*; ACI 440.2R-08; Farmington Hills, MI, USA, 2008.

2. Fédération internationale du béton (*fib*). *Externally Bonded FRP Reinforcement for RC Structures*; Bulletin No. 14; fib Bulletin: Lausanne, Switzerland, 2001.

3. National Research Council (CNR). *Guide for the Design and Construction of Externally Bonded FRP Systems for Strengthening Existing Structures (Materials, RC and PC Structures, Masonry Structures)*; CNR-DT200; Rome, Italy, 2004.

4. Realfonzo, R.; Napoli, A. Concrete confined by FRP systems: Confinement efficiency and design strength models. *Compos. B Eng.* **2011**, *42*, 736–755.

5. Realfonzo, R.; Napoli, A. Confining concrete members with FRP systems: Predictive *vs.* design strain models. *Compos. Struct.* **2013**, *104*, 304–319.

6. Iacobucci, R.D.; Sheikh, S.A.; Bayrak, O. Retrofit of square concrete columns with carbon fiber reinforced polymer for seismic resistance. *ACI Struct. J.* **2003**, *100*, 785–794.

7. Harries, K.A.; Ricle, J.R.; Pessiki, S.; Sause, R. Seismic retrofit of lap splices in nonductile square columns using carbon fiber-reinforced jackets. *ACI Struct. J.* **2006**, *103*, 874–884.

8. Harajli, M.H.; Dagher, F. Seismic strengthening of bond-critical regions in rectangular reinforced concrete columns using fiber-reinforced polymer wraps. *ACI Struct. J.* **2008**, *105*, 68–77.

9. Rutledge, S.; Seracino, R.; Kowalsky, M.; Witt, M. FRP repair of damaged large-scale circular reinforced concrete columns. In Proceedings of the 15th World Conference on Earthquake Engineering (WCEE), Lisboa, Portugal, 24–28 September 2012.

10. Bousias, S.N.; Fardis, M.N.; Biskinis, D. Retrofit of RC columns with deficient lap splices. In Proceedings of Fib Symposium "Keep Concrete Attractive", Budapest, Hungary, 23–25 May 2005.

11. Bournas, D.A.; Lontou, P.V.; Papanicolaou, C.G.; Triantafillou, T.C. Textile-reinforced mortar *versus* fiber-reinforced polymer confinement in reinforced concrete columns. *ACI Struct. J.* **2007**, *104*, 740–748.

12. Realfonzo, R.; Napoli, A. Cyclic behavior of RC columns strengthened by FRP and steel devices. *J. Struct. Eng.* **2009**, *135*, 1164–1176.

13. Realfonzo, R.; Napoli, A. Results from cyclic tests on high aspect ratio RC columns strengthened with FRP systems. *Constr. Build. Mater.* **2012**, *37*, 606–620.

14. *Nuove Norme Tecniche per le Costruzioni (NTC)*; Consiglio Superiore dei Lavori Pubblici: Rome, Italy, 2008. (In Italian)

15. Interbau Srl. Available online: http://www.interbau-srl.it (accessed on 30 June 2015).

16. Sikadur®30 Adhesive for bonding reinforcement. Available online: http://www.sika.com (accessed on 30 June 2015).

17. Mayes, R.L.; Clough, R.W. *State of the Art in Seismic Shear Strength of Masonry—An Evaluation and Review*; EERC Report: Berkeley, CA, USA, 1975.

Chapter 7

Flexural Strengthening of RC Slabs with Prestressed CFRP Strips Using Different Anchorage Systems †

José Sena-Cruz [1], Julien Michels [2,*], Yunus Emre Harmanci [2,3] and Luís Correia [1]

[1]*Institute for Sustainability and Innovation in Structural Engineering (ISISE), Department of Civil Engineering, University of Minho, Guimarães 4800-058, Portugal*

[2]*Structural Engineering Research Laboratory, Empa, Swiss Federal Laboratories for Materials Science and Technology, Überlandstrasse 129, Dübendorf 8600, Switzerland*

[3]*Department of Structural Mechanics, ETHZ, Swiss Federal Institute of Technology Zurich, Stefano-Franscini-Platz 5, Zürich 8093, Switzerland*

ABSTRACT

Externally Bonded Reinforcement (EBR) technique has been widely used for flexural strengthening of concrete structures by using carbon fiber-reinforced polymers (CFRP). EBR technique offers several structural advantages when the CFRP material is prestressed. This paper presents an experimental and numerical study on reinforced (RC) slabs strengthened in flexure with prestressed CFRP strips as a structural strengthening system. The strips are applied as an externally bonded reinforcement (EBR) and anchored with either a mechanical or a gradient anchorage. The former foresees metallic anchorage plates fixed to the concrete substrate, while the latter is based on an accelerated epoxy resin curing followed by a segment-wise prestress force decrease at the strip ends. Both anchorage systems, in combination with different CFRP strip geometries, were subjected to static loading tests. It could be demonstrated that the composite strip's performance is better exploited when prestressing is used, with slightly higher overall load carrying capacities for mechanical anchorages than for the gradient anchorage. The performed investigations by means of a cross-section analysis supported the experimental observation that in case a mechanical anchorage is used, progressive strip debonding changes the fully bonded configuration to an unbonded end-anchored system. The inclusion of defined debonding criteria for both the anchorage zones and free length between the anchorage regions allowed to precisely capture the ultimate loading forces.

Keywords: RC plate strengthening; CFRP strips; prestressing; mechanical anchorage; gradient anchorage; static loading tests; cross-section analysis; numerical simulations

1. INTRODUCTION

In recent decades, several research initiatives has been devoted to the development of Fibre Reinforced Polymer (FRP) materials and related strengthening techniques used by the construction industry [1,2,3,4,5,6,7,8]. Nowadays, these materials and strengthening techniques are relatively well-unknown by researchers, designers and constructors being, respectively, for the case of retrofitting concrete structures the Carbon FRP materials (CFRP) and the externally bonded reinforcement (EBR) technique the most used ones. Epoxy adhesives are typically used as bond agent. In some specific applications, the use of prestressed FRP materials is appropriate or even required [9,10,11,12,13,14,15]. This alternative technique combines the benefits of the passive EBR FRP systems with the advantages associated with external prestressing. El-Hacha *et al.* [16] pointed out the following main advantages: (1) superior durability since non-corrosive materials are used; (2) the existing deflections are reduced; (3) the crack widths reduction and the onset of cracking initiation is delayed; (4) internal steel reinforcement strains are relieved; (5) higher fatigue failure resistance; (6) more efficient use of the concrete and FRP; (7) opposes stresses due to both dead and live loads; (8) ultimate capacity can be further increased; (9) it can be worked as a substitute of internal pre-stress that has been lost; (10) shear capacity is increased by the longitudinal stresses induced by prestressed FRP laminates. Mainly three systems have been proposed to prestress the FRP materials [16]: (1) cambered prestressing systems; (2) prestressing against an independent element; and (3) prestressing against the element to be strengthened. Up to now the systems that apply the prestressing against the element to be strengthened are the most used, due to their main advantages in terms of practicality. At the ends of the prestressed FRP elements special end-anchorage systems are required in order to transfer the high shear stress developed from the reinforcing material into the concrete substrate in order to avoid a premature FRP peeling-off failure. From all the proposed systems, two of them have been mostly used, mainly [17]: the mechanical anchorage (MA) system fixing the ends of the FRP reinforcement to the concrete substrate by means of metallic plates and bolts and the gradient anchorage (GA). Detailed information about these two systems is given in Section 2.3.

To assess to the performance of the mechanical and gradient anchorage systems, an experimental program was carried out. In addition to the study of anchorage systems (MA and GA), the cross-section geometry of the CFRP laminate strip was also investigated. For that purpose, eight reinforced concrete (RC) slabs, strengthened with pre-stressed CFRP laminate strips and anchored according the to the MA and GA systems, were monotonically tested under displacement control up to failure by using a four-point bending test configuration. The observed performance of the tested RC slabs is presented and critically analyzed. Additionally, by using a cross-sectional analytical model, numerical simulations were carried out to better understand the obtained results with regard to the overall structural behavior. This calculation tool is able to predict the structural behavior of RC structures strengthened with an additional reinforcement. This external reinforcement can be implemented as either a fully bonded or an unbonded strengthening element.

2. EXPERIMENTAL PROGRAM

2.1. Test Program

The experimental program was composed of eight reinforced concrete (RC) slabs, as presented in Table 1: (1) two slabs were used as reference specimens (REF1 and REF2); (2) one slab (SL50_1.4_EBR) was strengthened with one laminate according to the externally bonded reinforcement (EBR) technique; (3) five slabs were strengthened with one EBR prestressed CFRP laminate strip, three by using the mechanical anchorage (MA) system and two with the gradient anchorage (GA) system, with distinct geometry types of CFRP laminate strips, mainly 50 mm × 1.4 mm, 50 mm × 1.2 mm, and 80 mm × 1.2 mm. The CFRP strip prestrain $\varepsilon_{f,p}$ was measured with strain gauges installed at its mid-length.

Table 1. Experimental program (b_f = strip width, t_f = strip thickness, $\varepsilon_{f,p}$ = strip prestrain, f_{cm} = average compressive strength on cylinder of concrete at slab testing day, E_{cm} = average Young's modulus of concrete at slab testing day; the values between parentheses are the corresponding coefficients of variation (CoV)).

Slab	CFRP Strip ($b_f \times t_f$)	Anchorage	$\varepsilon_{f,p}$ (‰)	f_{cm} (MPa)	E_{cm} (GPa)
REF1	-	-	-	53.4 (4.3%)	32.2 (7.5%)
REF2	-	-	-	57.4 (3.0%)	32.6 (0.1%)
SL50_1.4_EBR	50 mm × 1.4 mm	-	-	53.4 (4.3%)	32.2 (7.5%)
SL50_1.4_MA	50 mm × 1.4 mm	MA	3.98	53.4 (4.3%)	32.2 (7.5%)
SL50_1.4_GA	50 mm × 1.4 mm	GA	4.05	53.4 (4.3%)	32.2 (7.5%)
SL50_1.2_MA	50 mm × 1.2 mm	MA	4.19	49.5 (3.1%)	-
SL80_1.2_MA	80 mm × 1.2 mm	MA	3.99	57.4 (3.0%)	32.6 (0.1%)
SL80_1.2_GA	80 mm × 1.2 mm	GA	4.06	57.4 (3.0%)	32.6 (0.1%)

The specimens' geometry and test configurations are shown in Figure 1. The RC slabs have a total length of 2600 mm and rectangular cross section of 600 mm (width) by 120 mm (depth). In the tension region, the slabs are reinforced with 5 longitudinal steel bars of 8 mm of diameter (5Ø8), whereas in the compression zone 3Ø6 are used. Secondary and transverse reinforcement is also used composed of closed steel stirrups of Ø6 at 300 mm spacing.

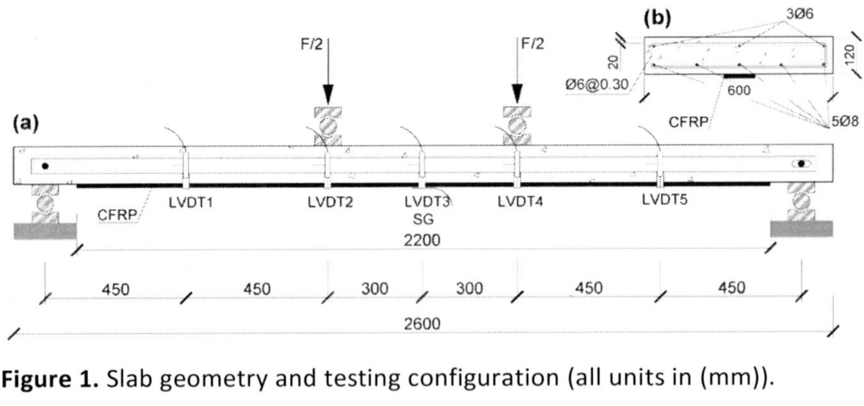

Figure 1. Slab geometry and testing configuration (all units in (mm)).

All the slabs were monotonically tested up to the failure, by using a servo-controlled equipment under midspan displacement control at a rate of 1.2 mm/min. Four point bending test configuration was used. Instrumentation includes: (1) 5 linear variable differential transducers (LVDT1 to LVDT5) to measure the displacements along the slab's longitudinal axis; (2) a TML BFLA-5-3 strain gauge sensor (SG) to record the CFRP strain at its mid-length; and (3) a load cell to measure the applied load F. Figure 1 shows the position of the five displacement transducers: three (LVDT2 to LVDT4) in the pure bending zone with the range of ±75 mm and a linearity error of ±0.10% and two (LVDT1 and LVDT5) between the supports and the applied load points with a range of ±25 mm and the same linearity error. The load cell, with a maximum measuring capacity of 200 kN and a linear error of ±0.05%, was placed between the actuator and the steel device that distributed the load into equal parts (see Figure 1).

2.2. Materials

The material characterization included the evaluation of the mechanical properties of concrete, steel, CFRP laminate strips and epoxy adhesive.

Three batches were used to cast the RC slabs. Six cylindrical concrete specimens with 150 mm of diameter and 300 mm of height of each concrete batch were used to evaluate the modulus of elasticity and compressive strength through the LNEC E397-1993:1993 [18] and NP EN 12390-3:2011 [19] recommendations, respectively. These tests were performed at the same age than for the static tests of the slabs. An average compressive strength of about 53.4 MPa was obtained, see Table 1. The tensile properties of the steel reinforcement were evaluated using the standard NP EN ISO 6892-1:2012 [20]. Three specimens with 500 mm of length were used for each bar type. To measure the strains during the linear branch of the stress-strain curve, a clip gauge with a 50 mm of initial length was used. Figure 2 shows the relationship between the tensile stress and displacement between grips, whereas Table 2 includes the obtained results in terms of Young's modulus E_s, yielding strength $f_{s,y}$, and ultimate tensile strength $f_{s,u}$. In spite of both bar types presented similar Young's modulus, Figure 2 shows a yielding plateau and lower values in terms of $f_{s,y}$ and $f_{s,u}$ for the case of a Ø8 bar.

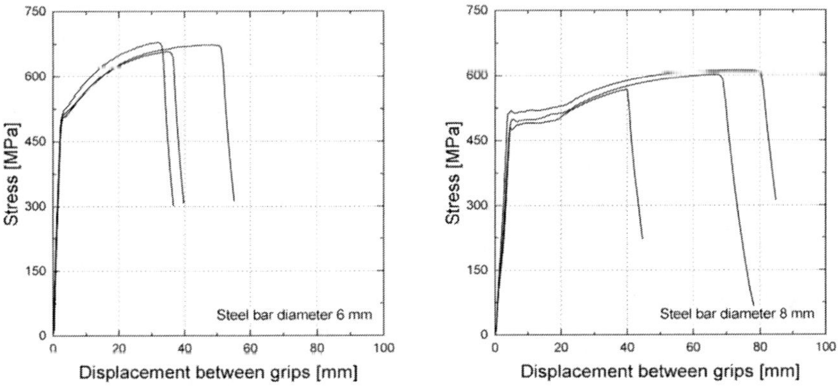

(a) (b)

Figure 2. Tensile stress *vs.* displacement between grips of (**a**) Ø6 steel bar; (**b**) Ø8 steel bars.

Table 2. Tensile properties of the reinforcing steel bars.

Bar Type (ø in (mm))	E_s (GPa)	f_y (MPa)	$f_{s,u}$ (MPa)
ø6	209.5 (8.5%)	579.3 (3.3%)	669.7 (1.7%)
ø8	212.8 (9.7%)	501.4 (5.9%)	593.9 (3.9%)

The CRFP laminate strips (Type: *S&P Laminates CFK*) used in the present experimental program are prefabricated (pultruded) and consist of unidirectional carbon fibers held together by an epoxy vinyl ester resin matrix. This composite material is provided in rolls of 150 m total length, has a smooth external surface and a black appearance. Tensile properties were assessed throughout ISO 527-5:1997 [21] recommendations. Four samples were used for each CFRP section type. A clip gauge with 50 mm of initial length was used to measure the strains. Table 3 presents the obtained results in terms of Young's modulus E_f and ultimate tensile strength $f_{f,u}$.

Table 3. Mechanical properties of the CFRP strips.

Geometry (mm × mm)	E_f (GPa)	$f_{f,u}$ (MPa)
50 × 1.2	167.7 (2.9%)	2,943.5 (1.6%)
50 × 1.4	154.8 (4.6%)	2,457.1 (1.2%)
80 × 1.2	164.6 (0.2%)	2,455.3 (5.0%)

The two-component epoxy adhesive (Type: *S&P Resin 220*) was used to bond the CFRP laminate strips to the concrete substrate. This epoxy adhesive is a solvent free, thixotropic and grey two-component (Component A = resin, light grey color and Component B = hardener, black color). The tensile properties of this epoxy adhesive was assessed following the ISO 527-2:1993 [22]. The specimens were cured with similar conditions to the ones that are used with mechanical (MA) and gradient (GA) anchorage systems (Michels *et al.*, 2013 [17]). Six and four specimens were tested in each series, and were cured during 7 and 3 days, respectively, for the MA and GA. The difference in age of the epoxy at the testing date is explained by the different necessary curing time in order to reach a more or less stable tensile strength and stiffness. For the GA, the epoxy resin undergoes accelerated curing at high temperatures, and hence already develops significant strength and stiffness during the anchorage application. For the MA, however, the epoxy resin undergoes normal curing at room temperature and is hence delayed in strength and stiffness development compared to the resin with an initial accelerated curing. Detailed studies on strength development of epoxy resins under different curing conditions can be found in Moussa *et al.*[23]. Figure 3 shows the stress-strain relationships from the tests carried out. Secant modulus of elasticity was defined using the two points with 0.05% and 0.15% of strain, being the average values equal to 8.68 GPa (CoV = 5.7%) and 5.79 GPa (CoV = 9.5%) for the case of MA and GA, respectively. Finally, the obtained average tensile strength was equal to 20.72 MPa (CoV = 18.25%) and 14.85 MPa (CoV = 11.28%) for the MA and GA, respectively.

2.3. End-Anchorage

2.3.1. Anchorage Systems

As mentioned before, two distinct end-anchorage systems were investigated: (1) the mechanical anchorage (MA), which uses metallic plates at the ends of the FRP strips, and (2) the gradient anchorage (GA), which is produced by sector-wise heating of the adhesive and a step-wise releasing of the prestressed force F_p. The studied commercial systems, available from the same supplier as for the CFRP reinforcement and adhesive, present several common and specific equipment. These are presented in Figure 4. The MA system uses metallic anchor plates of 200 mm × 272 mm × 12 mm fixed with 6 M16 8.8 bolt anchors, guides, clamp units, frames, hydraulic hoses and cylinders, and a manual hydraulic pump. For the case of the GA system, guides, clamp units, frames, hydraulic hoses and cylinders, manometer and valves, and heating device for accelerated curing the adhesive, are required. Depending on the level of prestressing to be applied and the length of the CFRP laminate strip to be prestressed, the MA can have one or both active anchorages. In the case of the GA system both anchorages are active. Regarding the duration of the strengthening application, in the MA system all the steps are concluded after approximately 24 h, whereas for the case of GA only 3 to 4 h are required. In the MA a significant cure of the epoxy adhesive along the CFRP laminate is required, which takes in normal conditions at about 24 h (Granja *et al.* [24]). Since the GA system uses the adhesive's ability to cure fast at high temperatures, significant lower period of time is needed to conclude the strengthening application. More detailed information about the accelerated curing properties of commercially available adhesives are listed in Czaderski *et al.* [25] and Michels *et al.* [26]. Finally, at the end in the case MA system in addition to the CFRP laminate strips two metallic anchors will exist, while for the case of the GA system the strengthening solutions will be composed by a purely bonded CFRP strip/epoxy/concrete system.

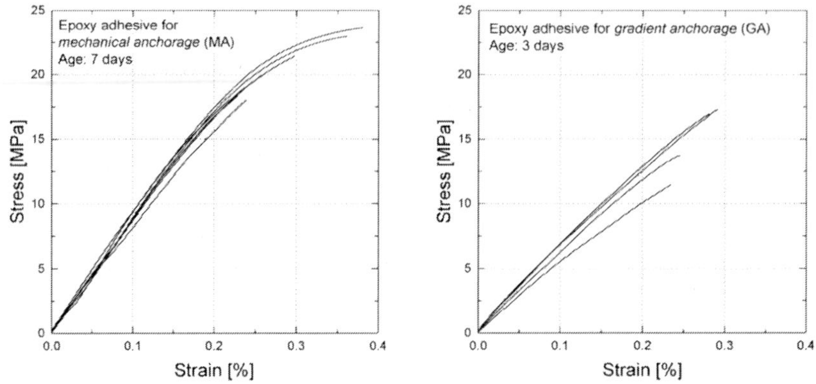

(a) (b)

Figure 3. Tensile stress *vs.* strain of epoxy adhesive used with (**a**) Epoxy adhesive for MA; (**b**) Epoxy adhesive for GA.

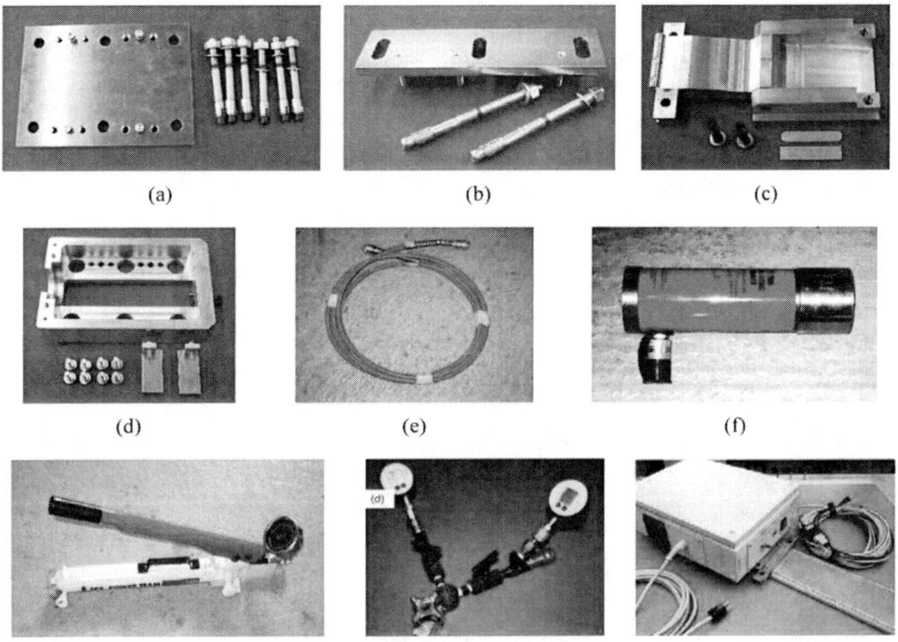

(a) (b) (c) (d) (e) (f) (g) (h) (i)

Figure 4. Equipment composing the anchorage systems: (**a**) metallic plate anchors (MA); (**b**) guides (MA&GA); (**c**) clamp unit (MA&GA); (**d**) frame (MA&GA); (**e**) hydraulic hoses (MA&GA); (**f**) hydraulic cylinder (MA&GA); (**g**) manual hydraulic pump (MA&GA); (**h**) manometer and valves (GA); (**i**) heating device (GA).

2.3.2. Strengthening Procedure

Figure 5a,b schematically present the main steps for the strengthening using the MA and GA systems, respectively, mainly:

(A) The first step consists on surface preparation of the slabs where the CFRP laminate strip will be applied. In the present case a grinding stone wheel is used. Subsequently, compressed air was used to clean the treated region.

(B) Several holes are drilled to accommodate temporary and permanent bolt anchors. GA system complies only temporary bolts, while for the case of MA system, six M16 8.8 permanent bolt anchors are used to fix each metallic anchorage plate. The HIT-HY 200-A® chemical bond agent was used to fix these bolts to concrete. Then, aluminium guides are placed in the right position to guide and fix the clamp units. Afterwards the clamp units are placed in its position, *i.e.*, in-between the guides at each extremity of the slab.

(C) The CFRP laminate strip is cleaned with a solvent and the epoxy adhesive is prepared according to the information given by the producer's technical datasheet. It should be remarked that before this step, the CFRP laminate strip had been already instrumented with a strain gauge at its mid-length (see also Section 2.1). Subsequently, the adhesive is applied on the surface of the CFRP laminate as well as on the concrete surface region in contact with the laminate. The CFRP laminate strip is then placed in its final position and slightly pressed against the concrete substrate. In the end a minimum of 2 mm of thickness of epoxy is assured. The clamp units are installed in-between the guides at each extremity of the slab and are closed to fix the CFRP laminate strip.

(D) For the case of MA system, metallic anchor plates are slightly grinded with sandpaper and cleaned with a solvent before they are installed in their predefined location. In the case of GA, heating devices are placed in the gradient zone.

(E) The aluminium frames are then placed on their predefined locations and fixed against the concrete with the bolt anchors in order to accommodate the hydraulic cylinder.

(F) Eventually, the hydraulic cylinder cylinders are installed in the aluminium frame and using a manual hydraulic pump, the prestress is applied to the CFRP laminate strip.

After prestressing the CFRP, distinct procedures are followed for the case of the MA and the GA systems. For the case of the MA system, in order to increase the confinement provided by the metallic anchor plates at the anchor's region and consequently reduce the probability of the CFRP laminate sliding at the ends, a torque of 150 N·m is applied in the six bolt anchors. Additional fixing screws mounted in-between the frame and the clamp units are used to block the prestressing system in order to avoid pre-stress losses during the curing of the epoxy. As referred before, the strengthening application is concluded after approximately 24 h. In the end, all the equipment is removed and the temporary bolt anchors and CFRP laminate outside of the anchor plates are cut off.

For the case of the GA system, in the ambit of the present experimental program, a 600 mm anchorage length was used, composed of three sectors (50/80 mm wide and 200 mm long each). During the application of the gradient, the specimens are always monitored in terms of applied force by the hydraulic cylinders and temperature at the distinct sectors composing the heating devices. Details about the evolution of the temperature inside the epoxy adhesive in the respective sectors during the application of the gradient can be found in Michels*et al.*, 2013 [17]. Generally, the aimed epoxy temperature T_a is about 90 °C. The releasing force in each sector was equal to about 1/3 of the total applied force F_p. The total heating duration was 35 min with the subjected temperature profile of the heating elements T_h as shown in Figure 6 for both strips 50 mm × 1.4 mm and 80 mm × 1.2 mm. Each force release ΔF was performed approximately 15 min after the initiation of the cooling phase of the corresponding sector. This waiting time was chosen in a way to ensure that the epoxy adhesive has cooled down to temperatures below 50 °C.

As previously mentioned, in each single prestressed slab a strain gauge was installed on the at midspan of the CFRP laminate strip to control the prestrain level $\varepsilon_{f,p}$ to be applied during the strengthening phase and to record the strains attained during the monotonic test up to the failure. The CFRP laminate strip was prestrained to an average value of more or less 0.4%. Table 1 includes the values of the registered prestrain at the middle of the laminate at the end of the strengthening. The specimens were kept in lab environment after strengthening at least one month prior to testing.

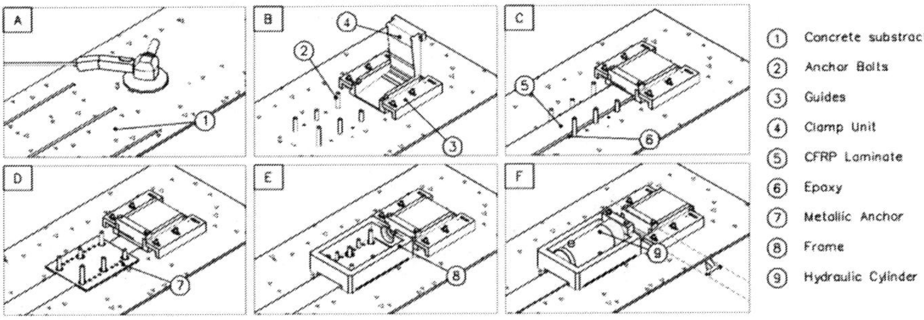

(a)

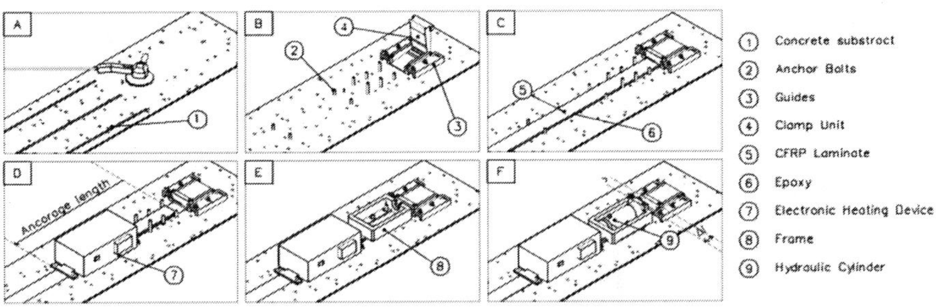

(b)

Figure 5. Application procedures for MA and GA systems, (**a**) Mechanical Anchorage; (**b**) Gradient Anchorage.

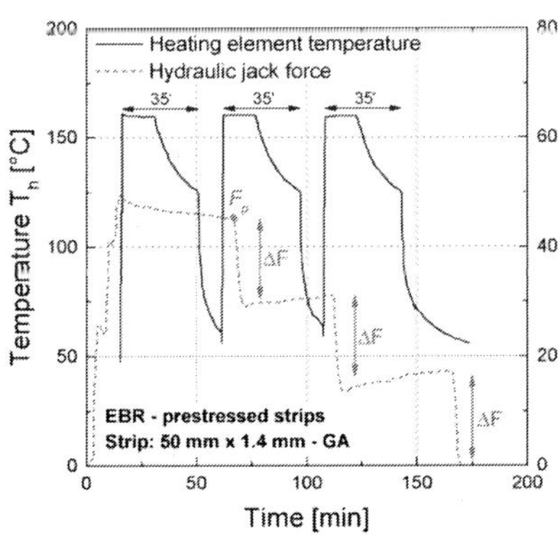

(a)

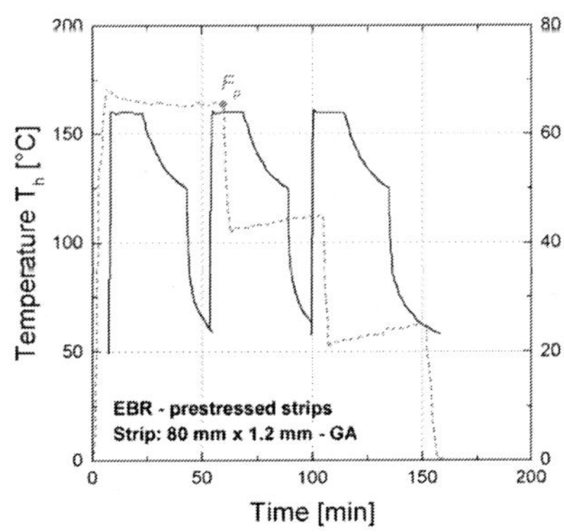

(b)

Figure 6. Heating element temperature T_h and gradient force F_{GA} evolution over time. (**a**) 50 mm × 1.4 mm strip; (**b**) 80 mm × 1.2 mm strip.

3. NUMERICAL INVESTIGATION

3.1. Cross Section Analysis (CSA)

Cross-Section Analysis (CSA) was performed with a Matlab-based calculation tool. The differences in strain distribution between *bonded* and *unbonded* external reinforcements are given in Figure 7. For *bonded*configurations, conventional force equilibrium and strain compatibility in the cross-section can be considered. After establishing the moment-curvature (M,κ)-relation, deflection values can afterwards simply be derived by double integration of the curvature over the span length. For *unbonded* configurations, however, strain compatibility between the CFRP tensile strain and the bottom concrete surface is not given anymore. In this case, CFRP tensile strain ε_f, which is for each loading step constant over the whole strip length L, has to correspond to the average concrete tensile strain $\varepsilon_{c,bot,avg}$ at the bottom for each loading step:

$$\varepsilon_f = \frac{1}{L} \cdot \int_0^L \varepsilon_{c,bot}(x) \cdot dx = \varepsilon_{c,bot,avg} \tag{1}$$

Such a condition requires a much more complex calculation procedure, implicating iteration procedures for the cross-section analysis on several locations over the whole span lengths and afterwards a verification of the CFRP tensile strain and average concrete bottom strain compatibility. Tension stiffening has been included through the*tension chord model* by Marti *et al.* [27]. The exact calculation procedure with a vectorized optimization approach can be found in Harmanci [28] and Harmanci *et al.* [29].

3.2. Debonding Failure Modes

In addition to the listed material failures in compression (concrete) and tension (concrete, steel, CFRP), several debonding criteria have been included in the cross-section analysis in order to precisely assess the ultimate load carrying capacity.

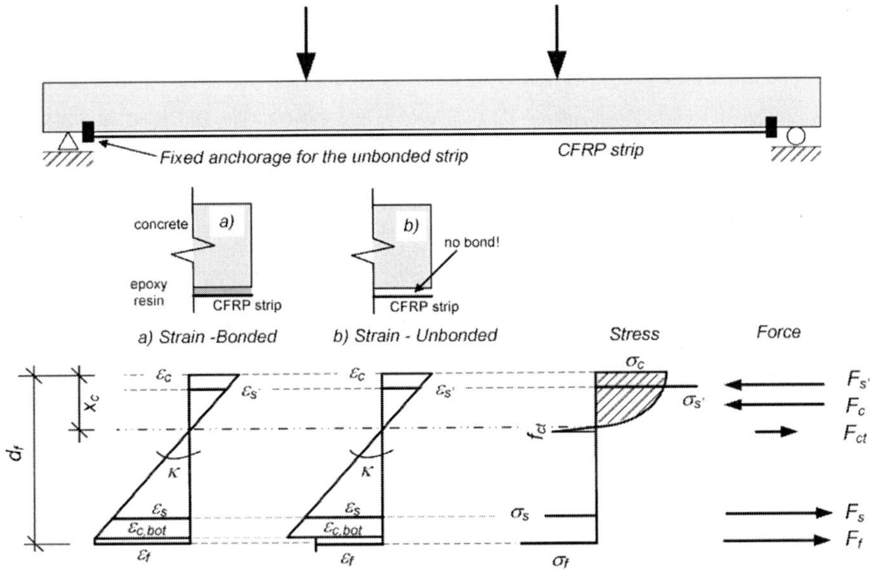

Figure 7. Differences in strain distribution for a bonded and unbonded CFRP/concrete system configuration.

3.3. Constitutive Material Laws

The different materials have been implemented as shown in Figure 8. Concrete was defined linear elastic up to failure in tension, and by a Hognestad parabola in compression [30]. Reinforcing steel was considered elastic-plastic with a simple bi-linear law, and eventually CFRP was modelled linear-elastic up to unidirectional tensile failure.

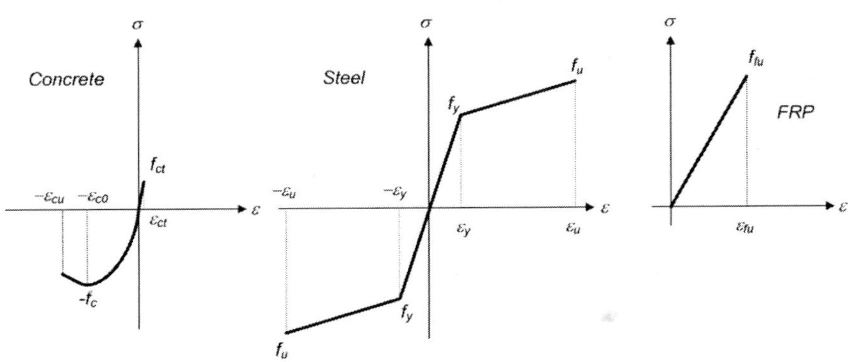

Figure 8. Constitutive material laws for concrete, steel, and CFRP.

1. For initially unstressed EBR, the debonding modes listed in the Swiss design recommendations SIA166 [2] have been used. These include:

   - *End-anchorage failure* (DM1, with: $F_{f,cr}$ being the CFRP strip force at the last crack, F_R being the anchorage resistance):

$$F_{f,cr} < F_R \tag{2}$$

   - *Intermediate strip debonding due to interfacial shear stress exceeding* (DM2, limit value for the shear strength τ_{lim} being a function of the concrete tensile strength):

$$\tau < \tau_{lim} \tag{3}$$

   - *Intermediate strip debonding due to CFRP strip strain exceeding* (DM3):

$$\varepsilon_f < 8‰ \tag{4}$$

2. For prestressed CFRP strips, design criteria for the gradient anchorage failure defined in the PhD thesis of Czaderski [31] have been included.

3. For prestressed CFRP strips with a mechanical anchorage (Type: S&P Clever Reinforcement Company AG), an upper limit for the CFRP strip strains ε_f of at maximum 10‰ at the anchorage start have been implemented. This value is based on suggestions given by Suter and Jungo [32].

4. For both prestressed configurations with either a gradient or a mechanical anchorage, the intermediate strip debonding as defined in the SIA166 [2] have also been incorporated. The strip strain limitation was based on the additional strip strain $\Delta\varepsilon_f$ (Verification: $\Delta\varepsilon_f < 8‰$). The applicability to prestressed systems has been demonstrated in Harmanci [33].

4. RESULTS AND DISCUSSION

4.1. General Comments

Experimental load-midspan deflection curves are presented in Figure 9, and in Figure 10 along with the numerical simulations. All characteristic results for the cracking *"cr"*, yielding *"y"*, and ultimate loading state *"u"* are presented in Table 4. Additionally, the structural ductility with the ductility index for deflection μ_δ as well as the strength factor μ_F as defined in Michels *et al.* [15] are given. As expected, the strengthened slabs (EBR, GA and MA specimens) exhibit higher cracking, yielding, and ultimate loads but lower deflections at failure compared to the reference slabs REF1 and REF2.

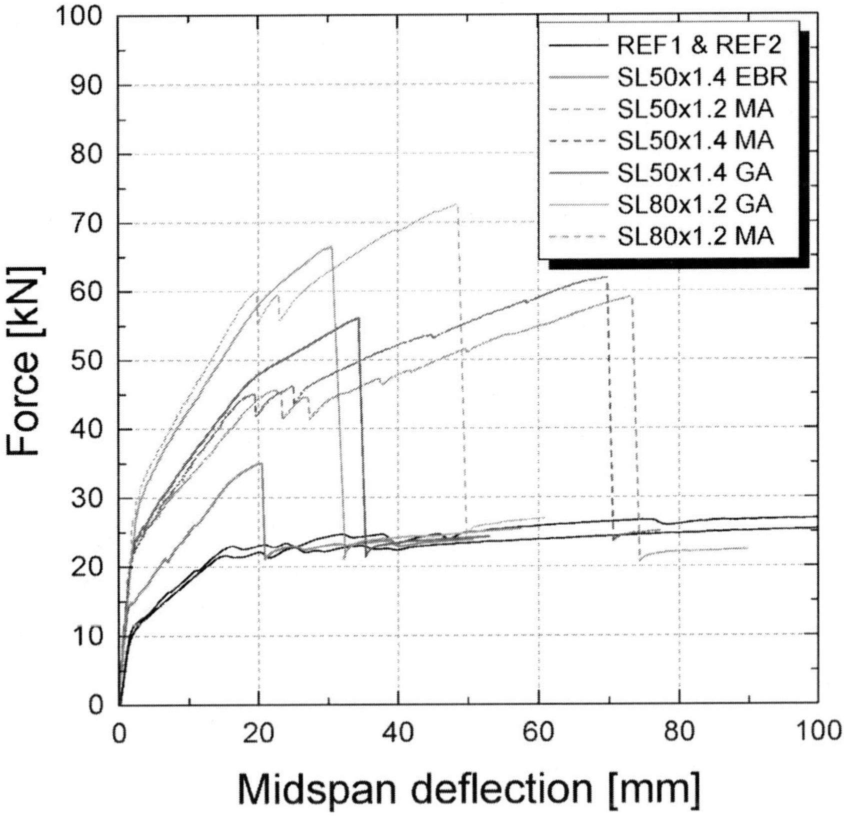

Figure 9. Experimental load-deflection curves.

4.2. Failure Modes

The reference slab tests were stopped at a midspan deflection of 100 mm due to the corresponding LDVT measurement range limitation. Failure mode would in this case most likely be concrete crusing on the top fiber. The strengthened slabs all exhibited CFRP strip debonding, which occured partially in the concrete substrate or in the CFRP/epoxy or epoxy/concrete interface. Debonding mostly initiated at one slab extremity and then propagated towards the other support. For the slabs with a mechanical end-anchorage, strip debonding did not implicate the immediate reaching of the

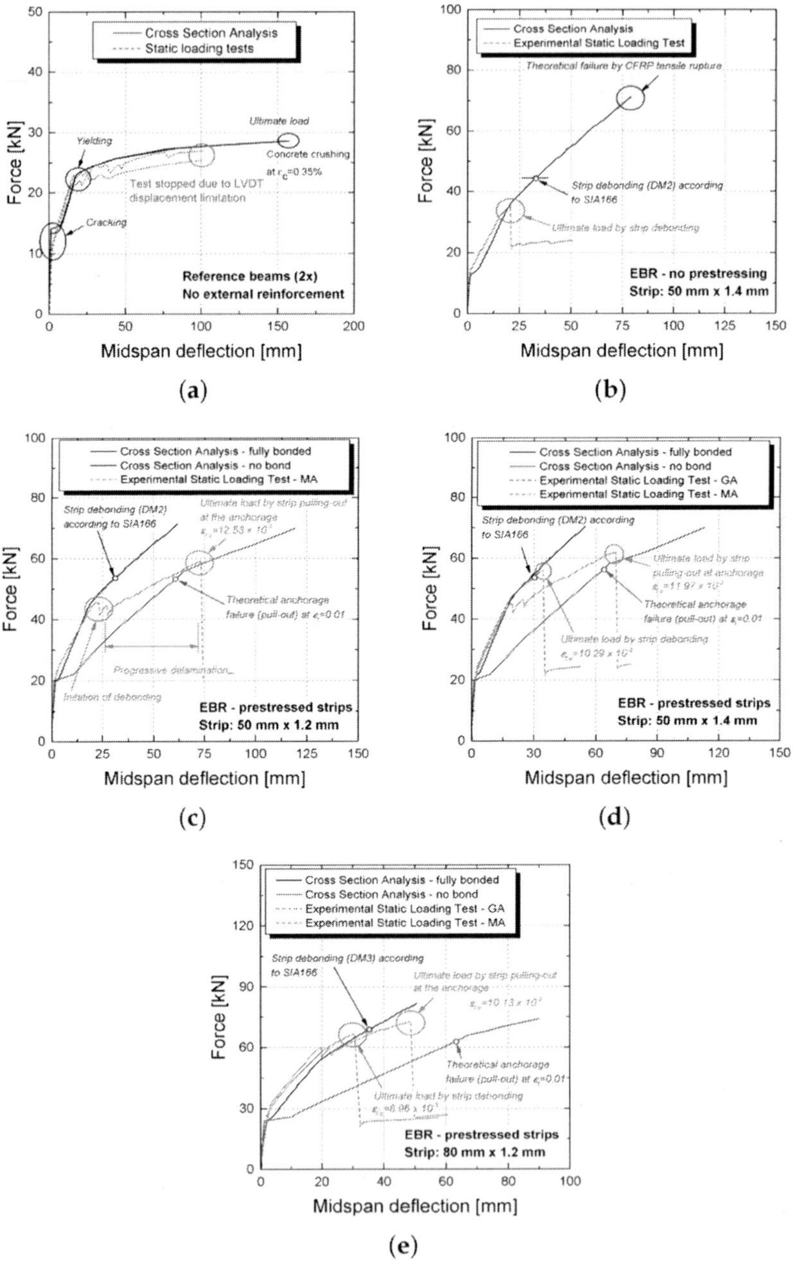

Figure 10. Experimental and Numerical force-deflection comparison for all slabs. (**a**) Reference slabs; (**b**) EBR—No prestressing; (**c**) Prestressed strip—50 mm × 1.2 mm; (**d**) Prestressed strip—50 mm × 1.4 mm; (**e**) Prestressed strip—80 mm × 1.2 mm.

Table 4. Key experimental results (δ_{cr} = deflection at crack initiation, F_{cr} = cracking load, δ_y = deflection at yielding, F_y = yielding load, δ_u = ultimate deflection, F_u = ultimate load, $\varepsilon_{f,u}$ = CFRP tensile strain at failure, μ_δ = ductility index, $\mu_F = F_u/F_y$ = strength factor).

Plates	δ_{cr} (mm)	F_{cr} (kN)	δ_y (mm)	F_y (kN)	δ_u (mm)	F_u (kN)	$\varepsilon_{f,u}$ (‰)	Failure Mode	$\mu_\delta = \delta_u/\delta_y$	$\mu_F = F_u/F_y$
REF1	2.47	11.04	15.74	21.5	-	-	-	-		
REF2	2.49	11.12	15.96	22.9	-	-	-	-		
SL50x1.4_EBR	1.64	14.73	17	33.3	20.47	35.06	4.64	Strip debonding	1.20	1.05
SL50x1.4_GA	2.25	23.84	18.86	48.35	34.39	56.02	10.29	Strip debonding	1.82	1.16
SL50x1.4_MA	2.25	22.07	17.8	44.32	69.84	61.76	11.97	Strip pulled out of mech. anchorage	3.92	1.39
SL50x1.2_MA	2.53	22.81	20.57	44.89	73.23	59.09	12.53	Strip pulled out of mech. anchorage	3.56	1.32
SL80x1.2_GA	2.88	28.56	20.31	58.31	30.61	66.21	8.96	Strip debonding	1.51	1.14
SL80x1.2_MA	2.51	28.71	18.43	58.67	48.62	72.58	10.13	Strip pulled out of mech. anchorage	2.64	1.24

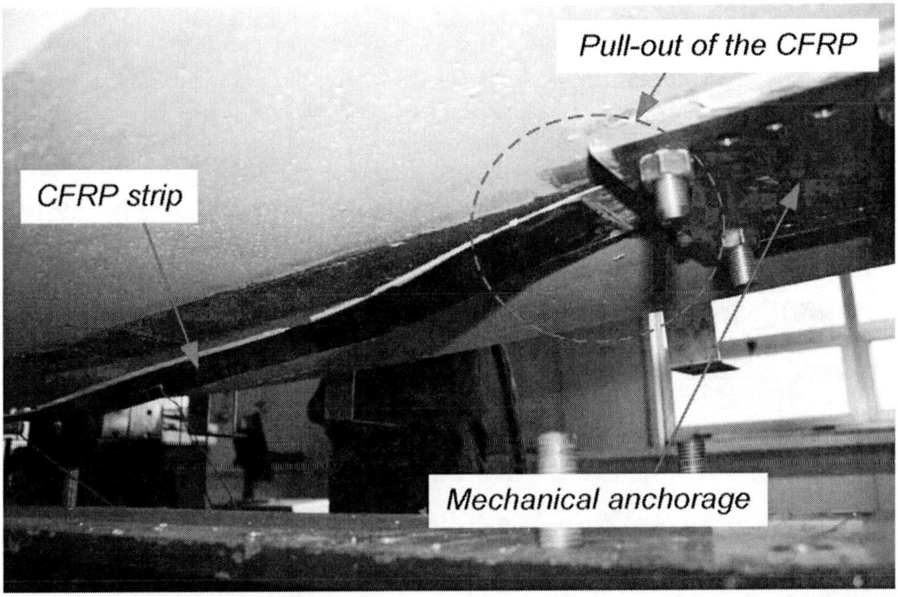

Figure 11. CFRP strip pull-out from the mechanical end anchorage.

ultimate load carrying capacity, since the strip was still held at its both ends. Further load increase was still possible. Eventually, the strip was pulled out of the mechanical anchorage as shown in Figure 11. Details regarding the structural behavior will be discussed in the next sections. The slabs with gradient anchorage exhibited a more sudden strip debonding, as in this case no mechanical end-fixing could provide additional anchorage resistance.

4.3. Prestressed versus Non-Prestressed

As expected, the use of prestressing improved the overall structural behavior of the slabs in terms of cracking and yielding initiation, stiffness, and load carrying capacity compared to a simple EBR solution (Figure 9 andFigure 10a,b). Although the initial stiffness in the uncracked stage did not change significantly, mainly due to the low level of strengthening, cracking load is significantly higher (almost doubled with a 50 mm × 1.4 mm prestressed strip compared to only 33% with an unstressed one). Similar observations can be made for overall stiffness prior to yielding. The ultimate load increase is between about 25% for en EBR strips and between 100 and 120% for the prestressed configurations with gradient and mechanical anchorage, respectively. Finally, it should be also referred that prestressing also better explores the CFRP's mechanical properties in tension. The measured ultimate CFRP strains $\varepsilon_{f,u}$ were 10.29 and 11.97‰ in case the laminates were prestressed, but only 4.64‰ could be used with an initially unstressed strip (Table 4). Additionally, as presented in Table 4, a better structural behavior in terms of ductility can be noticed when a prestressed EBR with a MA is applied. A relative improvement of about 68% in ultimate deflection was observed, while the deflection at yielding is almost identical. This also results in a much higher ductility index μ_δ. In terms of load, the prestressed EBR system with MA also exhibits a more distinct load carrying improvement after the yielding initiation.

4.4. Gradient Anchorage (GA) vs. Mechanical Anchorage (MA)

Up to yielding initiation, GA and MA systems exhibited similar behavior (Figure 10d,e). From this point on, additional contribution provided by the steel reinforcement is limited, the CFRP strip being now responsible to carry additional load. This is the reason why the force increment carried by CFRP material drastically increases at the onset of yielding initiation (Figure 12). For all MA slabs, two drop points can be observed just after yielding initiation. These two drops are related to a strip debonding initiation at the both extremities (in between the metallic plate anchors). After this point the slab still continues to carry load due to the existence of the metallic anchors that avoided the premature debonding of the CFRP laminate strip. For this reason, the slabs of MA series present a better performance in terms of ultimate load when compared with the GA series, for which the initial debonding process is rapidly transformed into the complete strip detachment. In Table 4, it can be seen that the ultimate loads reached with a mechanical anchorage are about 10% higher than for a gradient anchorage. Deflection at failure, is around 103% (50 mm × 1.4 mm) and 59% (80 mm × 1.2 mm) higher for a mechanical anchorage compared to a purely bond-based gradient anchorage. In terms of structural ductility, both times the

MA exhibit a more favorable behavior with ductility indexes of 3.92 and 2.64 for the strip section 50 mm × 1.4 mm and 80 mm × 1.2 mm, respectively. The same is valid for the additional force increase after yielding initiation.

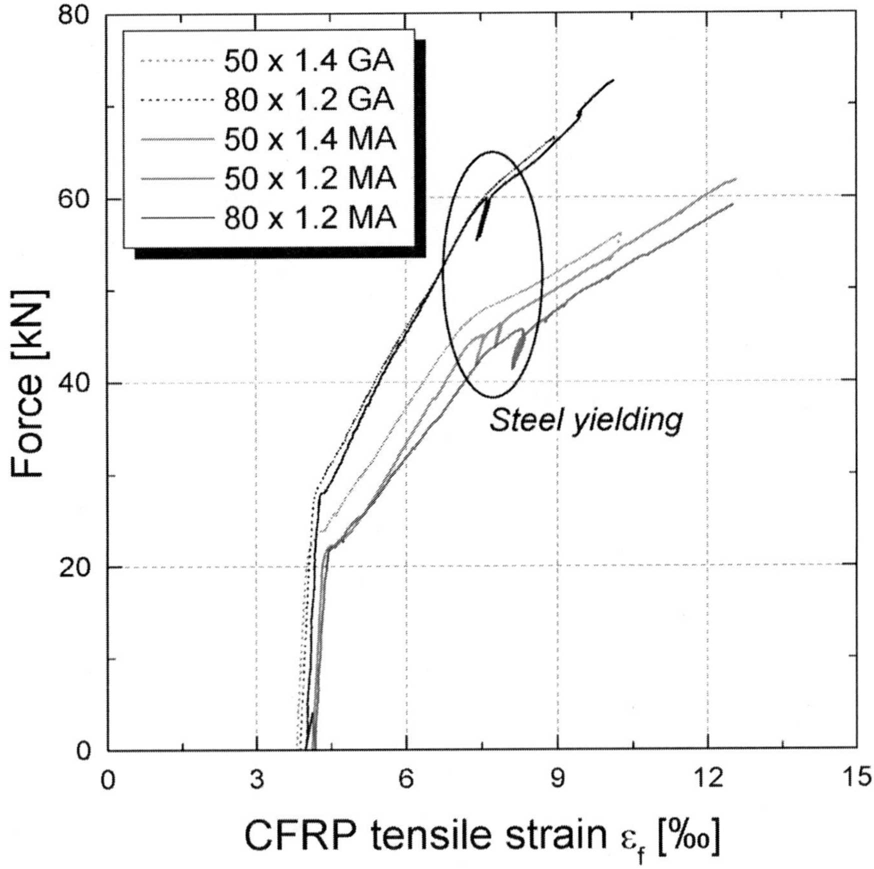

Figure 12. Experimental CFRP tensile strains in function of the applied load.

4.5. Influence of FRP Thickness t_f and Width b_f

As shown in Figure 10c,d, SL50x1.2_MA and SL50x1.4_MA slabs presented a relatively similar behavior, with eventually a slightly higher ultimate load for the thicker laminate (61.8 kN against 59.09 kN). Considering the measured strains at failure listed in Table 4 together with the elastic moduli for both strip thicknesses, the pull-out forces are in both cases very close (see Figure 13): 126 kN for t_f = 1.2 mm and 129.7 kN for t_f = 1.4 mm. Both strip geometries exhibited relatively similar values concerning ductility index and strength factor.

When switching to the clearly higher CFRP cross-section 80 mm × 1.2 mm, due to the higher prestress force, the overall behavior under loading is obviously stiffer compared to the more narrow strip with b_f of 50 mm. When slabs SL50x1.2_MA and SL80x1.2_MA are compared (Figure 10c,e,) in terms of pull-out load at the end-anchorage, a higher force for the laminate with a higher cross-section can be deducted (see Figure 13). A strip width b_f of 80 mm results in an anchorage resistance of about 160 kN, whereas a more narrow laminate with b_f = 50 mm implicates a pull-out force of 126 kN. When comparing an average shear stress over the anchorage length of 270 mm the narrow strip delivers a higher strength than the larger one (9.16 MPa against 7.40 MPa). However, the CFRP strip with a higher axial stiffness might also implicate higher peak shear stresses at the anchorage tip (location where the laminate exits the anchorage zone towards midspan), being again responsible for an earlier strip pull-out. All in all, structural ductility is reduced when a higher CFRP cross-section is prestrained to the same level. Whereas relative force increases after yielding are rather similar for both cross-sections (SL50x1.2_MA and SL80x1.2_MA), clearly higher ductility indexes can be notived for the more slender strip. This behavior is analogue to classic prestressed concrete, for which a higher prestress level results in an enhanced cracking and yielding load, which eventually goes together with a reduced structural deformation capacity.

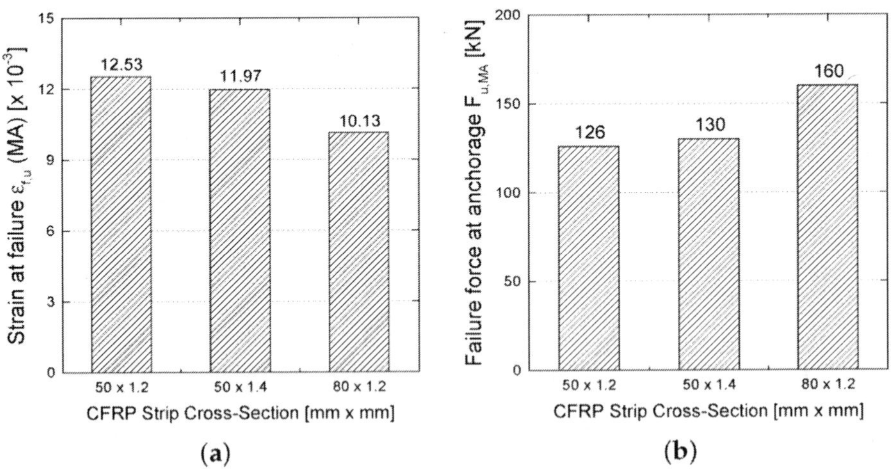

Figure 13. Measured CFRP tensile strain and deducted tensile force at pull-out. (**a**) CFRP tensile strain at failure; (**b**) CFRP force at failure.

4.6. Structural Behavior and Experiments/Simulation Comparison

Figure 10 presents the force-deflection predictions for the different configurations tested. In general, numerical predictions of the overall structural behavior fit well with the experimental results obtained from the static loading tests. Regarding the reference tests, it has to be stated that the experiments were stopped at a midspan deflections of 100 mm due to the LVDT displacement range limitation. This explains the larger deflections at ultimate load carrying capacity for the numerical simulations. However, cracking, yielding and ultimate load along with the different stiffnesses are captured very well.

All retrofitted plates have been calculated up to theoretical tensile rupture of the CFRP strip or concrete crushing in compression. For the unstressed EBR configuration, again cracking and yielding load as well as the initial two pre- and postcracking stiffness are simulated in an accurate manner. Strip debonding occurred at a load of around 35 kN, corresponding to a strip tensile strain of 4.64‰. The simulated ultimate load is in this case somewhat higher at 44 kN with a tensile strain of the CFRP strip of 6.45‰. Failure is numerically governed by a debonding mode DM2, exceeding the interfacial shear stress.

For all tests with a prestressed CFRP strips, whether with a mechanical or a gradient anchorage, cracking and yielding loads together with the structural stiffness are again simulated quite precisely. Both configurations with a gradient anchorage eventually fail by strip debonding, which after an initiation rapidly propagates over the complete strip length. Numerically, the implemented debonding criteria as listed in Section 3.2 allow calculating the ultimate load in an accurate manner. Interesting observations can be made when analyzing the results of the slabs with a mechanical anchorage. For all specimens, strip debonding initiates on one side at a certain force level, followed by a progressive debonding over the complete strip length. Due to the mechanical end-anchorage, however, the CFRP strip does not becomes inactive, but acts eventually as an outer tension chord without any remaining bond to the concrete substrate but in both anchorage regions. This gradual transfer from a fully bonded to a finally unbonded system can be well observed by comparing the experimental results to both simulated cases. Up to the initiation of debonding, both the experimental curves coincide with the simulated bonded configuration. Afterwards, a gradual decrease in stiffness can be noticed up to failure by pulling the CFRP strip out of the mechanical anchorage. For both the strip sections of 50 mm × 1.2 mm and 50 mm × 1.4 mm, the structural behavior is finally almost identical to the one of a fully unbonded tendon, which can be well observed via the comparison between the experimental data and the simulated "unbonded" curve. Additionally, it can be noticed that the applied failure criteria (3) allows in these cases also predicting the ultimate load very closely.

5. CONCLUSIONS AND OUTLOOK

Out of the presented results, several conclusions can be drawn:

- As expected, clearly higher CFRP strains in tension can be reached with an initial prestressing.
- The type of anchorage for prestressed CFRP strips does not affect the behavior at serviceability. Cracking and immediate post-cracking structural behavior is identical for both techniques.
- The application of a gradient anchorage results in a more or less sudden strip debonding. The failure mode is very similar to conventional externally bonded reinforcement without any end-fixations.
- A mechanical end-anchorage system does also not prevent strip debonding, but allows to transfer the structural behavior from a bonded to an external unbonded-tendon-like behavior. The numerical simulations support this observation. Eventually, failure is obtained by pulling the CFRP strip out of the mechanical anchorage.
- In the present case, the application of a mechanical end-anchorage system implicated a much higher structural ductility when comparing the deflection at failure to the one at yielding. A gradient anchorage was not able to deliver the same amount of deformability.
- The implemented debonding criteria for conventional EBR reinforcements as well as the gradient failure criteria allow to re-calculate the slabs' behavior in an accurate manner. Future investigations, both experimental and numerical, shoud aim at defining precise pull-out resistances for mechanical anchorages.

ACKNOWLEDGMENTS

This work is supported by FEDER funds through the Operational Program for Competitiveness Factors—COMPETE and National Funds through FCT—Portuguese Foundation for Science and Technology under the project FRPreDur-PTDC/ECM-EST/2424/2012. Additionally, the authors would like to thank all the companies that have been involved supporting and contributing for the development of this study, mainly: S&P Clever Reinforcement Ibérica Lda, S&P Clever Reinforcement Company (Switzerland), Tecnipor-Gomes & Taveira Lda., Vialam-Indústrias

Metalúrgicas e Metalomecânicas, Lda., Hilti Portugal-Produtos e Serviços, Lda. The fourth author wishes also to acknowledge the grant SFRH/BD/98309/2013 provided by FCT. This paper is dedicated to Tiago Teixeira (1988–2015), former doctoral student at the ISISE R&D Research Centre at the University of Minho.

AUTHOR CONTRIBUTIONS

José Sena-Cruz, Julien Michels and Luís Correia conceived and performed the experimental program; Yunus Emre Harmanci and Julien Michels developed the numerical tool; all the authors analyzed the data and wrote the paper.

REFERENCES

1. Bakis, C.; Bank, L.; Brown, V.; Cosenza, E.; Davalos, J.; Lesko, J.; Machida, A.; Rizkalla, S.; Triantafillou, T. Fiber-reinforced Polymer Composites for Construction—State-of-the-art Review. *J. Compos. Constr.* **2002**, *6*, 73–87.
2. *Klebebewehrungen (Externally Bonded Reinforcements)*; SIA 166; Schweizer Ingenieur-und Architektenverein (SIA): Zürich, Switzerland, 2004.
3. *Strengthening Reinforced Concrete Structures with Externally-Bonded Fibre Reinforced Polymers*; The Canadian Network of Centres of Excellence on Intelligent Sensing for Innovative Structures (ISIS Canada): Winnipeg, MB, Canada, 2007.
4. *Guide for the Design and Construction of Externally Bonded FRP Systems for Strnegthening Concrete Structures*; ACI440.2R; American Concrete Institute (ACI): Farmington Hills, MI, USA, 2008.
5. *Design Handbook for RC Structures Retrofitted with FRP and Metal Plates: Beams and Slabs*; Standards Australia: Sydney, Australia, 2008.
6. *Design Guidance for Strengthening Concrete Structures Using Fibre Composite Materials*; Technical Report No. 55 for the Concrete Society: Camberley, UK, 2012.
7. *Guideline—Strengthening of concrete members with adhesively bonded reinforcement*; Deutscher Ausschuss für Stahlbeton (DAfStb): Berlin, Germany, 2012.
8. National Research Council (CNR). *Istruzioni per la Progettazione, l'Esecuzione ed il Controllo di Interventi di Consolidamento Statico*

Mediante L'utilizzo di Compositi Fibrorinforzati; CNR-DT 200; CNR: Rome, Italy, 2013.

9. Czaderski, C.; Motavalli, M. 40-Year-old full-scale concrete bridge girder strengthened with prestressed CFRP plates anchored using gradient method. *Compos. Eng.* **2007**, *38*, 878–886.

10. Motavalli, M.; Czaderski, C.; Pfyl-Lang, K. Prestressed CFRP for strengthening of reinforced concrete structures: Recent developments at Empa, Switzerland. *J. Compos. Constr.* **2011**, *15*, 194–205.

11. Wight, R.; Green, M.; Erki, M.A. Prestressed FRP sheets for poststrengthening reinforced concrete beams. *J. Compos. Constr.* **2001**, *5*, 214–220.

12. Svecova, D.; Razaqpur, A. Flexural behavior of concrete beams reinforced with carbon fiber-reinforced polymer CFRP prestressed prisms. *ACI Struct. J.* **2000**, *97*, 731–738.

13. Kim, Y.; Wight, R.; Green, M. Flexural strengthening of RC beams with prestressed CFRP sheets: Development of nonmetallic anchor systems. *J. Compos. Constr.* **2008**, *12*, 35–43.

14. Pellegrino, C.; Modena, C. Flexural strengthening of real-scale RC and PRC beams with end-anchored pretensioned FRP laminates. *ACI Struct. J.* **2009**, *106*, 319–328.

15. Michels, J.; Martinelli, E.; Czaderski, C.; Motavalli, M. Prestressed CFRP strips with gradient anchorage for structural concrete retrofitting: Experiments and numerical modeling. *Polymers* **2014**, *6*, 114–131.

16. El-Hacha, R.; Wight, R.; Green, M. Prestressed fibre-reinforced polymer laminates for strengthening structures. *Prog. Struct. Eng. Mater.* **2001**, *3*, 111–121.

17. Michels, J.; Sena-Cruz, J.; Czaderski, C.; Motavalli, M. Structural strengthening with prestressed CFRP strips anchored with the gradient method. *J. Compos. Constr.* **2013**, *17*, 651–661.

18. Laboratório Nacional de Engenharia Civil (LNEC). *Concrete—Determination of the Elasticity Young Modulus under Compression*; LNEC-E397; LNEC: Lisboa, Portugal, 1993.

19. Instituto Português da Qualidada (IPQ). *Testing Hardened Concrete—Part 3: Compressive Strength of Test Specimens*; NP-EN-12390-3; IPQ: Caparica, Portugal, 2011.

20. Instituto Português da Qualidada (IPQ). *Metallic Materials. Tensile Testing. Part 1: Method of Test at Room Temperature*; NP-EN-ISO-6892-1; IPQ: Caparica, Portugal, 2012.

21. International Organization for Standardization (ISO). *Plastics—Determination of Tensile Properties—Part 5: Test Conditions for Unidirectional Fibre-Reinforced Plastics*; ISO-527-5; ISO: Geneva, Switzerland, 2009.

22. ISO. *Plastics—Determination of Tensile Properties—Part 2: Test Conditions for Moulding and Extrusion Plastics*; ISO-527-2; ISO: Geneva, Switzerland, 1993.

23. Moussa, O.; Vassilopoulos, A.; de Castro, J.; Keller, T. Early-age tensile properties of structural epoxy adhesives subjected to low-temperature curing. *Int. J. Adhes. Adhes.* **2012**, *35*, 9–16.

24. Granja, J.; Fernandes, P.; Benedetti, A.; Azenha, M.; Sena-Cruz, J. Monitoring the early stiffness development in epoxy adhesives for structural strengthening. *J. Adhes. Adhes.* **2015**, *59*, 77–87.

25. Czaderski, C.; Martinelli, E.; Michels, J.; Motavalli, M. Effect of curing conditions on strength development in an epoxy resin for structural strengthening. *Compos. Eng.* **2012**, *43*, 398–410.

26. Michels, J.; Czaderski, C.; El-Hacha, R.; Brönnimann, R.; Motavalli, M. Temporary bond strength of partly cured epoxy adhesive for anchoring prestressed CFRP strips on concrete. *Compos. Struct.* **2012**, *94*, 2667–2676.

27. Marti, P.; Alvarez, M.; Kaufmann, W.; Sigrist, V. Tension chord model for structural concrete. *Struct. Eng. Int.* **1998**, *8*, 287–298.

28. Harmanci, Y. Prestressed CFRP for Structural Retrofitting-Experimental and Analytical Investigation. Master's Thesis, ETH Zürich, Zürich, Switzerland, 2013.

29. Harmanci, Y.; Michels, J.; Czaderski, C.; Motavalli, M. Calculation technique for externally unbonded CFRP strips in structural concrete retrofitting. *J. Eng. Mech.* **2015**.

30. Hognestad, E. A study of combined bending and axial load in reinforced concrete members. *Univ. Ill. Eng. Exp. Stn.—Bull. Ser.* **1951**, *399*, 128.

31. Czaderski, C. Strengthening of Reinforced Concrete Members by Prestressed Externally Bonded Reinforcement with Gradient Method. Ph.D. Thesis, ETH Zürich, Zürich, Switzerland, 2012.

32. Suter, R.; Jungo, D. Vorgespannte CFK-Lamellen zur Verstärkung von Bauwerken. *Beton Stahlbetonbau* **2001**, *96*, 350–358.

33. Harmanci, Y. *Externally Bonded Reinforcement EBR for Flexural Strengthening of Reinforced Concrete Beams with Rectangular Cross Section*; Semester Project, ETH Zürich: Zürich, Switzerland, 2011.

Chapter 8

The Use of Fiber Reinforced Plastic for The Repair and Strengthening of Existing Reinforced Concrete Structural Elements Damaged by Earthquakes

George C. Manos[1] and Kostas V. Katakalos[1]

[1] *Laboratory of Experimental Strength of Materials and Structures, Department of Civil Engineering, Aristotle University of Thessaloniki, Greece*

1. INTRODUCTION

During the last fifty years various parts of the world have been subjected to a number of damaging earthquakes. Greece is one of the countries where such damaging earthquakes occur quite frequently ([1]). Some of these earthquakes, not necessarily the most intense, occurred near urban areas and thus subjected various types of structures to significant earthquake forces leading to damage ([2]). For some of these earthquakes, ground motion acceleration recordings were obtained at distances relatively close to the area of intense shaking, thus providing valuable information for correlating the observed damage with this ground motion recording and its characteristics. Moreover, following the most damaging of these earthquakes, studies were initiated that led to the revision of the provisions of Seismic Codes [3]. The damaged structures included old structural formations, sometimes older than one hundred years, which were not designed for seismic forces. They usually belong to cultural heritage and are under various forms of conservation status that does not allow all types of retrofitting but only retrofitting materials and techniques that are compatible with the existing materials; moreover, the applied retrofitting in these cases must also be reversible so that it can be easily removed in case it demonstrates undesirable effects with time. Apart from the cultural heritage structures, the damaged structures also include relatively contemporary structures that are usually less than fifty years old. The vast majority of these structures are multistory reinforced concrete (R/C) buildings. There are other types of structures apart from (R/C) buildings, such as structures forming the infrastructure or industrial facilities which can also develop earthquake damage. However, this chapter will be devoted to the usual R/C residential multi-story buildings, the earthquake damage of their structural elements and their strengthening.

1.1. Damage Observations For Contemporary R/C Structures

These structures are usually designed and built according to the provisions of a Seismic Code ([3]). The cause of damage may be due to:

The code provisions underestimating the severity of the shaking and thus underestimating the seismic demand posed upon the various structural elements, their connections and the foundation of the whole structure.

The code provisions together with the specification of the materials resulting in such strengths that are below the seismic demands placed upon the structural elements.

The detailing and the realization of the design during construction or alteration during the life time of the building resulting again in such strengths for the various structural elements that are below the corresponding seismic demands placed upon them.

In all cases the appearance of structural damage results from the above fundamental causes, either from one of them or from their combination; this is expressed by the following inequality (Eq. 1) between the strength and the demand put upon the various members of the structural system, whereby the demand is signified as S_d and the strength as R_d. Damage is expected to occur when this inequality is not satisfied [4].

$$S_d < R_d \tag{1}$$

A serious task, after a strong earthquake affects an urban area, is to appraise the severity of the structural damage for a large number of buildings. This is usually done in Greece by engineers specially appointed for this task by the Government. The screening process commences a few days after the natural disaster, provided that the seismic sequence has subsided. There are certain guidelines published by the Hellenic Organization of Earthquake Planning and Protection to facilitate the appointed engineers in this damage screening process [5]. In general terms the first round of inspection must lead to the classification of each building to one of the three main categories. The first damage category is that there is no structural damage and the mainly non-structural damage does not pose any danger; so these buildings can be reused immediately. The second damage category is that there is non-structural damage as well as some structural damage; the latter, although contained, may have led to a considerable decrease of the bearing seismic capacity of the damaged structural elements and the structure as a whole. There may be need for temporary shoring and the removal of dangerously damaged non-structural elements. These buildings will be subjected to a second round of inspection after these countermeasures have been accomplished. Their reuse will be decided after the second round of screening. The third category includes buildings with relatively extensive damage to its structural elements (slabs, beams, columns or shear walls) [5], [6]; the damage is relatively widespread in terms of story level. Permanent deformations of the structural elements are evident in the form of concrete cracking and crushing in certain critical areas of the structural elements indicating that these areas have been overstressed and inequality 1 has ceased to be valid. There is serious consideration that this bearing seismic capacity reduction may lead to partial collapse. There may be urgent need for temporary shoring and the removal of dangerously damaged parts in order to avoid, if possible, partial collapse (figure 1b). These buildings will also be subjected to a second round of inspection. If it is decided that they pose a public threat, due to the possibility of partial or total collapse, they should be demolished (figure 1a); otherwise, a special design should devise a feasible scheme for their repair and strengthening. The following sections will address the task of repair and strengthening of structural elements that have been subjected to such earthquake damage to their structural system and have been classified as belonging to the second or third seismic damage categories after the second round of screening process.

a b

Figure 1. a the old building of a pharmaceutical company that collapsed during the athens 1999 earthquake together with an upgraded building next to it that survived unscathed [2]. B. Temporary shoring of 4-story building (hardas) damaged during the pyrgos 1993 eart hquake [2].

Figure 2. a Typical damage of beam at the joint with the nearby column (6th story building, Aharnes, Athens earthquake 1995), [2]. b Typical damage of the toe of a column at ground floor (two-story building, Nea Kifisia,, Athens earthquake 1995), [2]. c Typical shear damage at the columns of a "soft story" (4-story Metamorfosi building, Athens earthquake 1995), [2].

1.2. STRUCTURAL DAMAGE DESCRIPTION AT THE LEVEL OF THE STRUCTURAL ELEMENT

It is usual to describe the structural damage at the level of each reinforced concrete structural member, e.g. slab, beam, column and shear wall, always having in mind inequality 1 and the fact that the axial (N), bending (M) and shear (Q) force demands in each one of these structural members from the combination of dead and live loads plus the earthquake forces are of a particular nature. Thus, for the slabs the demands are mainly flexural whereas for the beams the demands are flexural and shear. For the columns the demands are flexural and shear with the presence of considerable axial forces whereas for the shear walls the demands are flexural and shear with the presence of a relatively lower level of axial forces than that of the columns. Apart from the structural elements themselves, one should also consider critical areas of their connections (joints) as well as the foundation. Both, the structural connections and the foundation are very critical areas that require special consideration in both identifying the nature of the structural damage as well as proposing countermeasures. This presentation does not deal with either of these critical areas. The main flexural structural damage in slabs and beams develops in the areas of maximum bending moments. For the beams they usually develop near the joints with the columns and shear walls (figure 2a) where large bending moments are expected to develop from the seismic forces. Similarly, at the ends of the beams are the areas of large shear forces from the combination of earthquake forces with the dead and live loads; these will cause the appearance of shear damage in the form of diagonal cracks (figure 2a). The presence of large bending moments mainly from the seismic load together with large axial forces will cause the formation of flexural damage at the top and the toe of columns (see figure 2b) whereas the presence of shear forces from the seismic loads together with axial forces will lead to the formation of shear damage at the columns (see figure 2c). The presence of large shear forces from seismic loads together with relatively low level axial forces will lead to the development of shear damage in the shear walls (see figure 3a)whereas the presence of short columns will lead to the development of large shear forces from seismic loads and the development of shear damage as shown in figure 3b.

1.3. The Strategy For The Repair And Strengthening Scheme

In the previous discussion the emphasis in the description of damage was on attributing the various forms of structural damage to the nature of the demand (either flexural or shear with or without axial forces). However, as already stressed by inequality 1, structural damage is due to the fact that the said demand was not met by the existing strength. The strategy for retrofitting damaged structural systems, or structural systems that can be demonstrated by analytical methods prior to a strong earthquake to be prone to potential damage in the future, must be based on either somehow lowering the demands or increasing the corresponding strengths or both. In an effort to lower the demands, one can try to decrease the masses mobilized by the earthquake vibrations. This can be accomplished either by removing unnecessary mass or by changing the structural system in such a way that the resulting dynamic system combined with the design spectrum leads to a decrease in the earthquake loads (e.g. seismic isolation). Lowering the demands in the way described before is not always feasible. Thus, the retrofitting scheme is usually based on increasing the strengths of the structural members. In doing so, one must be aware that it is advisable to increase the deformability of the structural members thus increasing the ability of the structural system to dissipate the seismic energy through plastic deformations that are designed to develop at predetermined locations ([7], [8] [9], [10]). The location where these plastic deformations occur should be such that it does not lead to unstable structural formations. Sometimes, a partial objective of the strengthening scheme is to increase the stiffness to a moderate degree, especially for structural systems that can develop excessive torsional response. However, frequently a considerable increase in the stiffness of the strengthened structural system results as an indirect consequence of the adopted scheme whereas its main objectives were to increase the strength and deformability of the structural elements and of the structure as a whole. This increase in stiffness results in larger demands due to the dynamics of the strengthened structure, as it corresponds to higher amplitudes of design spectral accelerations than the un-strengthened structural system. This is usually the case when traditional strengthening schemes are employed utilizing reinforced concrete jacketing of the structural members (columns and beams) or the addition of shear walls by encasing reinforced concrete elements within the bays of the corresponding R/C frames [6]. The strengthening schemes that can be employed in order to increase the flexural or shear capacity of columns, beams or shear walls utilizing fiber reinforcing plastics (FRP) usually made of carbon, glass or steel do not result in this undesirable increase in stiffness. Moreover, due to their external application, they usually require less interference with or breaking of the volume of existing R/C structural elements. Finally, such strengthening schemes become effective in a much shorter time than traditional strengthening schemes ([11], [12], [13], [14], [15], [16]).

Figure 3. a Typical damage of shear walls at the "soft story" during The Kozani 1995 Earthquake in Greece. b. Typical damage of "short" columns during The Kozani 1995 Earthquake in Greece.

2. MAIN APPLICATIONS FOR DEALING WITH EARTHQUAKE STRUCTURAL DAMAGE UTILIZING FIBER REINFORCING PLASTICS

In section 1.2 a brief description of earthquake structural damage that usually develops in slabs, beams, columns or shear walls of multi-story R/C buildings was presented. In this section, the use of fiber reinforcing plastics (FRP) will be discussed in a way dealing with the corresponding damage. These FRP materials behave in tension almost elastically till their ultimate state that for the material itself is the breaking of the fibers in tension; they do not develop any forces in shear or compression. The value for the modulus of elasticity is approximately 240Gpa for carbon fibers, 200Gpa for steel fibers and 80Gpa for glass fibers ([8], [17]).

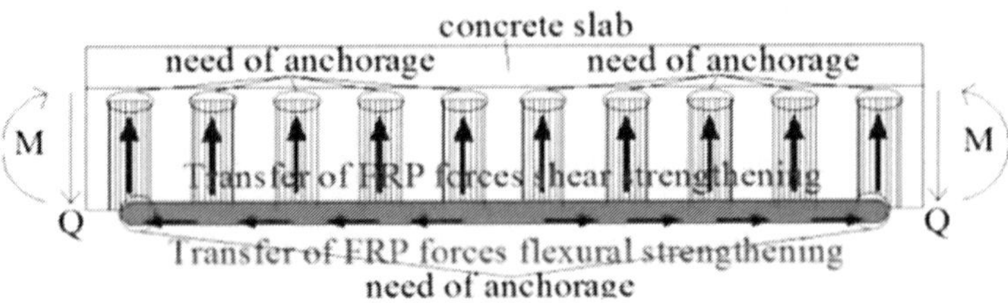

Figure 4. Flexural and shear FRP reinforcement for a T-beam.

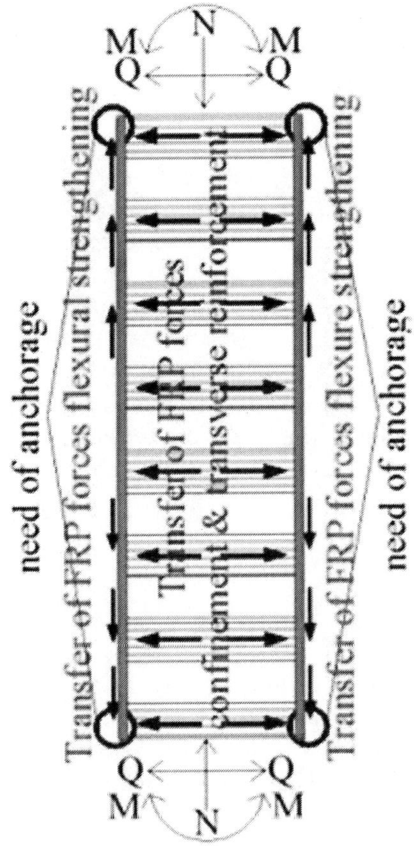

Figure 5. Flexural and shear FRP reinforcement for a column.

The ultimate axial strain values given by the manufacturers reach values in the range of 2%. Consequently, sheets made by these materials, despite their relatively small thickness which is usually below 0.2mm for one layer, can develop substantial tensile forces in the direction of their fibers (seefigures 4 and 5). This property accompanied with their low weight and the very easy external application to structural elements, by attaching them on the external surfaces by proper organic or inorganic matrices, results in their being used as effective longitudinal or transverse reinforcement for such structural elements that are in need of strengthening (see figures 4, 5 and [12], [15], [16]). However, the following limitations exist for this type of application. The first limitation springs from the fact that the ultimate axial strain value of the order of 2% for the material of the fiber cannot be reached for all the fibers together in a sheet due to the actual conditions of the attachment that results in non-uniform distribution of the strain field ([13], [18], [19]). Thus, for design purposes, the adopted limit axial strain value is usually below ½ of the ultimate axial strain value for the fiber material. The second limitation results from the way the tensile forces which develop on these FRP sheets can be transferred. When the transfer of these forces relies solely on the interface between the FRP sheet and the external surface of the reinforced concrete structural elements, the delamination (debonding) mode of failure of these sheets occurs, due to the relatively low value of either the ultimate bond stress at this interface or the relatively low value of the tensile strength of the underlying concrete volume. This mode of failure is quite common and it occurs in many applications well before the corresponding FRP sheets develop tensile axial strains in the neighborhood of values mentioned before as design limit axial strains (approximately of the order of 1%). Consequently, there is need of alternative ways in order to transfer these tensile forces apart from the simple attachment, in order to enhance the exploitation of the FRP material potential. This will be presented and discussed in the following paragraphs ([11], [13], [20], [21], [22], [23]).

2.1. Upgrading The Flexural Capacity of Slabs

The main cause of damage in this case is the fact that the flexural capacity cannot meet the demand. Fiber reinforced plastics either in the form of sheets or laminates are externally attached to the slab either at the top or at the bottom surface (positive or negative bending moment demand). Such a scheme was utilized in the upgrading of the seismic capacity of the slabs of a 4-story R/C building built in 1933 and subjected to the Kozani, 1995 earthquake in Greece ([2], [24]). For this strengthening scheme it was planned to make use of a certain type of Carbon Fiber Reinforcing Plastic laminates (CFRP) with a cross-section 50mm x 1.2mm. These laminates could be applied in-situ either on the upper or lower surface of the concrete slabs with the use of a special epoxy paste. An extensive experimental parametric study was performed on this type of attachment by testing a series of specimens prior to applying the best attachment detail to prototype slab specimens. The tests were performed at the laboratory of Strength of Materials of Aristotle University and utilized twin prismatic concrete specimens with dimensions 100mm x 100mm x 150mm each. These twin concrete prisms were joined with these CFRP laminates; they were attached on the two opposite sides of each twin concrete prism. These specimens were then subjected to such a loading as would force the laminates to be detached from the concrete surface. This is shown in figures 6a to 6d. A variety of attachments between the CFRP laminates and the twin concrete prisms were tried. The simplest form of attachment was the one employing only epoxy paste between the CFRP laminates and the concrete surface (figures 6a and 6b). Next, a variety of bolting arrangements were utilized with or without the epoxy paste.Figures 6c and 6d depict such an attachment of the laminates whereby the epoxy paste is combined with one bolt on each side of the twin concrete prism. The bond strength in the first case was found equal to 2.70Mpa whereas in the second case equal to 4.40Mpa, which represents an increase of 63%. The maximum bond strength that was achieved throughout these tests reached the value of 7.20Mpa, which represents an increase of 167% from the simple attachment of the laminates only with epoxy paste. These findings were utilized with slab specimens that were taken from the actual mezzanine slab of the prototype structure; they were cut from parts of the slab where an opening would be formed for a new staircase.

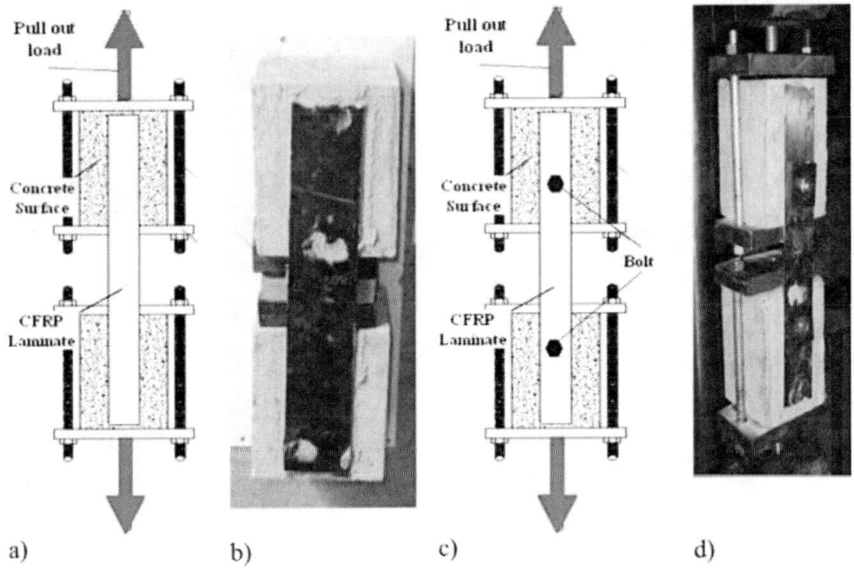

Figure 6. Bond tests of the used cfrp laminates

Figure 7 depicts the loading arrangement that was used to subject these slab specimens to four-point flexure. Initially, this was done for a specimen without any retrofitting that reached a maximum bending moment value equal to 8.84KNm. Then two laminate strips, each having a cross section of 50mm x 1.2mm, were attached on the bottom surface of this slab specimen, which was reloaded to flexure as before. Figure 8 depicts the obtained flexural behavior of this retrofitted slab specimen. At this stage, the CFRP laminates were attached to the specimen only with the use of epoxy paste. This time the specimen reached a maximum bending moment value equal to 20.89KNm, which represents an increase equal to 136% when compared to the maximum

Figure 7. Loading arrangement of the slab specimens in flexure.

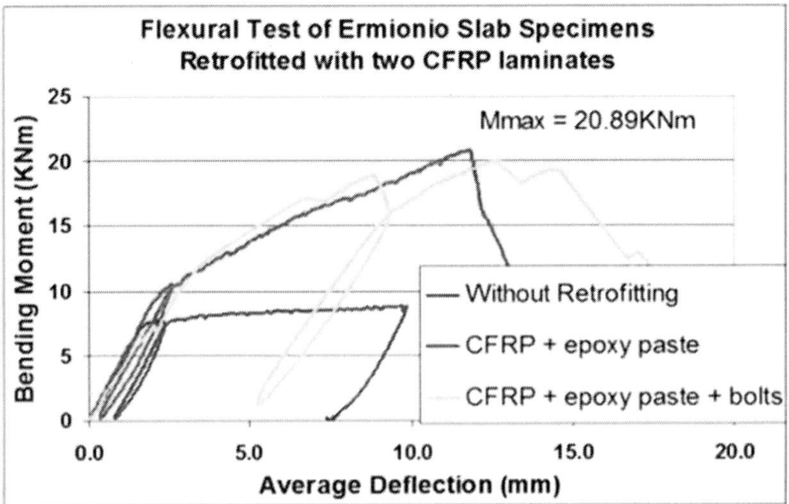

Figure 8. Flexural behavior of ermionio slab specimen retrofitted with two cfrp laminates either bonded only with epoxy paste or with epoxy paste and bolts.

Figure 9. Flexural failure of the ermionio slab specimen retrofitted with two cfrp laminates bonded only with epoxy paste. The cfrp laminates were debonded from the concrete slab in this case.

bending moment value of the un-retrofitted slab. This retrofitted slab failed in the form of the debonding of the CFRP laminates (figure 9). The debonding of the two CFRP laminates was followed by a sharp decrease in the flexural bearing capacity together with the formation of a plastic hinge at mid-span in the form of excessive plastic deformations of the old steel reinforcing bars at the bottom of the slab as well as crushing of the concrete at the top of the slab. Next, two new CFRP laminates were reattached to the same slab specimen; this time epoxy paste was used again together with bolts penetrating through the slab and securing the attachment of these CFRP laminates. The flexural bearing capacity of the slab this time reached the same maximum bending moment value as the one observed when only epoxy paste was used for the attachment of the laminates; they exhibited satisfactory performance without any signs of failure either in the form of debonding or fracture. This is shown by the bending moment versus mid-span deflection curve plotted for this test in figure 8. Despite the satisfactory performance of the CFRP laminates, no increase in the flexural bending capacity could be achieved due to the slab damage from its previous loading history; however; the performance of the anchors for the attachment of the CFRP laminates was satisfactory and the flexural behavior of the slab exhibited a much larger range of deformability (figure 8).

2.2. Upgrading The Flexural And Shear Capacity Of Beams

The upgrading of the shear / flexural capacity will be presented in this section together with the corresponding modes of failure of reinforced concrete (R/C) beam specimens including repair schemes with Fibre Reinforcing Plastics (FRP) ([13], [25], [26], [27], [14], [28], [29]). The specimens were constructed and tested at the laboratory of Strength of Materials and Structures of Aristotle University. The applied load and the deflections of the specimen at three points were monitored throughout the experiments. Moreover, strain gauges were applied at selective locations of the FRP strips employed in the repaired specimens in order to monitor their state of stress during the loading sequence. In what follows, the most important experimental results are presented together with a discussion of the observed performance in order to highlight the role of the FRP strips in the observed behavior. It must be stressed that in all the studied beam specimens closed loop FRP hoops could be applied; this cannot be done in T-beam sections [19].

2.2.1. The initial reinforced concrete beam specimen and its observed behavior.

The virgin beam specimen: The virgin specimen BEAM-1 is a typical R/C beam, shown in figure 10, of rectangular cross-section of 200 mm x 500 mm and length l = 3700mm, having longitudinal reinforcement 3Φ20 at the top and 3Φ20 at the bottom side with closed stirrups as transverse reinforcement Φ8/250 ([14]). The selected longitudinal and transverse reinforcement together with the loading arrangement, shown in figure 11, will lead the behavior of this virgin specimen to be dominated by the shear rather than the flexural mode of failure.

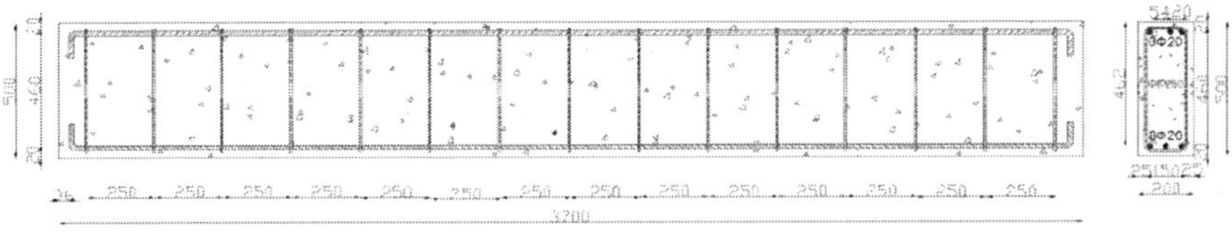

Figure 10. Dimensions (in mm) and structural details of specimen *beam-1*. Longitudinal reinforcement: 3ϕ20 at the top and 3ϕ20 at the bottom. Transverse reinforcement: ϕ8/250.

The behavior of BEAM-1: The specimen was simply supported at two point supports and was subjected to loads at two points symmetrically spaced at a distance 1150mm from each support; the mid-part had a length of 700mm. This loading was gradually increased, being monitored all the time. Moreover, throughout the loading sequence till the failure of the specimen, the cracking pattern was monitored together with the deflections of the beam at three points (mid-span and under each point load). As can be seen in figure 11, the shear cracking pattern is predominant at the two sides of the beam, whereas the middle part, as expected, is dominated by the flexural behavior. Finally, the dominant failure mode was from shear, as was planned during the design of this specimen. Figure 12adepicts the observed shear cracking patterns. In order to facilitate the repair of this specimen the loading was stopped after a decrease in the bearing capacity of the specimen was observed. In this way, the permanent deformations of this specimen at this stage were rather limited and the shear mode of failure was not allowed to fully develop.

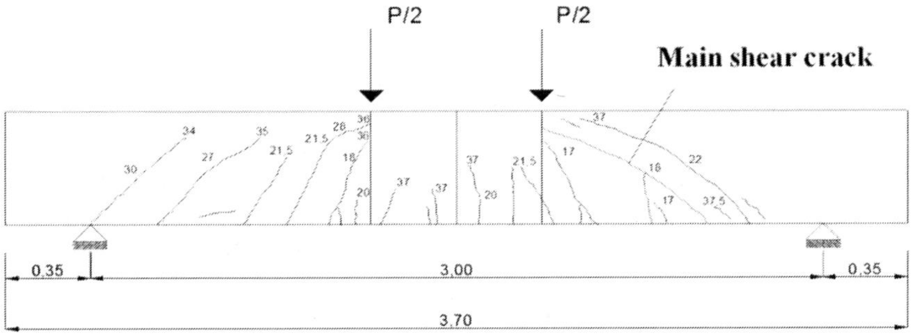

Figure 11. Flexural And Shear Cracking Patterns For Specimen *Beam-1*. Prevailing Final Mode Of Failure The Shear.

Evaluation of the observed behavior for specimen BEAM-1: Figure 12b depicts the observed behavior in terms of shear force (Q). In the same figure the horizontal lines of different colours indicate the shear capacity of this particular specimen, as was predicted following the procedure suggested by various researchers ([30], [31], [32]) as well as by either the Greek Code [9] (without safety factor) or by the Euro-code [7] (with safety factor). Figure 12c depicts the observed behavior in terms of bending moment (M). Again, the horizontal lines of different colours in this figure indicate the flexural capacity of this specimen as predicted by well established formulae as well as by the corresponding code procedures (without safety factor). Bearing in mind that the mode of failure is dominated by shear and not flexure the comparison between the shear and flexural limit states predicted in this way confirms the observed behavior.

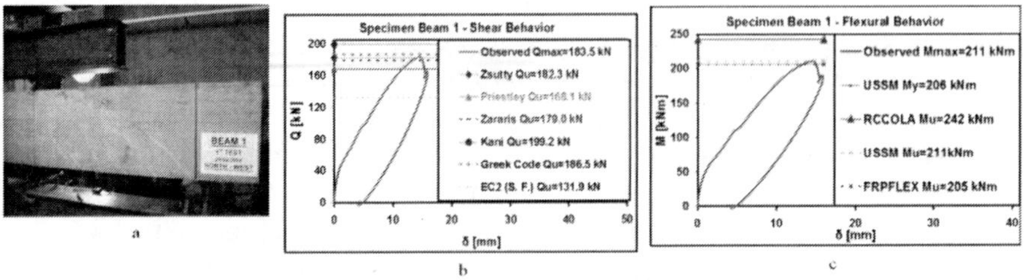

Figure 12. A cracking patterns of beam-1. B evaluation of shear behavior (beam-1). C evaluation of flexural behavior (beam-1).

2.2.2. REPAIRED BEAM SPECIMEN BEAM 1R-1 AND OBSERVED BEHAVIOR.

Description of Specimen BEAM 1R-1: After the end of the loading sequence of specimen BEAM-1, this specimen was repaired against shear failure in the way shown in figure 13. Closed loop CFRP (type C1 – 23) hoops of two layers (of nominal thickness 0.133mm per layer), having width b_{slice} = 75mm, were attached externally on the specimen spaced at distances of s=275mm. This was very easy to apply to this specimen and the only necessary treatment was to round up the corners of the specimen. However, in order to apply exactly such a repair scheme in prototype conditions one is faced with the considerable difficulty in drilling the appropriate holes at the top R/C slab that is usually monolithically connected to the beam at its top part. Moreover, the attachment of the CFRP layers in closed loop hoops will cause considerable difficulty. Alternatively, the CFRP shear reinforcement may not be of the closed loop hoop type but of the open loop U-shaped strips with or without anchors. The shear and flexural cracking that developed in specimen Beam-1 from the previous loading sequence was left without any repair (by epoxy resins or other means). An external layer of resin was used to paint the areas of the cracking in order to be able to monitor the activation of these cracks during the new loading sequence [14]. The mechanical properties of the employed CFRP layers (type C1 – 23), as they are given by the manufacturer, are : Ultimate stress f_u = 3800 Mpa, Young's Modulus E_f = 230 Gpa, Limit strain ε_u = 1.8 %, thickness t_f = 0.133 mm.

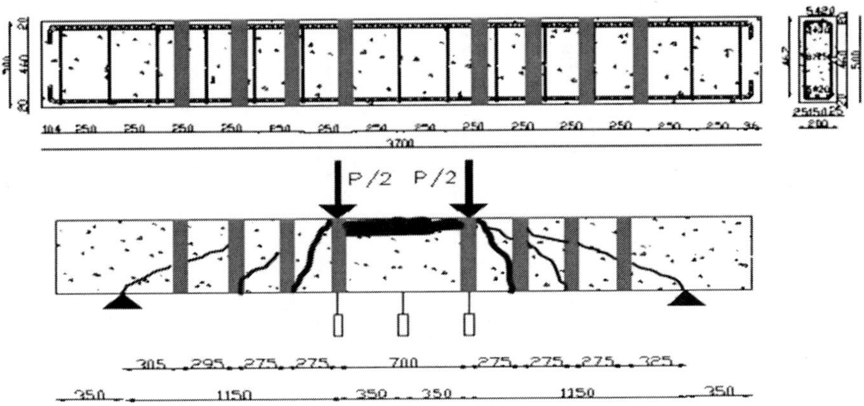

Figure 13. Repaired specimen BEAM 1R-1 The newly activated cracks are shown with bold lines. Dominant mode of failure the one at the compression zone of the mid-part of the specimen governed by flexure.

Observed Behavior of Specimen BEAM 1R-1: The main shear crack at the right side gave signs of becoming active for vertical load of 177KN (Q=88.5KN) whereas the same occurred for the left side main shear crack for vertical load of 216KN (Q=108KN). This meant that the corresponding CFRP shear type hoops started to develop stresses resisting part of the shear. Crushing was observed at the compressive upper side of the mid-part of this specimen undergoing flexure, when the vertical load reached 373KN. This was accompanied by the activation of the flexural cracks starting from the bottom side of the beam. These vertical flexural cracks developed even further for a maximum load of 414KN extending all the way from the bottom side to the crushed upper side, with the specimen reaching its flexural capacity. The shear mode of failure of the previous specimen was contained without further development by the applied CFRP hoops. The final condition of this specimen is depicted in figure 13. The observed behavior in terms of shear force (Q) and bending moment (M) is depicted in figures 14a and 14b. The maximum observed shear force was Q = 207 KN, and the maximum observed bending moment was M= 238 KNm.

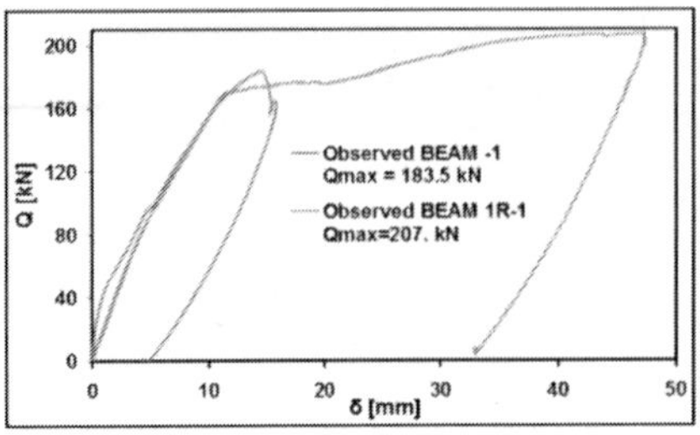

a

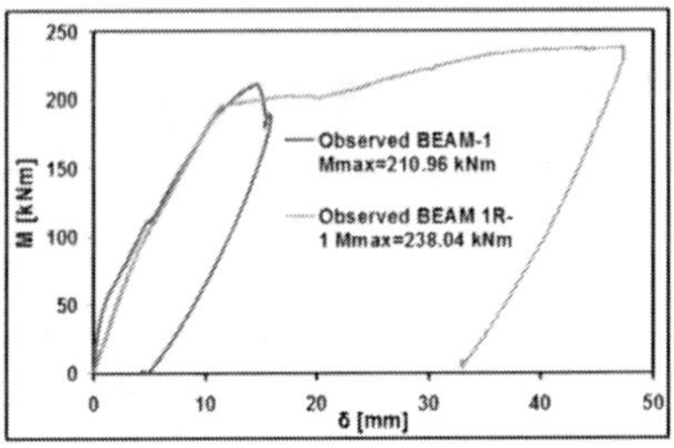

b

Figure 14. Observed shear behavior of BEAM 1R-1. 14b. Observed flexure of BEAM 1R-1.

In these figures 14a and 14b, the observed behavior before the repair of this specimen is also plotted for comparison with maximum observed shear force Q = 183.5 KN, and maximum observed bending moment M= 211 KNm. From this comparison it can be concluded that the applied CFRP hoops resisted a shear force of 23.5KN without signs of any distress. The successful increase (13%) in the shear bearing capacity of this specimen by this applied CFRP repair scheme, which leads to the exhausting of the specimen's flexural capacity and to reaching the desired flexural mode of failure, demonstrates the potential of such repair schemes for prototype conditions. However, there are construction difficulties in prototype conditions which were already pointed out before. Moreover, as was remarked before, the shear capacity of this scheme was not reached as the specimen survived the applied loading sequence in terms of shear. A subsequent reloading of specimen BEAM 1R - 1, with additional repair elements, namely specimen BEAM 2R -1, is presented in the following. The objective is to define by measurements the shear bearing capacity of the applied CFRP-hoops in this repair scheme.

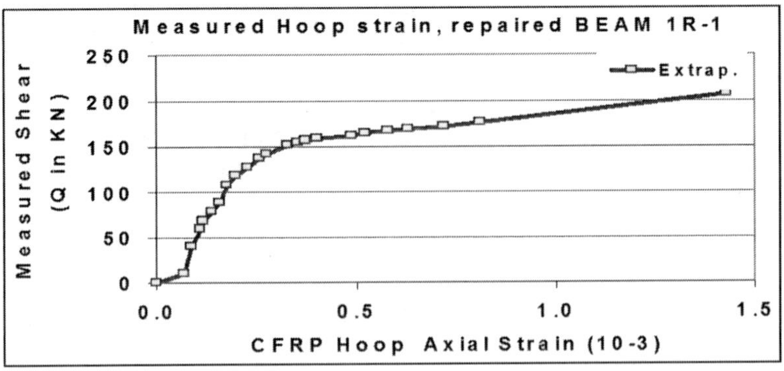

Figure 15. The variation of the CFRP-Hoop strain with the measured shear force for BEAM 1R-1.

Evaluation of the contribution of the CFRP for specimen BEAM 1R-1: During the loading sequence of Beam 1R-1, described above, the strains that developed at the mid-height of the third CFRP-hoop from the right end of the specimen (see figure 13) were monitored throughout the test by a strain gauge attached in this location. This strain gauge remained operational up to a total vertical load of 353KN (Q=177.6KN) measuring for this load an axial strain equal to 0.081 %. Unfortunately, beyond this load, the strain gauge became inactive due to a crack going through it. The variation of the axial (vertical) strain of CFRP-hoop with the applied shear force is depicted in figure 15. As can be seen in this figure, during the loading of specimen Beam 1R-1 the variation of the measured CFRP-hoop strain with the applied shear force is almost linear with different slope in two distinct ranges; the first range for shear force values up to Q=150KN and the second range for shear force values larger than Q=150KN. By extrapolating the second linear strain trend up to the maximum measured shear force value of 207KN a maximum strain value equal to 1.43 10^{-3} is found for the CFRP-hoop, which is also depicted in figure 15 (see also figure 14a). By employing the relationship:

tionship: $V_f = A^{hoop}(h/s)E_f \varepsilon_f^{eff}$ where A^{hoop} *=39.9mm2* the cross-section of one CFRP hoop, *h=500mm* the height of the beam,*s=275mm* the spacing of the hoops, E_f*=230Gpa* the Young modulus of the applied CFRP, $\varepsilon_f^{eff} = 1.43\ 10^{-3}$ the effective CFRP strain as was mentioned before, the value of the shear resisted by the applied CFRP-hoops can be found equal to V_f*= 23.86KN.*

When this value, which was found from direct strain measurements of the CFRP-hoop, is compared with the value of V_{fa} = 23.5KN, found by subtracting the shear resisted by the virgin specimen BEAM -1 from the shear resisted by the repaired specimen BEAM 1R-1 with the CFRP-hoops, a very good agreement is observed, which further validates the shear resisted by the applied CFRP-hoops. Using the software developed by Triantafillou [33] a prediction for the maximum shear force equal to 44KN is found to be resisted by the CFRP-hoop arrangement applied in BEAM 1R-1 (figure 13). The measured value of approximately 24KN and the observed shear performance of this specimen are in agreement with this predicted upper limit (44KN) for the CFRP-hoop shear resistance of the employed repaired scheme.

2.3. Repaired Specimen Beam 1r-2

As described in section 2.2.2. before, a middle part compressive zone crushing was observed as a limit state for specimen BEAM 1R-1. After unloading this specimen the crushed concrete was removed and the whole area cleaned till the compressive longitudinal reinforcement became visible. The reshaping of the orthogonal form of this part of the beam was achieved with a paste from a mixture of cement and epoxy resin, which becomes hard in one day attaining a compressive strength of 30 MPa. The thickness of this layer of paste was 20mm above the compressive longitudinal reinforcement and the corners were given a smooth curvature. This upper zone was next reinforced with layers of CFRP (C 1-23), as shown in figure 16, in order to increase its compression capability through confinement. In addition, three layers of GFRP (G-60 AR) were attached at the bottom side of the specimen to enhance through tension its flexural capacity beyond the one provided by the existing longitudinal reinforcement (figure 16).

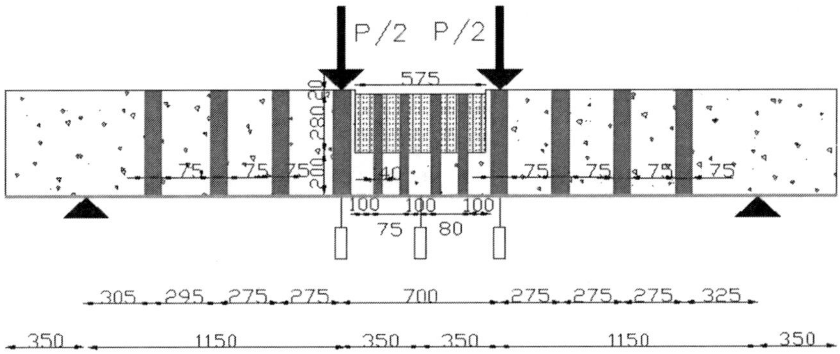

Figure 16. Specimen BEAM 1R-2 with CFRP layers attached in a way as to contain the compressive zone of the middle part as well as additional GFRP layer at the bottom side of the specimen extending the whole span.

2.3.1. OBSERVED BEHAVIOR OF SPECIMEN BEAM 1R-2

This specimen was reloaded in a way similar to the one employed for the preceding specimens BEAM -1 and BEAM 1R-1. Figures 17a and 17b depict for all three specimens the shear and flexural behavior, respectively. In figure 17b the predictions of the yield moment for specimen BEAM-1 and the ultimate moment of specimen BEAM 2R-1 are also indicated. Due to an increase in the flexural capacity for specimen BEAM 2R-1 when compared to the preceding specimen BEAM 1R-1, resulting from the described repair scheme for specimen BEAM 2R-1, the shear failure mode prevailed this time. This mode of failure was accompanied by the rupture of the CFRP hoops as well as by crushing of the concrete at the compressive shear transfer zone neighboring the point of loading (figure 18). As expected, this specimen (BEAM 1R-2) exhibited an increase in its flexural capacity when this is compared with the flexural limit state attained during the preceding test of specimen (BEAM 1R-1). This fact demonstrates that the repair scheme described before, aimed to increase the flexural capacity of specimen BEAM 1R-2 proved to be successful. The observed maximum shear force during this test was $Q_{\alpha v}$ = 215.8KN whereas the maximum bending moment was $M_{\alpha v}$ = 248.2KNm. Comparing these values with the ones observed during testing the original virgin BEAM-1 it can be concluded that the applied repair scheme resulted in an increase of the order of 18% for the shear capacity and more than 18% for the flexural capacity; the exact increase in the flexural capacity could not be measured as the shear failure for BEAM 1R-2 prohibited this specimen from reaching its flexural limit state. The predicted flexural capacity of BEAM 2R-1 is listed in Table 1.

TABLE 1. PREDICTIONS OF FLEXURAL CAPACITY FOR BEAM 2R-1.

METHODOLOGY	MU [KN m]	Mode of failure
Ultimate stress strain method	**301.26**	Failure from compression of the top zone and of tension of the GFRP
FRPFLEX	**319.02**	Failure of compression zone

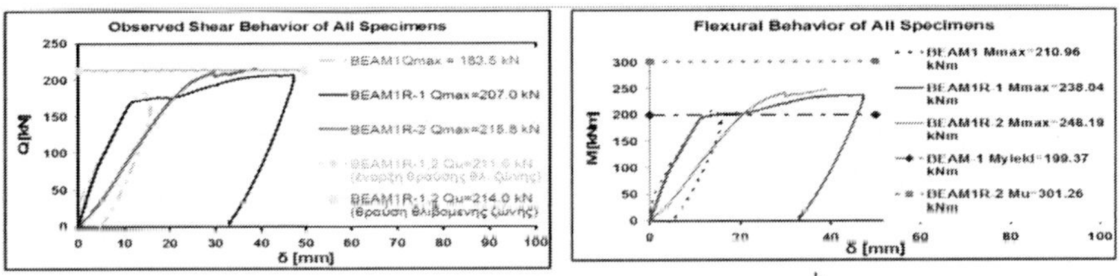

a b

FIGURE 17. Shear behavior of all specimens. b. Flexural behavior of all specimens.

Table 2. Important developments in the behavior of specimen BEAM 1R-2 during loading.

P [KN]	Description of the observed cracking
73.6	Partial activation of the shear cracks that were formed from past loading
166.8	Activation of the previously formed flexural cracks
235.4	Formation of main flexural crack next to the CFRP stirrup at loading point (left side)
343.4	Formation of main flexural crack next to the CFRP stirrup at loading point (right side)
421.8	Initiation of crushing of the concrete compressive zone at the shear transfer area neighboring the left loading point.
423.8	Rupture of a CFRP stirrup
431.6	Crushing of the concrete at the compressive shear transfer zone neighboring the left loading point. Rupture of the CFRP hoops that were crossed by the main shear crack (figure 18).

Figure 18. Shear failure accompanied by the rupture of the CFRP stirrups as well as of crushing of the concrete compressive zone at the shear transfer area neighboring the point of loading.

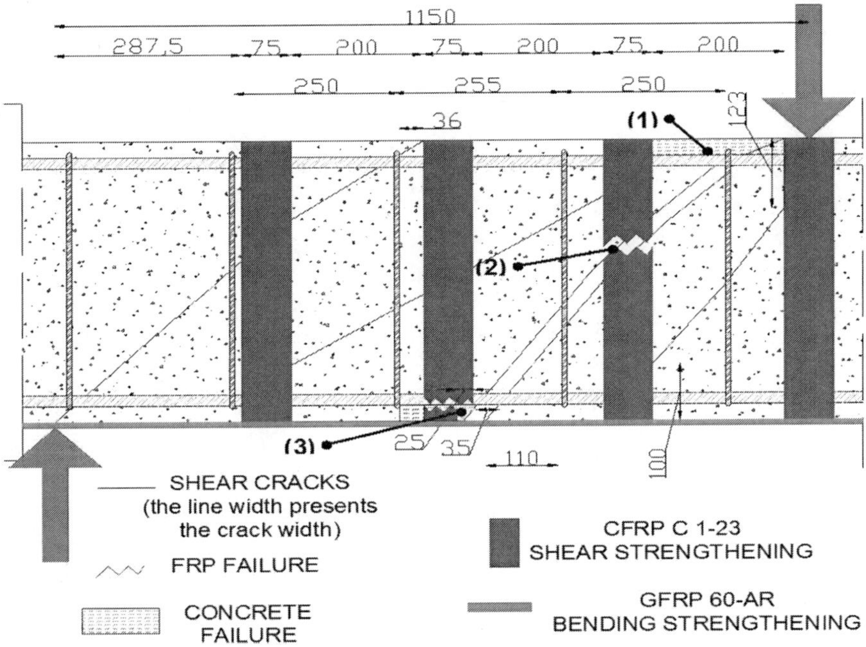

Figure 19. Mechanism of shear failure during the limit state of specimen BEAM 1R-2.

It must be pointed out here that the way the enhancement of the compressive zone capacity was achieved through the compressive zone confinement, attained from the CFRP layers, cannot be applied so easily in prototype conditions due to the presence of an R/C slab at this part of a prototype beam.Table 2 lists the total vertical load values which were measured during important developments in the observed behavior of specimen BEAM 1R-2. Moreover, figure 19 depicts the observed mode of failure in a detailed way.

2.3.2. EVALUATION OF THE CONTRIBUTION OF THE CFRP-HOOPS FOR BEAM 1R-2

As was done for specimen BEAM 1R-1, during the loading sequence of Beam 2R-1 the axial (vertical) strains that developed at the mid-height of the third CFRP hoop from the right end of the specimen (see figure 13) were again monitored employing a strain gauge attached in this location. This strain gauge remained operational throughout the whole load sequence measuring a maximum axial strain value ε_f^{eff} = 0.67 % when the total vertical load reached its maximum value of 435.6KN (Q=215.8KN). By subtracting the shear resisted by the virgin specimen BEAM -1 (Q=183.5KN) from the shear resisted by the repaired specimen BEAM 2R-1 (Q=215.8KN), a value of V_{fa} = 32.3KN is found as being resisted by the used repair scheme. As mentioned before, using the software developed by Triantafillou [33] the maximum shear force of 44KN is found to be resisted by the CFRP hoop arrangement applied in BEAM 2R-1. Moreover, employing the maximum strain measurement and the relationship $Vf = A^{hoop} (h/s)\ Ef \varepsilon f^{eff}$ the shear resisted by the applied CFRP-hoops can be found equal to $Vf = 111.8KN$. This discrepancy dictates further investigation on the shear resisting mechanisms that developed during the limit state of specimen BEAM 2R-1 (see figure 19).

The maximum axial strain measurement equal to 0.67 % was recorded at this particular CFRP-hoop location during maximum applied load just before the rupture of this CFRP-hoop and the final development of the shear failure for specimen BEAM 2R-1. The variation of the applied shear force with the measured CFRP hoop strain is depicted in figure 20; it is almost linear with different slope in two distinct ranges; the first for a shear force value from zero to Q=205KN and the second for a shear force value larger than Q=205KN up to $Q_{\alpha v}$ = 215.8KN, when the rupture of the CFRP occurred.

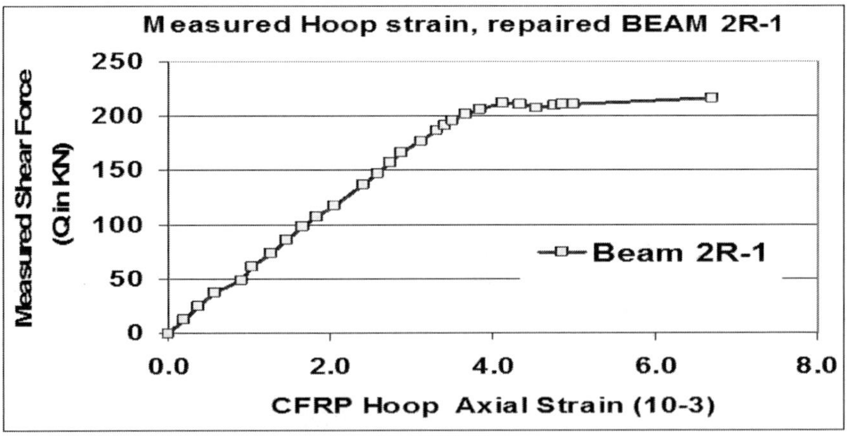

FIGURE 20. The variation of the CFRP-hoop strain with the measured shear force for BEAM 2R-1.

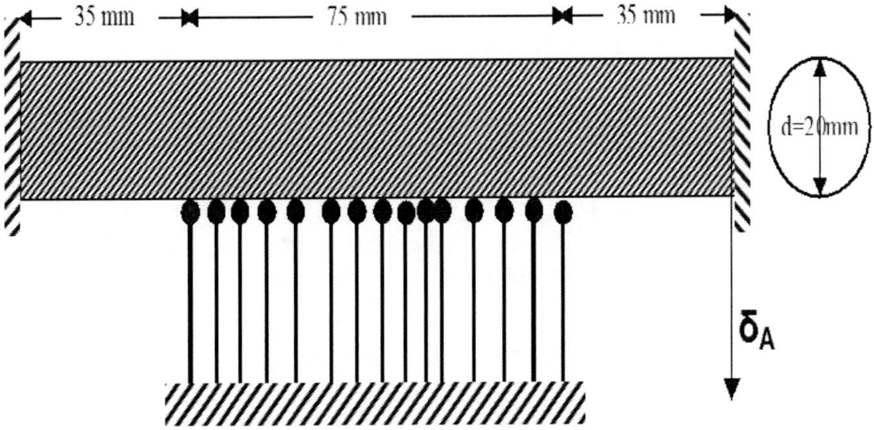

Figure 21. Numerical simulation of the dowel action.

2.3.3. Shear capacity of specimen beam 2r-1 based on the observed failure mechanism

Based on the observed shear failure mechanism, which is depicted in figure 19, the shear capacity of BEAM 2R-1 will be estimated as will be described below. It is assumed that the shear is resisted in this case by the following parts.

- One closed steel stirrup.
- The compressive zone for shear transfer next to the point of load application at the top of the main shear crack.
- Two CFRP-hoops.
- The contribution of the bottom side longitudinal reinforcement (dowel action).

The contribution of the compressive zone is found from the relationship: $V_c = 0.3\ (fc)^{2/3}\ (2)^{1/2}\ x\ b$. Where fc the concrete compressive strength, x the height of the compressive zone (as estimated from the flexural behavior), b the width of the beam. This leads to a shear force equal to V_c = 46.42 kN at the initiation of concrete compressive zone crushing.

The contribution of the assumed two CFRP-hoops is equal to V_f = 18234 $\varepsilon_f^{\,eff}$ [KN]. According to the strain measurements this contribution at the initiation of crushing of the compressive zone is equal to V_f = 73.54KN.

The contribution of the dowel action of the longitudinal reinforcement will be based on the geometry of this particular failure mechanism which was measured in detail at the laboratory. For this purpose, a simple numerical simulation of this dowel action was formed, which is depicted in figure 21. The longitudinal reinforcement is supported by a steel stirrup (left support) and by the CFRP-hoop (multiple mid-supports). All the supports are simulated by springs with the corresponding elastic / post-elastic and cross-sectional properties the actual steel stirrup and CFRP-hoop possess.

Due to the volume of concrete the rotations at the two ends of this dowel action numerical model were restrained; moreover, the employed simulation provided the capability for the longitudinal reinforcement to develop plastic hinges with parameters based on the cross-sectional area and yield stress of the employed longitudinal reinforcement, which were measured at the laboratory.

The limit state contribution of the dowel action of the longitudinal reinforcement, as resulted from this numerical simulation, was found equal to V_d = 37.67 KN. The contribution of the steel stirrup (V_w) is based on the assumption that its strain is the same as the measured strain of the neighboring CFRP hoop. The various contributions based on all the above assumptions are listed in Table 3 together with the corresponding total measured value. As can be seen a very good comparison is reached between the total estimated shear resistance and the corresponding measured value. It must be pointed out that the estimated value was based on the exact observation of the failure mechanism as well as on the measurements of the CFRP hoop strains. In the same table a predicted shear resistance value is also listed based on the prediction of the maximum shear resisted by the CFRP-hoops by the software developed by Triantafillou [33] (maximum predicted V_f value equal to 44KN is found).

Table 3. Shear resistance contributions of the various parts of specimen BEAM 2R-1 at the initiation of the compressive zone crushing.

Steel Stirrup V_w [KN]	2 CFRP Hoops V_f [KN]	Dowel action of longitudinal reinforcement (3Φ20) V_d [KN]	Compressive Zone of Concrete V_c [KN]	Total estimated shear resistance V_R [KN]
53.94	**73.54**	**37.67**	**46.42**	**211.57**
Measured total shear resistance **210.9KN**			Predicted total shear **227.5KN** [ref. 33]	

The variation of the estimated shear contributions of the CFRP-hoops (Vf), the steel stirrup (Vw), the dowel action of the longitudinal reinforcement (Vd) as well as the sum of all these (Vf+Vw+Vd) are plotted in figure 22 together with the measured total shear force (V_R) for specimen BEAM 2R-1. The common basis for the plotted displacements between the estimated contributions and the measured values is found from the correspondence of the load and strain measurements during testing. In addition, in the same figure, the variation of the contribution of the compressive concrete zone is plotted by subtracting from the total measured shear force (V_R) the contributions of the steel stirrups and the CFRP hoops ($V_c=V_R-V_w-V_f$). After the formation of the shear cracks this contribution includes the dowel action of the longitudinal reinforcement. This results in a maximum value Vc = 84.14 KN, which is in agreement with predictions based on formulas suggested by various researchers. Moreover, it can be seen that the summation of the various contributions is in good agreement with the total measured shear force if the contribution of the compressive zone is added, after the formation of the shear cracks and up to the initiation of the crushing of the compressive zone concrete.

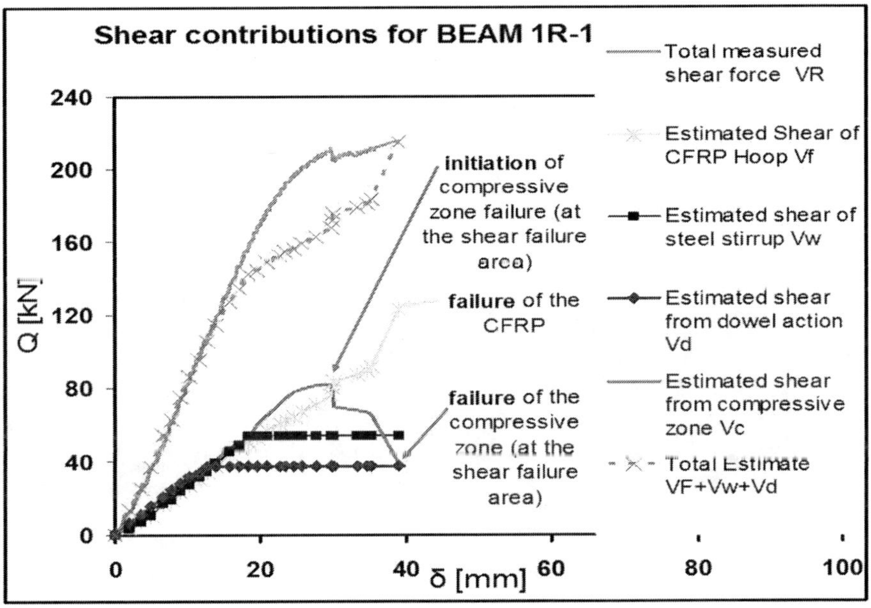

Figure 22. Shear resistance contributions, extracted from the mechanism of shear failure for specimen BEAM 2R-1.

2.4. Concluding Observations

a1. The technique of repairing reinforced concrete beam elements by attaching CFRP layers has been examined in the laboratory. It was found that using CFRP-hoops, spaced at the areas of the beam that resist the shear forces, was a quite successful repair technique in upgrading the shear capacity of beam specimens by inhibiting the shear mode of failure thus favouring the development of the flexural mode of failure.

b1. The repairing of the examined specimens by attaching GFRP layers to complement the tensile longitudinal reinforcement together with CFRP confinement of the compressive zone at these parts of the beam that undergo flexure was again a quite successful technique in upgrading the flexural capacity of the beam specimens. However, this must be properly controlled in order to prohibit the development of the undesired shear mode of failure.

c1. In order to be able to successfully control the shear and flexural capacities and modes of failure, after the application of CFRP repair schemes, reliable tools are necessary that successfully predict the bearing capacities of the repaired structural elements. It was demonstrated by this work that whereas this is possible for the flexural capacity it is not so straightforward for the shear capacity unless the basis of the shear capacity estimate is founded on a realistic failure mechanism that takes into account the realistic contributions of all the parts that are mobilized through this failure mechanism.

d1. It must be pointed out that the examined repair schemes for upgrading the shear resistance through CFRP stirrups will not be so easily applied in prototype conditions where the presence of a reinforced concrete slab, cast monolithically with the upper part of the beam, must be confronted.

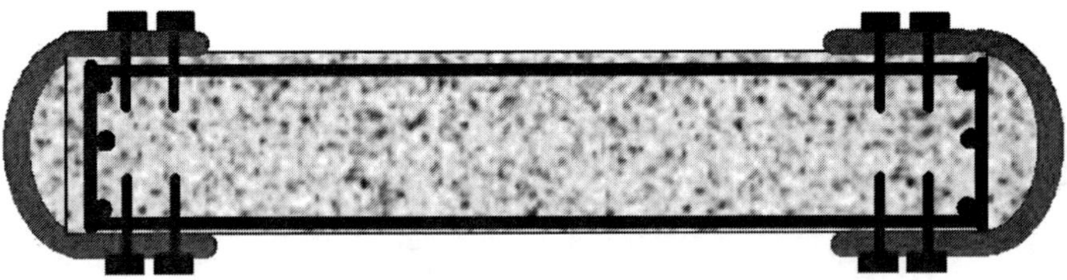

Figure 23. R/C cross-sections with one side longer than the other (h/b>1.5) and CFRP partial confinement.

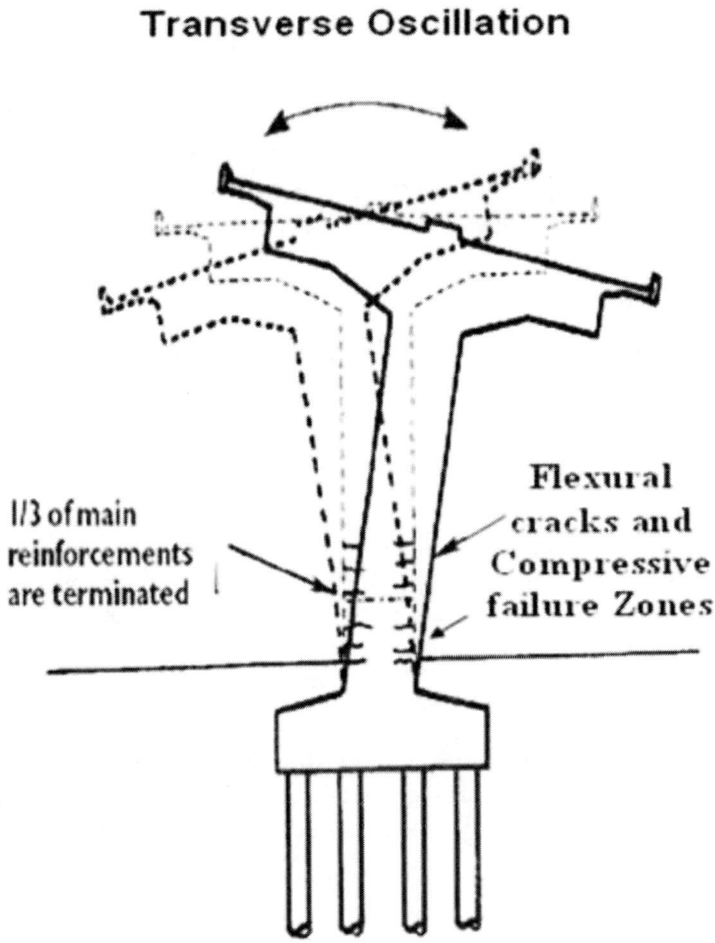

Figure 24. Bridge Pier compression failure mode.

3. PARTIAL CONFINEMENT

3.1. The Use Of Confinement For Upgrading The Flexural And Shear CAPACITY OF COLUMNS.

A very common application of externally applied fiber reinforced plastics (FRP) is confining parts of structural members. This can be easily envisaged for columns with their cross-section approximating a square shape [34]. This is done by wrapping around the structural element one or more that one layers of FRP sheets together with the proper resine in-order to bond them on the surface of the structural member as well as to bond the FRP sheets between themselves. The main points of attention here are the following [35]:

- Preparation of the surface of the existing structural member in order to prohibit premature delamination (debonding) of the attached FRP sheets.
- Localized rupture of the FRP jacket due to abrupt change of curvature at the corners, to buckling of reinforcing bars, or to excessive dilation of concrete.
- Proper application of the resin in order to achieve sufficient bond between the FRP sheets and the surface of the existing structural member.
- Proper application of the resin and enough overlapping length of the wrapped layers of the FRP sheet around the existing member in order to prohibit any unwrapping mode of failure.

The proper wrapping of the FRP multilayer sheet around the existing structural member (column or beam) introduces a passive confinement in the same way as steel stirrups do on such concrete members. Whereas the wrapping of FRP sheets around columns or beams of rectangular shape with width / height ratio no greater than 2 can be easily materialized this is not so easily applicable in the following cases:

- In columns with rectangular cross-sections having a width / height ratio greater than 2. It will be discussed later on in this section how this difficulty was confronted in the laboratory.
- In T- beams whereby the presence of the slab prohibits the wrapping of the FRP sheets around the cross-section.

Table 4. Characterization of the effectiveness of the partial confinement.

Grade	Compression failure mode	Increase in the compressive capacity (compared to the unconfined specimens)	Effectiveness of the partial confinement
1	Of the weak part as without the partial confinement	Small increase	**low**
2	Of the weak part as without the partial confinement	Considerable increase	**Considerable**
3	Of both the weak part and the strong parts	Substantial increase	**Very effective**

The beneficial effect of such a confinement on the compressive strength of axially loaded structural members is well known and will not be repeated here. Moreover, the function of these externally applied FRP sheets as transverse reinforcement for both columns and beams is also evident in the same way closed steel stirrups function in reinforced concrete sections. Acting as external transverse reinforcement FRP sheets fully wrapped around beams or columns can be utilized to upgrade their shear capacity (figure 5). This has been already discussed in the beginning of section 2. Again, for T-beams the presence of the slab prohibits the formation of closed loop transverse reinforcement by the FRP sheets. If the FRP sheets are applied as transverse reinforcement in an open loop formation they are susceptible to a premature delamination type of failure (see figure 4). This will be further discussed in section 6 together with the way this difficulty was confronted in the laboratory.

The upgrading of reinforced concrete (R/C) cross-sections, with one side rather longer than the other (h/b > 1.5), by partial application of CFRP (Carbon Fiber Reinforcing Plastic) confinement is investigated here (figure 23). This partial application of CFRP confinement is aimed at the retrofitting of bridge-pier type R/C cross-sections in order to prohibit, up to a point, the development of premature compressive failure at the base of the pier due to combined compression and flexure from seismic loads (see figure 24 and [36], [37]). The performance of such structural elements was studied extensively in the past ([38], [39]). This type of partial confinement may also be applied to upgrading vertical structural members with non-accessible sides. Design guidelines for rectangular FRP jackets applied on rectangular columns have been proposed with the limitation that the cross-sections have aspect ratio h/b < 1.5 [37]. For higher aspect ratios it is recommended to design a circular or oval jacket. However, it is expected that for rectangular cross sections with aspect ratios larger than 1.5 the radius of a circular or rectangular jacket will be too large and will result in ineffectual confinement and will prove costly and impractical. For this reason it is desirable to investigate alternative schemes for increasing the confinement of rectangular cross-sections with relatively large aspect ratio without resorting to complete circular or oval jackets. Such a scheme is studied here using CFRP layers that do not extend all around the cross-section (figure 23 "*partial confinement*").

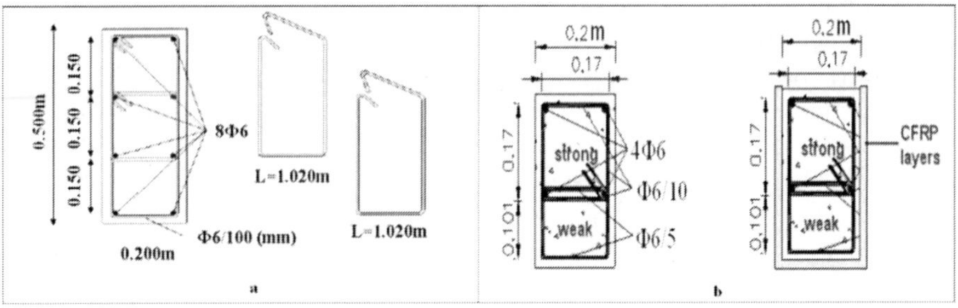

Figure 25. Initial cross-section with h/b ratio = 2.5. 25b. Test specimens without and with partial CFRP confinement.

To compensate for the fact that the CFRP layers do not enclose the cross-section entirely, anchorage of these layers must be provided, as shown schematically in figure 23. To this end, a laboratory investigation was carried out to study the effectiveness of such partial confinement together with alternative anchorage schemes. As will be explained, this effectiveness was tested by subjecting the specimens only to compressive loads. Despite this limitation, as will be demonstrated from the results of this investigation, the most significant aspects of the critical factors for this "*partial confinement*" scheme were brought to light.

3.1.1. THE FORMATION OF TEST SPECIMENS

The initial cross-section, which formed the basis of the tested specimens, is depicted in figure 25a with an aspect ratio h/b equal to 2.5. This is a rectangular cross section of a bridge pier model structure, which was tested both at the laboratory (figure 25a) and at the Volvi-Greece European Test Site in the framework of the European project Euro-Risk [36]. This cross section was intentionally designed to develop flexural mode of failure at the base of the pier; moreover, it was desirable to find ways to retrofit such specimens by prohibiting premature compression failure at the base by means of partial CFRP confinement. The effectiveness of the partial CFRP confinement is studied by subjecting the tested specimens only to pure compressive loads. This type of stress field is expected to develop at the base of such vertical members under combined vertical loads and seismic actions, where undesired compression failure may develop. In order to limit the maximum level of compressive loads required to bring to failure such a cross-section with the loading arrangements available, the tested specimens had a cross section (figure 25b) of 200mm by 300mm instead of 200mm by 500mm of the initial cross-section (figure 25a and figure 25b) for the bridge pier specimens tested both at the laboratory and at the test site under combined compression and flexure [36]. Moreover, in order for the tested specimens to form compression failure at the same part of the cross-section where such failure would develop at the base of the initial bridge pier model, one part of the tested cross-section was left identical to the initial cross-section (the one that is marked in figure 25b as weak) whereas the remaining part was strengthened both with longitudinal and, in particular, with transverse reinforcement (the one marked in figure 25b as strong).

In this way, with the compression capacity of the weak part being smaller than that of the strong part, the compression failure was expected to develop at the weak part. This proved to be correct during the experiments, as will be shown in the following sections. The CFRP partial confinement was applied at the weak part, as is shown at the right hand side of figure 25b. Then, by studying the resulting bearing capacity and mode of failure under compression of the tested specimens (with or without partial confinement) the effectiveness of such a repair scheme could be demonstrated and classified as listed intable 4.

3.1.2. CONSTRUCTION OF TEST SPECIMENS

Ten identical specimens were constructed and eight of them were used in the current experimental sequence (see table 5). All specimens were reinforced in the same way and were cast at the same time with the same mixture aiming for similar plain concrete strength values. The reinforcing details are depicted in figure 26. As can be seen, the total height of the specimens was 1600mm. They had a mid-part height of 580mm that was left to develop the compression failure (figure 26, part denoted as CROSS SECTION A). The cross section of this mid-part was the one shown in figure 25b, including the two distinct parts (the weak and the strong). The two edges of the specimens, with a height of 510mm each, were confined during the experiment with strong steel brackets covering these parts from all sides thus prohibiting any compressive failure developing at those two edges (see figure 27).

The partial CFRP confinement was attached only on the three sides of these specimens covering the weak part and leaving the fourth side (of the strong part) free without any CFRP layers (figure 25b). There were 8 virgin specimens, namely 1, 1a, 3, 3a, 4, 4a, 5, and 5a (Table 5). These specimens were tested in their virgin state in which some of them were without partial CFRP confinement whereas the rest had the partial CFRP confinement applied to them from the beginning. The second column oftable 5 indicates the partial confinement condition of the virgin specimens. The testing sequence of these virgin specimens is signified as Test 1. Most of these specimens were repaired after they had reached their limit state during their previous test. In all the repaired specimens the CFRP partial confinement was applied. Throughout these series of experiments the nominal thickness of the employed CFRP layers was 0.176mm with a given Young's Modulus E=350GPa. The measured maximum axial CFRP strain was approximately 1%.

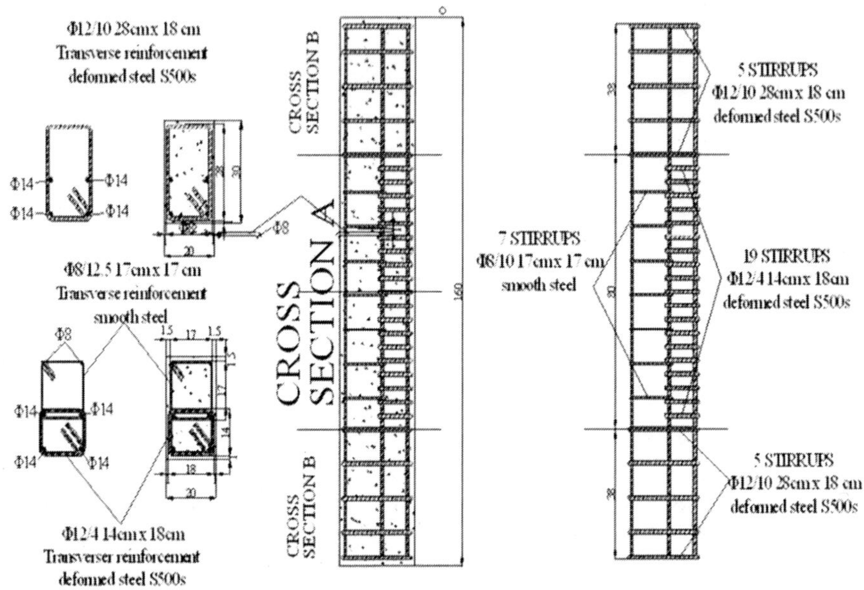

Figure 26. Reinforcement details common to all specimens.

Table 5. Test specimens with their corresponding concrete strength.

Virgin Specimens	CFRP Confinement	Repaired Specimens	Plain Concrete Strength (Mpa)
Specimen 1 Test 1	No	Test 2 3+(2) CFRP	28.0
Specimen 1a Test 1	3 CFRP layers	Test 2 3 GFRP	25.8
Specimen 3 Test 1	No	Test 2 3+(2) CFRP	27.6
Specimen 3a Test 1	5 CFRP layers	Test 2 5 CFRP	27.6
Specimen 4 Test 1	5 CFRP layers		27.7
Specimen 4a Test 1	5 CFRP layers		27.7
Specimen 5 Test 1	5+(2) CFRP layers	Test 2, 5+2 CFRP Test 3, 5+2 CFRP Test 4 7 CFRP	27.6
Specimen 5a Test 1	No	Test 2, 7 CFRP Test 3 7+4 CFRP	25.8

Figure 27. Specimen with confining steel brackets at the edges and partial CFRP confinement at the mid-part

This sequential test number of the repaired specimens is signified as Test 2 (for the 1st repair), Test 3 (for the 2nd repair) etc. (see column 3 of Table 5). In the same column the number of CFRP layers used in the partial confinement for the repaired specimens is also indicated. The total number of specimens, virgin and repaired, was seventeen. In Table 5 the unconfined concrete compressive stress is also listed, found from cylinders with diameter 150mm and 300mm height; these cylinders were obtained during the casting of each virgin specimen. The anchorage of the CFRP layers was applied along the two long sides, which were common to both the weak and the strong parts. The main load that was applied was axial compression, although in limited specimens the axial compression was combined with bending, which is not reported here. From the observed behavior, the effectiveness of the applied partial confinement could be deduced. As shown in table 4, this judgment was based on the level of the bearing capacity combined with the type of compression failure that was formed (at the weak or strong part). Moreover, the observed behavior of the various parts of the test specimens, such as the CFRP layers and their anchorage, helped to identify the factors that bear an adverse or beneficial influence on these aspects of the behavior.

3.1.3. Critical parameters and their variation:

The following parameters were critical (see results contained in Table 6) .

1 st. The type of anchorage of the CFRP layers, 2 nd. The number of the CFRP layers.

The second parameter becomes critical only if the first parameter performs satisfactorily. A large number of specimens (1, 1a, 3, 3a, 4 and 4a) exhibited failure of the anchorage of the partial confinement thus limiting the effectiveness of the partial confinement to low levels. Consequently, the increase in the compressive bearing capacity in this case was relatively modest. The compressive load was monitored by continuous sampling throughout the test. The "*Average Stress at failure*" listed intable 6 is found by dividing the compressive load by the gross cross-sectional area of each specimen. Listed in table 6 are certain particulars of the anchorage of the CFRP layers. Anchors-1 signifies bolts of 7mm diameter and 60mm length that do not penetrate the cross-section in its whole width, whereas Anchors-2 signifies bolts of 10mm diameter that penetrate the whole section (220mm length).For specimens 5 and 5a, due to the improvement of the anchoring details of the partial confinement, the increase in the compressive bearing capacity of these specimens was 50% higher than the observed bearing capacity of similar specimens with no partial CFRP confinement. When the maximum compressive bearing capacity achieved with the partial confinement is compared with the corresponding plain concrete strength value, a 230% increase can be observed. Moreover, the compressive failure in this case involved both the weak and the strong part thus improving substantially the effectiveness of the partial confinement and allowing the employment of a large number of CFRP layers.

Table 6. Summary of test results together with the basic specimen characteristics.* As % of the corresponding plain concrete strength.

Specimen No	Virgin	Repaired	Number of CFRP layers	Anchors-1	Anchors-2 + washers	Average Stress at failure (Mpa)	Failure Mode
1 Test 1	Yes	No	No	No	No	41.12 (146.9%)*	Weak part
1 Test 2	No	Yes/ EMACO	3 (+2)	Yes	yes	41.69 (148.9%)	Bolts
1a Test 1	Yes	No	3	Yes		45.78 (177.4%)	Pull out Anchors-1
3 Test 1	Yes	No	No	No		40.79 (147.7%)	Weak part
3 Test 2	No	Yes	3 (+2)	Yes	Yes	45.78 (165.9%)	anchorage
3a Test 1	Yes	No	5	Yes	Yes	47.09 (170.6%)	anchorage
3a Test 2	No	Yes/ EMACO	5	Yes	Yes	42.51 (154.0%)	anchorage
4 Test 1	Yes	No	5	No	Yes weak	46.60 (168.2%)	anchorage
4a Test 1	Yes	No	5	No	Yes weak	45.53 (164.4%)	anchorage
5 Test 1	Yes	No	5 (+2)	No	Yes strong	53.96 (195.5%)	Steel bracket
5 Test 2	Yes	-	5 (+2)	No	Yes strong	55.26 (200.2%)	Steel bracket
5 Test 3	-	Top / EMAKO	5 (+2)	No	Yes Strong	53.96 (195.5%)	CFRP mid height
5 Test 4	No	Yes/ EMAKO	7	No	Yes Strong	58.86 (213.3)	Strong stirrups
5a Test 1 bending	Yes	No	No	No	No	40.88 (158.4%)	Weak part
5a Test 2	No	Yes	7	No	Yes Strong	57.23 (221.8%)	CFRP mid-height
5a Test 3	No	Yes	7 +4	No	Yes Strong	60.17 (233.2%)	Strong part

3.1.4. Instrumentation to obtain the average stress-strain behavior

Apart from monitoring the compressive load, the deformations of the mid-part were also continuously recorded throughout each experiment with displacement measurements taken at each side of the cross-section. Eight displacement sensors (two at each side) were employed to record the deformations of the mid-part. Although the deformation of this mid-part was far from uniform, as can be seen from the obtained displacement measurements of the weak and strong parts (figure 29), the average axial displacement, which was found by averaging the measured displacement values at all four sides of each specimen, is mostly used here as an indication of the deformability of each specimen. By dividing this average axial displacement by the height of the mid-part an average axial strain could also be obtained in this way. The following discussion of the observed behavior of each specimen is based on diagrams of average axial stress versus average axial strain found from the previously described averaging process. More detailed study on the obtained non-uniform deformability for each specimen will be carried out at a future stage. An additional measurement that was obtained during the experimental sequence was the axial strain that developed at the CFRP layers of the partial confinement of the mid-part. These CFRP strain measurements are an additional indication of the effectiveness of the partial confinement.

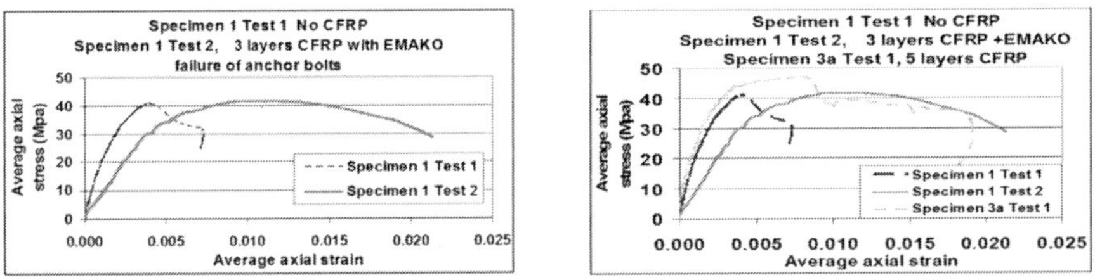

a b

Figure 28. a Partial confinement of low effectiveness. b. Partial confinement of low effectiveness.

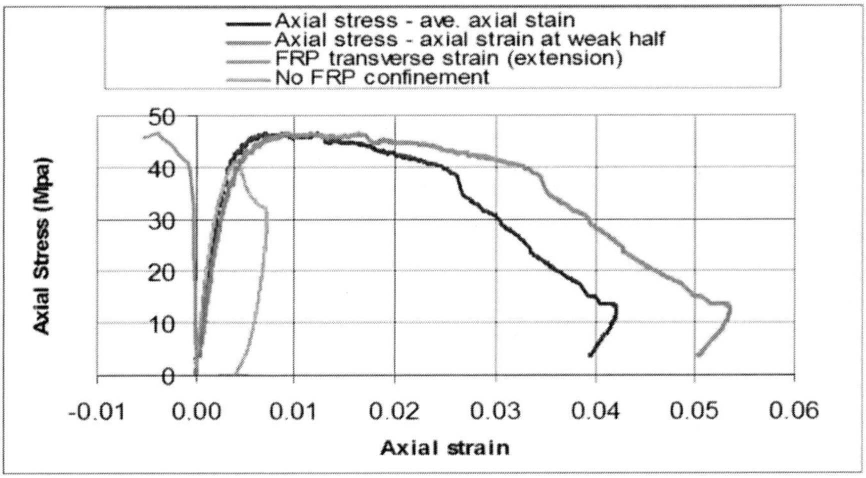

Figure 29. Specimen 4, Test 1 (5 layers CFRP) Comparison with Specimen 1 Test 1 (no confinement).

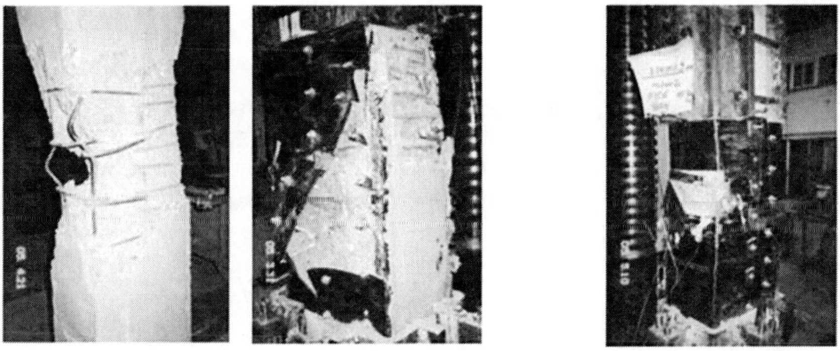

a b c

Figure 30. a Virgin Specimen without confinement. b. Failure of anchor bolts. 30c Low effectiveness of partial confinement. c. Tensile failure of CFRP Layers. Considerable effectiveness of partial confinement.

3.2. Discussion Of The Results

3.2.1. Partial confinement of low effectiveness

In figures 28a and 28b the obtained behavior of specimen1 (Test 2) with 3 CFRP layers and specimen 3a (Test 1) with 5 CFRP layers is compared with specimen

1 test 1 (no partial confinement). Specimen 1 test 2 was formed from specimen 1 test 1 by repairing the failed specimen 1 test 1 with special (low shrinkage) concrete as well as with 3 layers of CFRP forming the partial confinement. This repaired specimen failed in compression with almost the same capacity as the previously-tested virgin test with no partial confinement, but with larger deformability. The effectiveness of the partial confinement is low and it is due to the failure of the anchor bolts of the applied confinement. In figure 28b the observed behavior of specimen 3a Test 1 is also included. This is a virgin specimen that had a 5-CFRP layer partial confinement. Despite the increase in the CFRP layers, the observed effectiveness of the partial confinement is low as was for specimen 1 test 2, again because of the failure of the anchor bolts. In figure 29 the observed behavior of specimen 4 test 1 is depicted. This was a virgin specimen with 5 CFRP layers partial confinement. In this case, a certain alteration was applied in the anchor bolts, by increasing their length. However, this was not sufficient to improve accordingly the effectiveness of the partial confinement, which was again linked to the failure of the anchor bolts. The failed virgin specimen without the partial confinement is shown in figure 30a whereas figure 30bdepicts the failure of the anchor bolts for the "low effectiveness" partial confinement.

3.2.2. Partial confinement of considerable effectiveness

In figures 31a and 31b the obtained behavior of specimen 5 (Test 3) with 5 (+2) CFRP layers is compared to specimen 1 test 1 (no partial confinement). Specimen 5 test 3 was formed by repairing a previously failed virgin specimen with special (low shrinkage) concrete as well as with 5 layers of CFRP forming the partial confinement. An additional two (+2) CFRP layers were applied at the part of the section where the anchor bolts were placed. This repaired specimen failed in compression with a modest increase (31%) in its capacity when compared with the capacity of the virgin unconfined specimen. The effectiveness of the partial confinement in this case was classified as considerable. This was due to an alteration in the anchoring of the partial confinement which proved to be relatively successful. The limit state for this specimen commenced with the tensile failure of the CFRP layers at the central zone and was accompanied, as expected, by a consequent compressive failure of the neighboring weak part of the section. This is depicted in figure 30c where the anchor bolts, which were left intact, are also shown. In figure 31b, the comparison is extended to include specimen 4 (Test 1), in which the partial confinement exhibited low effectiveness due to the failure of the anchor bolts.

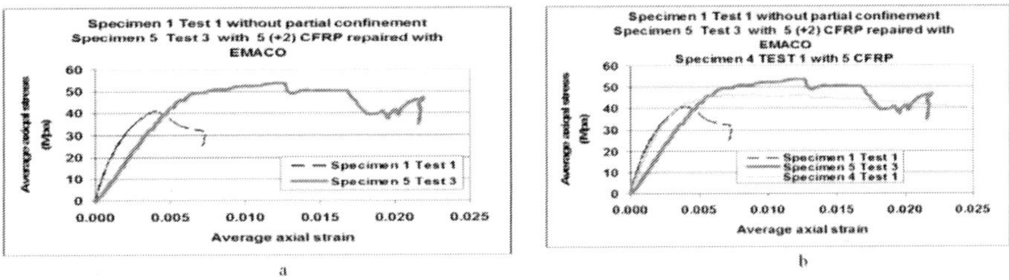

Figure 31. Partial confinement of considerable effectiveness. 31b. Partial confinement of considerable effectiveness.

3.2.3. Very effective partial confinement

In figure 32a, the behavior of specimens 5 (Test 4) and 5a (Test 2) is compared with the behavior of specimens 1 test 1 (with no partial confinement) and 5 (Test 3) discussed before. Specimens 5 (Test 4) and 5a (Test 2) were formed by repairing previously failed specimens with partial confinement of 7 layers of CFRP. Moreover, all the anchoring of their partial confinement was made with bolts going through the whole width of the strong part of the repaired section (see table 6 and figure 25). As can be seen in figure 32a, a substantial increase (40%), in the bearing capacity as well as in the deformability, resulted from the described partial confinement for these two specimens. Their behavior was in this way better than the behavior of specimen 4 (Test 1) which was classified before as one of considerable effectiveness of the partial confinement. Based in this increased capacity and deformability of specimens 5 (Test 4) and 5a (Test 2) they were classified as specimens with very effective partial confinement.

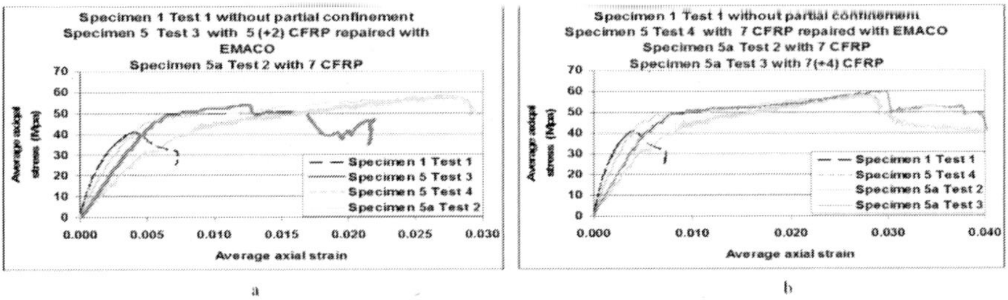

Figure 32. Very effective partial confinement. 32b. Very effective partial confinement.

a b

Figure 33. Strengthening stirrups - strong part. 33b. Failure of the CFRP layers of the mid-part.

The limit state of specimen 5 (Test 4) was accompanied with the failure of the stirrups of the strong part whereas the limit state of specimen 5a (Test 2) was accompanied with the tensile failure of the CFRP layers at the central zone. In figure 32b the observed behavior of specimen 5a (Test 3) is compared with that of specimens 5 (Test 4) and 5a (Test 2) previously discussed. Specimen 5a Test 3 had a partial confinement formed by 7 layers of CFRP. Moreover, the area sustaining the anchor bolts was strengthened by additional four (+4) layers of CFRP. Apart from the anchor bolts going through the whole width of the strong part of its section, the stirrups of this strong part were also strengthened (figure 33a). This time the increase in the bearing capacity was 46%; the increase in the deformability was also quite substantial. Thus, the behavior of this specimen is also one with very effective partial confinement. The observed limit state was accompanied with the compressive failure of the CFRP layers at the mid-part near the steel brackets (figure 33b).

3.3. Concluding Remarks

a2. The undesired compression failure expected to develop in the base of vertical members with reinforced concrete cross sections having h/b ratio larger than 1.5 under combined vertical loads and seismic actions is studied through specially formed specimens subjected to uniform compression. The retrofitting of such specimens with partial CFRP confinement is aimed at prohibiting, up to a point, such compression failure. This type of partial confinement may also be applied to upgrading vertical structural members with non-accessible sides.

b2.. From the results of the experimental investigation with identical specimens, with or without this type of partial CFRP confinement, the successful application of such partial confinement was demonstrated. An increase of almost 50% was observed in the compression bearing capacity of some of the tested specimens. Moreover, the deformability of these specimens was substantially increased, demonstrating the effectiveness of this type of partial confinement.

c2. It was found from the experimental sequence that critical factors for this increase were the type of anchorage of the CFRP partial confinement and the number of CFRP layers. Successful anchoring of the CFRP layers allowed this partial confinement to become effective and to permit the use of a larger number of CFRP layers. In the present study alternative anchoring schemes were tried with limitations imposed by the geometry of the model cross-section. Similar limitations imposed by the geometry and the reinforcement of the cross-section will also dictate the design of such an anchoring scheme for a prototype cross-section. Further investigation on the performance of such prototype anchoring arrangements may be necessary.

4. UPGRADING THE FLEXURAL CAPACITY OF (R/C) VERTICAL MEMBERS

4.1. Flexural Upgrading

The upgrading of the flexural capacity of reinforced concrete (R/C) vertical members with externally applied CFRP layers as means of tensile reinforcement is investigated here (see figure 34 and [4], [6], [11], [16], [22], [34], [40], [41])

The CFRP layers are applied at the two opposite sides of R/C specimens which were constructed for this purpose (figures 35a, 35b). They were designed with such longitudinal and transverse reinforcements shown in figure 35b that the flexural limit state would prevail. The specimens were tested to develop the ultimate flexural behavior with the corresponding damaged region concentrated near their joint with the foundation. These specimens had the foundation block fixed at the strong reaction frame of the laboratory of Strength of Materials and Structures of Aristotle University and were subjected to simultaneous constant vertical load as well as horizontal cyclic imposed displacements approximating thus the seismic actions. This loading sequence was applied prior to the application of any CFRP layers and was repeated again after these CFRP layers were attached to the damaged specimens. The observed upgrading of the flexural capacity in terms of ultimate overturning moment is presented and discussed.

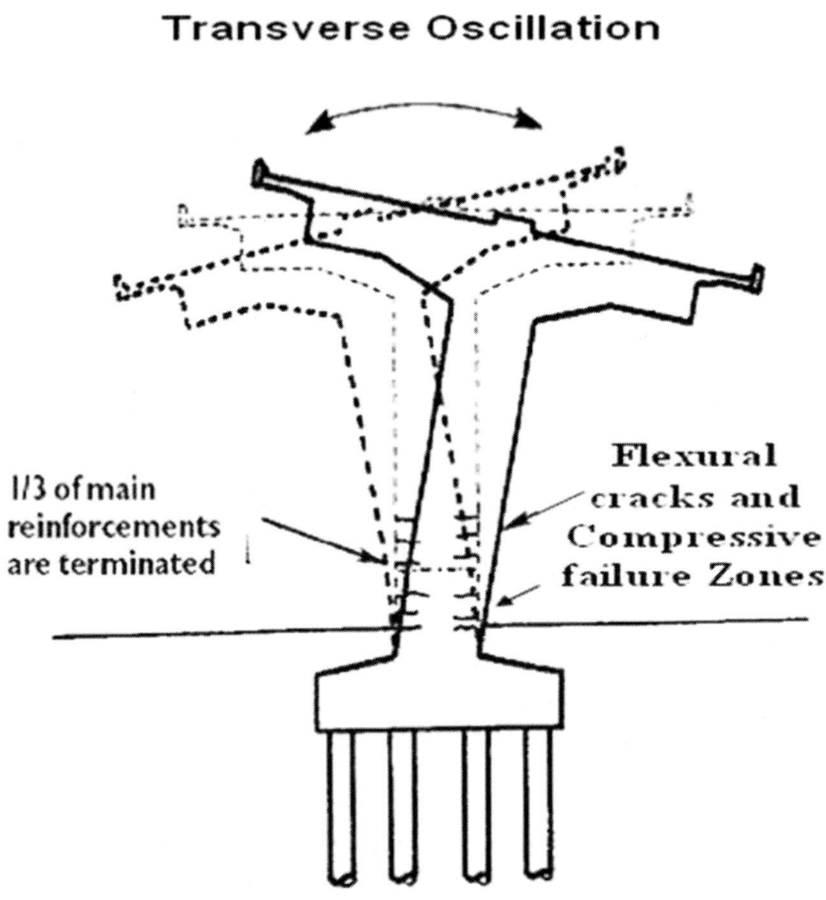

Figure 34. Flexural Damage

4.1.1. The virgin beam specimen

This specimen is shown in figure 35a with a rectangular cross-section of 200 mm x 500 mm and height h = 1815mm (from its top to the upper surface of its foundation). 8Φ6 was the longitudinal reinforcement that extended with no splices and was anchored to the foundation block and Φ6/100 closed stirrups the transverse reinforcement. The detailing of the cross-section is shown in figure 35b. The selected longitudinal and transverse reinforcement together with the loading arrangement, will cause the behavior of this virgin specimen to be dominated by the flexural rather than the shear mode of failure. More details are given in references [40] and [41].

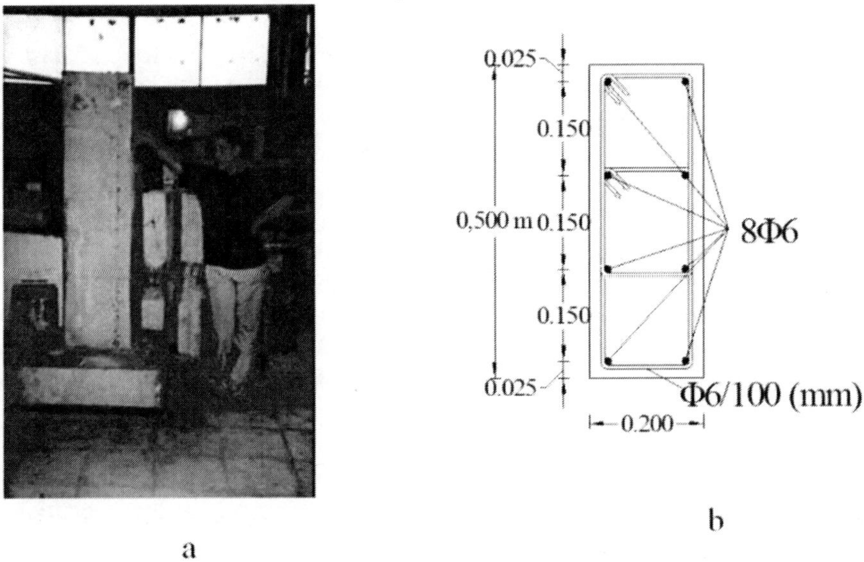

Figure 35. a Pier A. b. Cross-section and reinforcing details of Pier A.

4.1.2. The loading sequence

The foundation of the specimen was anchored at the strong floor of a reaction frame. The specimen was then subjected to a simultaneous constant vertical load and a cyclic horizontal displacement with increasing amplitude in time, utilizing servo-electronically-controlled dynamic actuators. The frequency of this cyclic displacement was in some tests 0.1Hz whereas in other selected tests it became 1.0Hz. The vertical load was kept constant at 95KN. The horizontal imposed cyclic displacement was applied in 13 groups of continuously increasing amplitude. Each group included 3 full cycles of constant amplitude. This time history of the imposed displacement is depicted in figure 36a. The amplitude in this picture is non-dimensional and is given as a percentage of the final maximum amplitude of this cyclic loading. In the experimental sequence the maximum displacement horizontal amplitude was initially relatively small (2mm); as testing progressed, in subsequent cyclic loading sequences it reached values of 20mm to 25mm. The imposed displacements were measured at the location of the horizontal actuator which was placed at a height of 1400mm from the top of the foundation (figure 36b). Figures 36b and 36c depict this experimental set-up. As can be seen the specimen was placed in the strong reaction frame, having its foundation block, with dimensions in plan 1000mm by 1000mm and 300mm height, fully anchored to the strong floor. The horizontal and vertical actuators, as part of this strong reaction frame, applied the loading sequence described above. Instrumentation was provided in order to measure the variation of the applied horizontal and vertical loads as well as the most important aspects of the displacement field that resulted from the application of these loads to the specimen. The horizontal displacements of the specimens at the top were monitored as well as the displacements of the specimen at the region near the foundation block in order to identify the flexural, and shear deformations as well as the plastic hinge behavior at this part of the specimen. In figure 36b the two vertical sides of the specimen where the CFRP layers were attached are indicated together with the region where the anchoring arrangements were placed as part of the current investigation. Throughout these series of experiments the nominal thickness of the employed CFRP layers was 0.176mm with a given Young's Modulus E=350GPa. The measured maximum axial CFRP strain was approximately 1%. These CFRP layers were placed at these locations

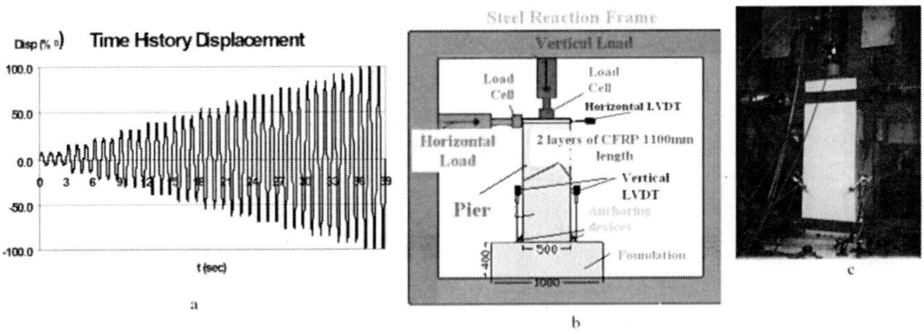

Figure 36. Imposed horizontal displacement. 36b. Experimental set-up. 36c Pier A placed at the strong reaction frame.

4.2. Observed Behavior - Results

4.2.1. The behavior of pier a

The yield stress of the longitudinal reinforcements was equal to 344.8MPa whereas the concrete strength was equal to 21.2MPa. Based on these mechanical characteristics and the cross-section reinforcing detailing,

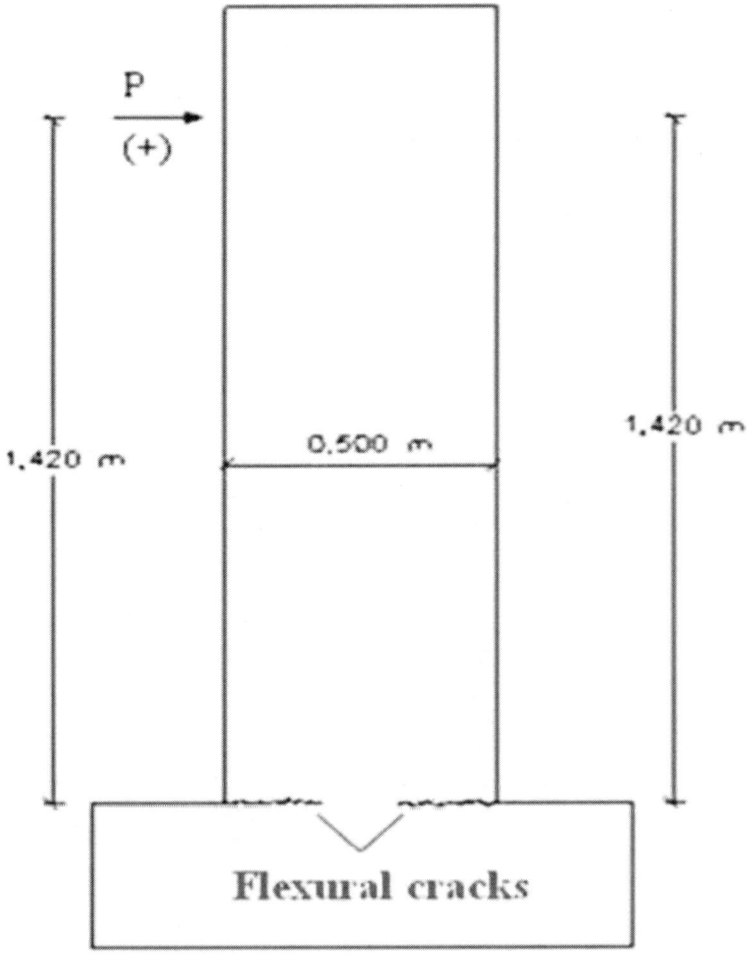

Figure 37. Observed flexural cracks for Pier A.

predictions of the limit-state flexural behavior for this virgin specimen (without CFRP) were obtained in terms of the M-N interaction. The limit-state flexural behavior of a specimen with 2 layers of CFRP attached to both vertical sides of a repaired specimen is also obtained, based on the assumption that the cross-sections under flexure remain plane even with the attachment of the CFRP layers. The development of flexural cracks took place as shown in figure 37. At this horizontal cross-section the measured flexural behavior of Pier A is depicted in figure 38, which presents the variation of the resulting bending moment at this cross-section of the specimen against the rotation of the same cross section,. The rotation was obtained from measurements made by displacement transducers that were placed at the two sides of the pier monitoring the relative vertical displacement between the upper part of the specimen and its foundation block. Additional instrumentation was provided to monitor the rocking or the sliding of the foundation block itself, which proved to be non-significant. The observed maximum bending moment value compares quite well with the corresponding predictions of the limit-state flexure for this cross-section. The red line in these plots represents the corresponding envelope curve.

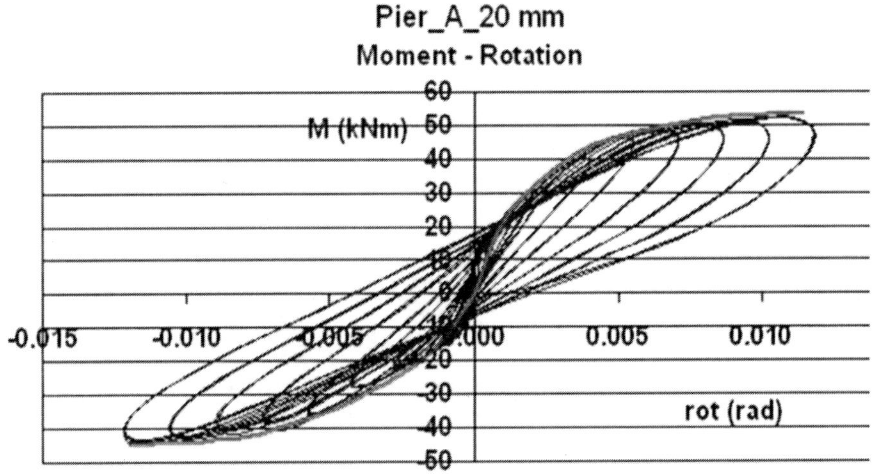

Figure 38. The measured flexural response of Pier A in terms of bending moment - plastic rotation.

4.2.2. Repaired specimen rep-1 pier a and observed behavior

The damaged Pier A was repaired in a way depicted by figures 39a,b. The resulting specimen is designated as Rep-1 Pier A. Two layers of CFRP were attached at each vertical side of the pier's cross-section. This attachment extended to a height of 1100mm from the foundation surface as well as at the top surface of the foundation block depicted in figure 39b. Moreover, in order to increase the bond strength between the CFRP layers and the top surface of the foundation block two double T steel sections were placed at these locations. A force normal to the bond surface at this location was applied by bolting these double T steel sections to the foundation block utilizing pre-stressing rods. Due to space limitations no further details are given here.

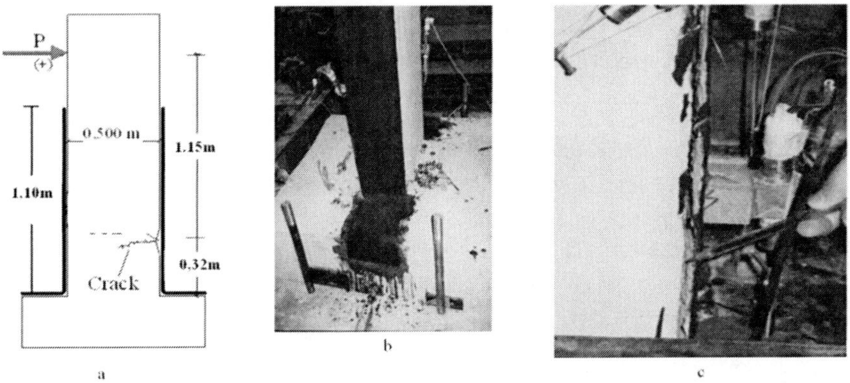

Figure 39. Specimen Rep-1 Pier A. Specimen 39b Rep-1 Pier A, attachment of CFRP layers. 39c Debonding of the CFRP layers from the right side of the repaired specimen Rep-1 Pier A.

4.2.3. Observed behavior of specimen rep-1 pier a

The loading sequence, described before, was applied to this repaired specimen. During increasing of the imposed horizontal displacement the debonding of the CFRP layer at the right side of the pier occurred as depicted in figure 39c. The obtained flexural behavior is depicted in figure 40 in terms of applied horizontal load against the imposed horizontal displacement at the top of the pier. If this behavior is compared with the corresponding behavior of the virgin specimen Pier A, a modest increase in its bearing capacity of approximately 20% can be observed as a result of the applied CFRP repair scheme, despite the debonding of the CFRP layers. The red line in this figure represents the corresponding envelope curve.

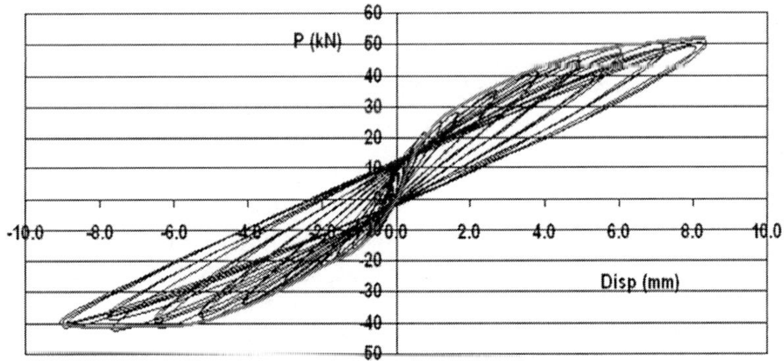

Figure 40. The measured flexural response of Rep-1 Pier A in terms of Horizontal load - Horizontal Displacement.

4.2.4. Repaired specimen rep-2 pier a and observed behavior

The damaged Rep-1 Pier A was repaired again in a way which is shown in figures 41a,b. The CFRP layers that were separated from the concrete at the right side of Rep-1 Pier A were reattached, both with the appropriate resin and additional 2mm diameter small bolts of 35mm length; these bolts were spaced at regular 50mm intervals along all the height of the CFRP layer at both sides. Moreover, the edge of the double T steel section neighboring the location of the CFRP layer at the pier-foundation joint was machined to form a curvature so that it provided a relatively uniform contact between the steel section and CFRP layers at this location. The locations of the bolts were first marked and drilled and then the bolts were applied including the corresponding plastic anchoring inserts. The specimen repaired in this way is designated as Rep-2 Pier A.

Figure 41. a The repair of specimen Rep-1 Pier A. b. The repair of specimen Rep-1 Pier A. c. Fracture of CFRP for specimen Rep-2 Pier A.

4.2.5. Observed behavior of specimen rep-2 pier a

The loading sequence, described in section 2, was also applied for this repaired specimen. During the increasing of the imposed horizontal displacement the fracture of the CFRP layers at the right side of the pier occurred as depicted in figure 41c. This fracture occurred at the location where the pier joins the foundation block. Figure 42a presents the variation of the measured horizontal load versus the horizontal displacement, whereas the resulting bending moment against the plastic hinge rotation is plotted in figure 42b. The plastic hinge which was formed during testing specimen Rep-1 was not repaired in any other way, but with the means described before. The fracture of the CFRP layer depicted in figure 41c took place at the same cross section. As was mentioned before the plastic rotation was measured by displacement transducers that were placed at the two sides of the pier. Because of the dislocation of one of these transducers the rotation angle after that was not monitored. This is noted at the bottom left hand side of figure 42b. Despite this lack of rotation measurements after the dislocation, the bending moment capacity remains approximately at the same levels, as deduced from the variation of the horizontal load (figures 42a and 43a). The effectiveness of the applied repair schemes for specimens Rep-1 and Rep-2 can be seen in figures 43a and 43b in terms of envelope curves. In figure 43a the comparison is made in terms of horizontal load – horizontal displacement whereas in figure 43b this is done in terms of bending moment-rotation. As can be seen, the effectiveness of the CFRP layers is inhibited by the debonding of the CFRP layers; thus although a considerable increase is achieved in the bearing capacity, this is not sustained in terms of displacement because of the premature debonding of the CFRP. The improvement of the bonding with the use of the employed bolting scheme for Rep-2 resulted in an increase of the bearing capacity both in terms of load and displacement and resulted in the fracture of the CFRP.

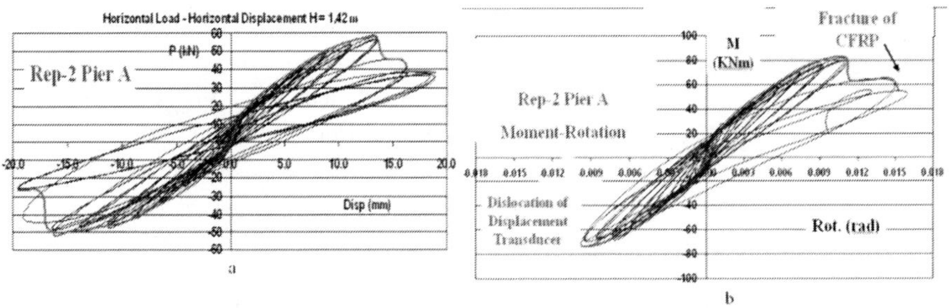

Figure 42. The measured flexural response of Rep-2 Pier A in terms of horizontal load - horizontal Displacement. 42b. The measured flexural response of Rep-2 Pier A in terms of bending moment - plastic rotation.

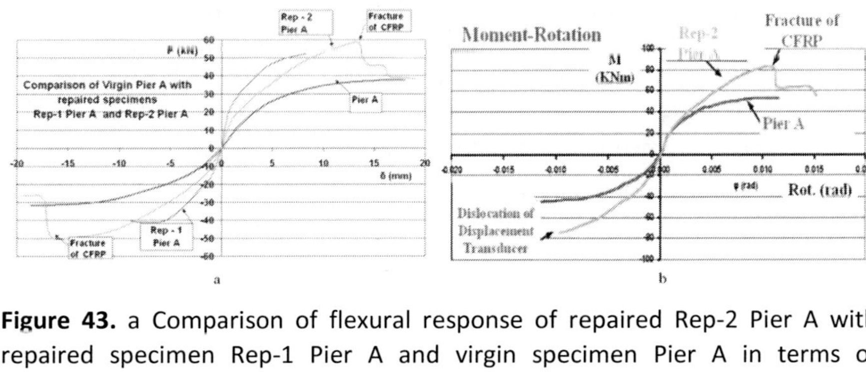

Figure 43. a Comparison of flexural response of repaired Rep-2 Pier A with repaired specimen Rep-1 Pier A and virgin specimen Pier A in terms of Horizontal load – Horizontal displacements. b. Comparison of flexural response of repaired Rep-2 Pier A with virgin specimen Pier A in terms of bending moment - plastic rotation.

From figure 43b the effectiveness of the repair scheme utilized in specimen Rep-2 Pier A can be deduced compared with the virgin specimen Pier A in terms of bending moment - plastic hinge rotation. The achieved increase in bending moment is almost 50% and the flexural behavior in terms of rotation is satisfactory up to the fracture of the CFRP layer at the joint between the pier and the foundation. Comparing the maximum value of the bending moment (83KNm) that was resisted by specimen Rep-2 Pier A with the one predicted (130KNm), as an ultimate bending moment that the cross-section reinforced by the applied CFRP layers can ideally resist, it can be concluded that the fracture of the CFRP did not allow this ideally maximum bending moment value to be reached. One reason for this is the way the anchoring of the CFRP layers at the foundation block is achieved with the utilization of the double T steel sections described before. Despite the machining of the edge of these double T steel sections the development of high stress levels concentrated at the CFRP layers at these locations is the cause of the observed CFRP fracture. Moreover, such an anchoring scheme is quite impractical for prototype conditions. For these reasons a different anchoring scheme was investigated in order to address both these shortcomings that were found to be critical in the repair schemes investigated so far. This is presented next in a summary form.

4.2.6. Repaired specimen rep-3 pier a and observed behavior

Figure 44a depicts the outline of this anchoring scheme and figure 44b depicts the application of this anchoring scheme for repair specimen Rep-3 Pier A, whereby the CFRP layers are folded around a cylinder that is anchored to the foundation [40]. Moreover, in order to avoid the separation (detachment) of the CFRP layers from the sides of the specimen, as observed in specimen Rep-1, the bolting scheme that was used in specimen Rep-3 was further improved by using bolts and washers of larger size (due to space limitations no details are given here). In addition, to avoid the development of shear failure at the bottom part of the pier, three horizontal CFRP closed hoops were attached in this region of the specimen (see figures 44c, 44d and [40], [41], [42]).

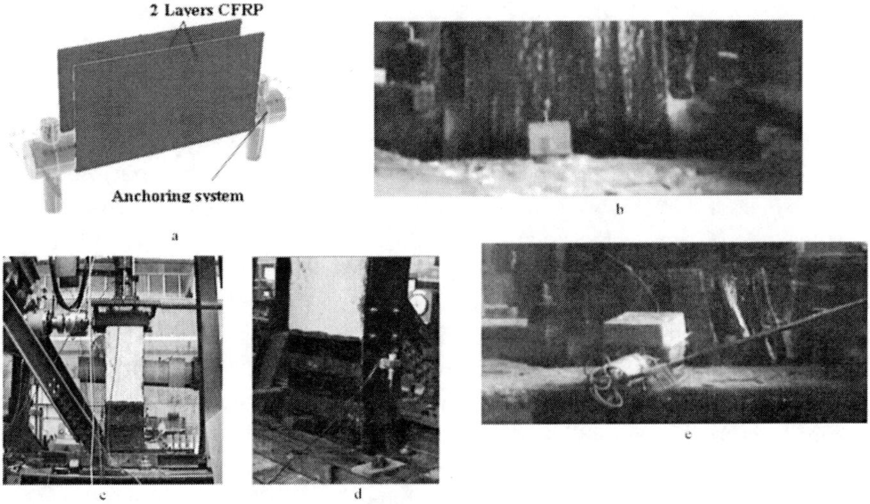

Figure 44. a New anchoring device. b. New anchoring device. c. Specimen Rep-3 Pier A. Figure d. Specimen Rep-3 Pier A. Figure e. Partial fracture of CFRP layers at the cylindrical part of anchoring device

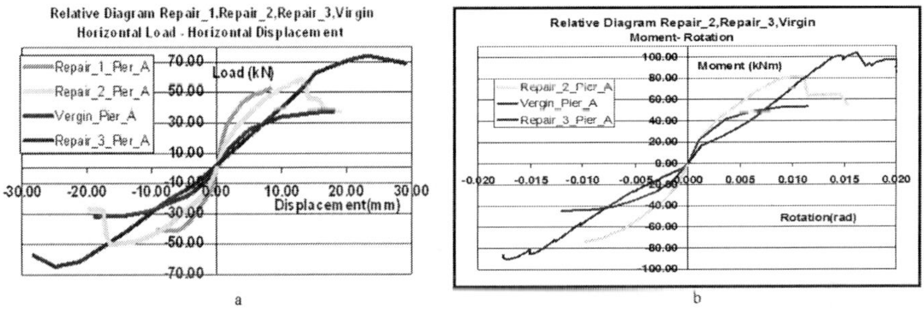

Figure 45. Comparison of flexural response of repaired specimens Rep-3, Rep-2, Rep-1 and virgin specimen. 45b. Comparison of flexural response of repaired Rep-3, Rep-2 and virgin specimens.

4.2.7. Observed behavior of specimen rep-3 pier a

The loading sequence, described before, was also applied for this repaired specimen. During increasing of the imposed horizontal displacement the partial fracture of the CFRP layer at the right side of the pier occurred as depicted in figure 44e. This fracture occurred at the location where the CFRP layers are folded around the cylindrical part of the anchoring device of figures

44a,b. In figures 45a and 45bthe measured flexural behavior in terms of horizontal load-horizontal displacement and bending moment-rotation envelope curve diagrams for repaired specimen Rep-1, Rep-2 and Rep-3 is compared with the coresponding flexural behavior of the virgin specimen Pier A. Due to space limitations the measured full cyclic response for specimen Rep-3 is not included in this presentation. The following observations can be made on the basis of these diagrams. a) The horizontal load and maximum bending moment bearing capacity of the latest repair scheme, namely specimen Rep-3, exhibits an increase of almost 100% when compared with the corresponding bearing capacity of the virgin specimen. b) When this comparison is made between the latest repair scheme Rep-3 and repair scheme Rep-2 the increase is of the order of 25%. Moreover, the new anchoring scheme does not cause an abrupt fracture of the CFRP at the anchoring device. This is reflected in figures 45a and 45b by the fact that the obtained response of specimen Rep-3 extends to relatively large displacement and rotation values. c) A reduction in the initial stiffness appears to be present in specimen Rep-3 when its obtained cyclic response in terms of the envelope curve of figures 45a,b is compared to the corresponding curves of specimen Rep-2 and the virgin specimen. This must be attributed to the fact that most parts of specimen Rep-3 have been subjected to three full loading sequences except the new parts of the CFRP. d) Rep-1 corresponds to the least effective repair scheme due to the debonding of the CFRP layers.

4.3. Concluding Observations

a3. The upgrading of the flexural behavior of vertical R/C structural elements was investigated utilizing the possibility of attaching longitudinal CFRP layers that can be stretched developing tensile forces externally at the opposite sides of these elements. Critical aspects for their satisfactory performance are the bonding of the CFRP layers as well as the effective transfer of the tensile forces to the foundation block.

b3. The bonding of the CFRP layers at the sides can be improved by bolting schemes that, in this way, increase the cooperation between the CFRP layers and the concrete part.

c3. The transfer of the tensile forces that develop at the CFRP layers to the foundation is a difficult technical problem. Two schemes were tried. The 2nd scheme is more practical and was successful in mobilizing sufficiently the available capacity of the applied CFRP layers in such a way that it resulted in approximately 100% increase in the flexural capacity when compared to the initial flexural capacity of the virgin specimen before the application of the CFRP layers.

d3. The applicability of such an anchoring system must be validated in a more general way before final practical conclusions can be reached.

e3. Such an upgrading of the flexural capacity may lead to the appearance of the shear mode of failure. However, the shear capacity can be increased with relative ease by closed CFRP hoops.

5. RELEVANT CODE PROVISIONS - EMPHASIS IN THE APPLICATION OF FRP STRIPS FOR SHEAR STRENGTHENING

During the last fifteen years numerous researchers ([12], [18], [29], [34]) have proposed different approaches for predicting the behavior of FRPs when they are used for strengthening structural members. Several design guidelines and code provisions adopted some of the proposed approaches. More specifically, the approach of Triantafillou and Khalifa [33] has been adopted by the American Code ACI 440 [11] whereas the model of Chen and Teng ([18], [25]) is utilized in the Greek Code of Interventions [6]. Eurocode 8 part 3 ([7]) follows the guidelines of FIB [8]. In the following paragraphs the aforementioned three Codes are discussed with particular emphasis being given to the application of FRP strips for shear strengthening. All codes recommend a limit-state tensile force that a cross-section (A_{FRP}) of an FRP sheet can sustain. This limit-state tensile force is obtained throughequation 2:

$$V_{FRP} = \frac{A_{FRP} \cdot \varepsilon_{FRP} \cdot E_{FRP}}{\varphi} \qquad (2)$$

Where: V_{FRP}: Limit-state tensile force of the FRP strip, ϕ: safety factor, A_{FRP}: cross section area of a strip of FRP

ε_{FRP}: developed axial strains on the FRP strip, E_{FRP}: modulus of elasticity of the FRP strip

The various codes recommend the application of Equation 2 by defining various values for the safety factor (ϕ) and the limit-state axial strain value (ε_{FRP}). These values vary according to the application depending on whether the predicted FRP contribution is for shear, flexure or confinement. In doing so, the basic factors that are varied among the Codes are the safety factor and the allowable developed strain on FRP. Table 7 presents the maximum allowable ε_{FRP} values and the recommended safety factor (ϕ) values according to ACI 440, Eurocode part 3, and the Greek Code for structural interventions, when anchoring is used in the transfer of FRP strip forces.

Table 7. Maximum allowable developed strains ε_{FRP} and proposed safety factors ϕ when anchoring is used.Where: ε^{*}_{nom}: max strain from manufacturer

Code	Max allowable ε_{FRP}		φ
	Flexure	Shear	
ACI 440 [11]	$0.9\,\varepsilon^{*}_{nom}$	0.4 %	1.24 - 1.47
Eurocode 8 (part3) 7	-	0.6 %	1.5
Greek Code of Interventions 6	$\varepsilon^{*}_{nom} / 2 < 1.5\%$	$\varepsilon^{*}_{nom} / 2 < 1.5\%$	1.25

The allowable limit-state axial strains (ε_{FRP}) value defined by the various codes also depend upon the way the FRP sheets are attached upon the structural elements in need of strengthening. According to all code provisions, there are two basic ways of attaching FRP sheets on structural elements. First, they can be attached using an anchoring device or they can be fully wrapped around the structural element. Secondly, they can simply be attached on the surface of the structural element through an organic or inorganic matrix in an open loop U-shaped without any anchoring. When the simple attachment is used without anchoring the debonding mode of failure prevails, whereas when FRP sheets are combined with an anchoring device or when they are fully wrapped around a column or a beam, fracture of FRP's occurs. The debonding mode of failure poses a limitation to the FRP axial strains and stresses. This limitation is reflected in all three Codes through a factor (usually designated as k_v), that takes values less than 1. This reduction factor depends on the recommended value of the attachment FRP length (effective attachment length). Table 8 lists the basic formulae included in the various codes for calculating this effective attachment length. The most important factor in calculating the effective attachment length is the determination of bond strength, which is not easily obtained since it depends on the actual concrete tensile strength and the state of the surface where the FRP is attached; both these parameters can easily vary even for the same structural element.

When the attachment of FRP sheets is combined with an anchoring device a more reliable transfer of forces can be achieved thus resulting in an equally reliable design of the relevant strengthening scheme. Moreover, since the prevailing mode of failure is the fracture of the FRP's, when anchoring of the FRP strips is used, the exploitation of the material of the FRP sheets is enhanced by reaching higher values of axial strains than for the cases governed by the debonding mode of failure. Under all circumstances the maximum allowable strains should not exceed the value presented in table 7. All codes demand the use of a safe anchoring device that would allow the fracture of FRP sheets without proposing any calculations for the safe design of these anchoring devices. In any case, equation (3)describes this condition for the bearing capacity of a permissible anchoring device

TABLE 8. Calculation of effective length. Where V_{FRP}: total force received by FRP, $V_{anchoring\ device}$: total strength of the anchoring device.

Code	Effective Length (Le)	
ACI 440 [11]	$L_e = \frac{23300}{(nt_f E_f)^{0.58}}$	where n: number of FRP layers t_f: thickness of FRP E_f: Modulus of elasticity
Eurocode 8 (part3) [7]	$L_e = \sqrt{\frac{E_f \cdot t_f}{\sqrt{4 \cdot \tau_{max}}}}$	where E_f: Modulus of elasticity t_f: thickness of FRP τ_{max}: bond strength
Greek Code of Interventions [6]	$L_e = \sqrt{\frac{E_j t_j}{2 f_{ctm}}}$	where E_j: Modulus of elasticity t_j: thickness of FRP f_{ctm}: tensile concrete strength

$$V_{FRP} \le V_{anchoring\ device} \tag{3}$$

6. SPECIAL STUDY FOR ANCHORING FRP STRIPS

In this section results from a recent research effort conducted at the Laboratory of Strength of Materials and Structures of Aristotle University will be briefly presented and discussed ([13], [42]). This research aimed to investigate the effectiveness of a specific FRP strip anchoring device by utilising a number of small concrete prismatic specimens, that can house such an FRP strip with sufficient width and length. In all, twelve (12) specimens were investigated. For six specimens no surface preparation of the concrete specimen was applied when attaching the FRP layers, whereas six other specimens had their surface treated according to construction guidelines. CFRP layers were attached to all specimens ([28], [43]). The use of the anchoring device was utilised on three (3) specimens with treatment preparation and on three specimens without such treatment. All the concrete prisms were fabricated using the same concrete mix and the same internal reinforcement, which was used to prohibit any accidental failure. The measured cylinder strength of the concrete was equal to 22 MPa. The properties of the used CFRP are listed in Table 9, as given by the manufactures.

Table 9. Properties of the used FRP sheets.

Material	Type / Name	Modulus of Elasticity (GPa)	Thickness of Layer (mm)	Ultimate strain
CFRP	SikaWrap 230 C/45	234	0.131	0.018

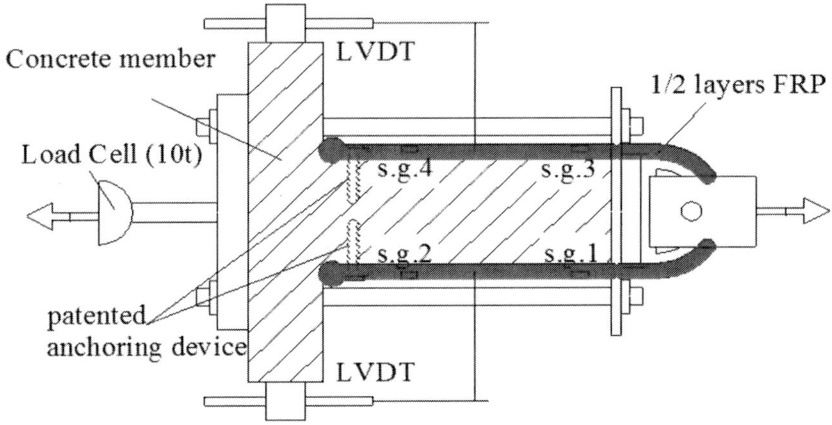

Figure 46. Experimental Set-up.

The loading arrangement is depicted in figure 46, whereby the tensile force is directly applied in the axis of symmetry at the right part of the FRP strips that forms an open hoop at this location; the other two sides of the FRP strip are bonded in a symmetric way on the top and bottom side of the concrete prism, as shown in this figure. When anchors were employed they were added at these locations (ends of the FRP strips). Despite the symmetry of this test set-up, instrumentation was provided that was able to record symmetric as well as asymmetric response of the specimen, especially during the initiation and propagation of the debonding process. During testing, the applied load is measured together with the longitudinal (axial) strains at four different locations of the external surface of the FRP strip, as indicated in figure 46 (s.g.1 to s.g.4), in order to calculate the stress field that develops at the FRP layer before and during the debonding.

Moreover, the relative longitudinal displacement between the concrete prism and the FRP surface is also monitored using four displacement transducers that are properly attached to the specimen, as indicated in this figure, in order to record the initiation and propagation of the debonding of the FRP. The used anchoring device was developed at the Laboratory of Strength of Materials and Structures of Aristotle University of Thessaloniki in Greece and it is patented with patent number WO2011073696 [42]. Figure 47 presents some details of this device. The tested specimens with their details are listed intable 10 together with their code names. The first letter C in the code name denotes a carbon fiber reinforcing polymer strip.(CFRP). The type of surface preparation is denoted by the second letter of the code name (S for smooth surface, R for rough surface; the type of anchor is denoted by the third letter of the code name (N for no anchor and P for patented anchoring device). Moreover, the number of layers of these strips is denoted by the fourth character of the code name (1 for one layer and 2 for two layers). The tests were conducted using a 1000 kN capacity hydraulic piston. The measurements of load, displacements and strains were recorded using an automatic data acquisition system.

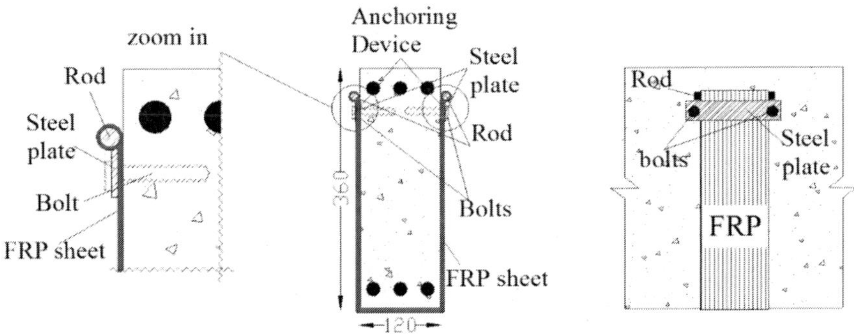

Figure 47. Patented Anchoring Device (WO2011073696).

Table 10. Details of Specimens.

Spec. Name	Material Type	Number of Layers	Surface Type	Anchor Type	No. of specimens	Bolt Type
CSN1	CFRP	1	smooth	no	3	no
CRN1	CFRP	1	rough	no	3	no
CSP2	CFRP	2	smooth	patented	3	2XHUS by Hilti
CRP2	CFRP	2	rough	patented	3	2XHUS by Hilti

Experimental Results and Discussion: The summary of the experimental results is listed in table 11. In this table, the observed failure mechanism is listed together with the corresponding value of the ultimate measured load as well as the value of the load recorded at the initiation of debonding. Moreover, the maximum strain values measured by the strain gauges at locations 1 and 3 on the FRP strip surface are also listed. The average value of these maximum FRP strains was utilised to calculate indirectly the load sustained by the FRP strips taking into account their total cross-sectional area and the value of the Young's modulus, listed in table 10. Finally, the material exploitation indicator (M_e) is given in the same table as the ratio of the maximum measured strain by the ultimate strain value provided by the manufacturer (Table 10).

Table 11. Summary of experimental results. * Specimens with only one CFRP layer. If for these specimens a second CFRP layer was added without anchor, due to the debonding failure, no increase in the maximum load and maximum strain can be achieved. Consequently, in such a case the exploitation ratio value would be half the values listed in this table.

Spec. Name	Anchor Type	Max Load (kN)	Failure Mechanism	Load at Debonding (kN)	Max Strain (μStrain) s.g.1	Max Strain (μStrain) s.g. 3	Material Exploitation M_e	Load from Strain (kN)
CSN1	no	27.9	debonding	27.9	5400	5400	0.30*	32.5
CRN1	no	42.7	debonding	42.7	6335	7215	0.38*	40.8
CSP2h	patented	112.8	CFRP fracture	30	9810	8320	0.50	109.2
CRP2h	patented	103.0	CFRP fracture	40	7220	9330	0.46	99.7

For specimens CSN1 having non-prepared surfaces, an average ultimate load was found equal to 27.9 kN. The equivalent ultimate load value when prepared contact surfaces were used, specimens CRN1 and SRN1, was 41 kN. Thus, the proper preparation of the contact surface resulted in a 46% increase of the ultimate load. This increase that is attributed to the surface preparation was observed when no anchoring device was utilized. The employed surface preparation needs at least twice as much time as when no surface preparation is made.

When an anchoring device is employed, it can be seen that preparation of the concrete contact surface is of no significance (see the ultimate load values of the anchored specimens of table 11 and those of debonding). In all tested cases the ultimate load is greater than the load at debonding. Thus, when using an effective anchoring device, the cost of properly treating the contact surface can be avoided.

The load at debonding for those specimens whose surfaces were not specially treated had a value approximately equal to 30 kN with small deviations. Similarly, the load at debonding for those specimens whose surfaces were specially treated had a value approximately equal to 40kN with small deviations. When an anchoring device is utilized the ultimate capacity increases from 28 kN to 112.8 kN for the set of specimens strengthened with CFRP. This represents a fourfold increase in the value of ultimate load.

Finally, when the patented anchoring device was applied for specimens CSP2h, CRP2h, a significant increase in the bearing capacity was observed. This time the performance of the anchoring device was very satisfactory and the observed failure was that of the fracture of the FRP strips for all these specimens. Figure 6 depicts such a failure mode for specimen SSP2b (see also table 11).

In order to discuss the observed behavior in terms of exploitation of the high strength of the FRP materials the following procedure was used. As already mentioned, a material exploitation indicator was found (M_e) as the ratio of the maximum measured strain values (average of the two sides, Table 3)for each specimen over the ultimate strain values as they are measured for the used FRP materials (seetable 9). These material exploitation indicator values are also listed in Table 11, having ideally as an upper limit the value of 1. As can be seen in this table, the highest M_e value during the present experimental sequence reaches the value 1 and this was achieved by the specimen that utilises the patented anchoring device together with two layers of CFRP strips. As expected, debonding of the FRP strips or failure of the anchoring device results in relatively low values of the material exploitation indicator M_e. (0,60 and 0,79) This research effort is in progress experimenting with various alternative anchoring details.

Figure 48. Detail of the anchoring device and the fracture of the CFRP strip when such an anchoring device is employed.

6.1. Concluding Observations

a4. Proper preparation of the contact surface between the FRP strips and the concrete face resulted in a 46% increase of the ultimate load. This increase that is attributed to the surface preparation was observed when no anchoring device was utilized. The employed surface preparation needs at least twice as much time as when no surface preparation is done.

b4. When an anchoring device is employed, it can be seen that preparation of the concrete contact surface is of no significance. With the proper anchoring device the ultimate capacity increases four times.

c4. It is important to properly detail the anchoring device in order to drive the mode of failure to the fracture of the FRP strip rather than the failure of the anchor thus exploiting the high tensile strength FRP potential.

d4. The highest value of the FRP material exploitation indicator was achieved in the specimen that utilises the anchoring device patented by Aristotle University together with two layers of CFRP strips.

e4. As expected, debonding of the FRP strips or failure of the anchoring device results in relatively low values of the material exploitation indicator M_e. This research effort is in progress experimenting with various alternative anchoring details.

7. BASIC QUALIFICATION TESTS FOR FIBER REINFORCING POLYMERS (FRP) SHEETS TO BE USED IN DEALING WITH EARTHQUAKE STRUCTURAL DAMAGE.

In what follows, a brief description is given of basic qualification tests to be performed with FRP sheets used in repair / strengthening schemes of R/C structural elements in the framework of earthquake structural damage. As already discussed in sections 5 and 6, the main critical parameters for such use of FRP sheets are: a) the modulus of elasticity, b) the maximum axial strain and c) the effective thickness of these sheets. These properties are usually provided as technical specification data by the manufacturers of these materials. Moreover, the manufacturers of the FRP sheets for this type of application also provide technical information on the organic or inorganic matrices that are compatible with the relevant FRP sheets and the type of material surface of the structural element to which these sheets are to be externally applied. Finally, the technical information of the manufacturers as well as the relevant code guidelines include preparatory actions which must be taken before the FRP sheets are applied, such as concrete surface preparation, rounding of corners and proper application of the matrices together with the FRP sheets prohibiting the formation of any air pockets. In case the tensile characteristics of the FRP sheets must be confirmed, specimens of the FRP material must be taken in order to define the modulus of elasticity (E), the Poisson's ratio (v), the tensile strength and the maximum tensile failure strain (ε_m). These tests are performed following the appropriate specifications of the relevant standard (e.g. European Standard EN ISO 527-5: 1997).

During the various experimental sequences that were presented in the preceding sections 2, 3 and 4 the employed FRP sheets were tested in order to verify their basic tensile characteristics. Selective measured values are reported in the relevant sections in brief (see subsections 2.2.2, 3.1.2 and 4.1.2). For all tests reported in sections 2, 3 and 4 the FRP sheets were attached to the reinforced concrete specimens using the same organic matrix. This was a two part, solvent free, thixotropic epoxy based impregnating resin / adhesive. The density of this epoxy resin is 1.31 kg/l and its average viscosity is approximately 7.000 mPas. The Thermal Expansion Coefficient is equal to 45 x 10-6 per °C, its tensile strength according to DIN 53455 is 30 Mpa, the Young's Modulus is 4500 Mpa and the ultimate elongation is 0.9%. There is a manufacturers warning that this product is not suitable for chemical exposure.

7.1. Bond Strength Between CFRP Layers And Concrete Substrate.

Another property of interest is the bond strength between the FRP sheets and its matrix with the surface of the structural member, especially when the FRP sheets are not accompanied by the appropriate anchoring, as discussed in the preceding sections. There are certain bond strength tests aiming to insure that the debonding mode of failure does not occur between the matrix and the FRP sheet (e.g. European Standard EN ISO 1542: 1999). However, when designing for the debonding limit-state use is made of the tensile strength of the concrete substrate instead of this bond strength. A special investigation was performed at the Laboratory of Strength of Materials and Structures of Aristotle University aiming at measuring directly this bond strength of the employed CFRP sheets attached with the named above epoxy resine to the concrete substrate. Figure 49a depicts this simple test aimed at measuring the bond strength between the employed CFRP layers and the concrete surface (ref. [40], [41]). Figure 49b depicts the obtained results together with a best-fit linear variation of the bond strength (τ) versus the applied normal stress ($\sigma_{\varepsilon\gamma\kappa}$). As expected, a modest increase can be obtained in the bond strength (τ) between the CFRP layers and the concrete surface if a normal stress ($\sigma_{\varepsilon\gamma\kappa}$) increase is applied at the bond surface. This was partly made use of in the various anchoring schemes employed in sections 2, 3 and 4.

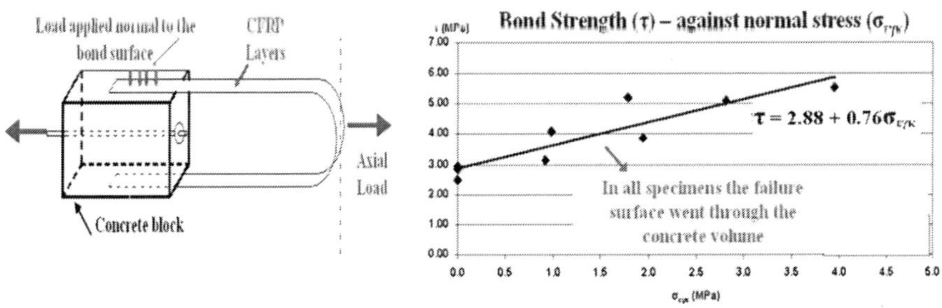

a b

Figure 49. a Test for estimating the bond between CFRP layers and concrete surface and b obtained results.

7.2. Cost Estimates

From a comparative study of data dealing with typical cases of repair and strengthening of R/C structural elements with conventional methods (jacketing with gunite or cast-in-place R/C concrete) versus CFRP based strengthening schemes the following constitute the average findings. The cost of a fully wrapped CFRP strengthening scheme when compared to a cast-in-place jacketing, represents an increase of approximately 20% to 30%. However, when a jacketing strengthening scheme with gunite (shotcrete) is applied instead of FRP's the cost increase is of the order of 30% to 40%. If the CFRP repair / strengthening solution does not include full wrapping the above cost increase is expected to be reduced.

8. CONCLUSIONS

This presentation dealt with repair and strengthening schemes of earthquake damaged reinforced concrete (R/C) structural elements utilizing externally attached fiber reinforcing plastics (FRP's). Such strengthening schemes were studied when applied to slabs, beams and vertical structural members. The success of such an upgrading scheme was discussed together with its limitations on the basis of a series of relevant experimental results.

One of the main limitations results from the way the tensile forces which develop on these FRP sheets can be transferred. When the transfer of these forces relies solely on the interface between the FRP sheet and the external surface of the reinforced concrete structural elements, the delaminating (debonding) mode of failure of these sheets occurs, due to the relatively low value of either the ultimate bond stress at this interface or the relatively low value of the tensile strength of the underlying concrete volume. This mode of failure is quite common and it occurs in many applications well before the corresponding FRP sheets develop tensile axial strains in the neighborhood of values mentioned before as design limit axial strains (approximately of the order of 1%).

Alternative ways of transferring these tensile forces, apart from the simple attachment, in order to enhance the exploitation of the FRP material potential have been also presented and discussed based on experimental evidence from ongoing research at Aristotle University.

9. ACKNOWLEDGEMENTS

Partial financial support for this investigation was provided by the Hellenic Earthquake Planning and Protection Organization (EPPO), which is gratefully acknowledged.

REFERENCES

1. B. C. Papazachos, 1990 Seismicity of the Aegean and the Surrounding Area Tectonophisics 178 287 308
2. G.C. Manos, 2011 Consequences on the urban environment in Greece related to the recent intense earthquake activity Int. Journal of Civil Eng. and Architecture Dec 5 12Serial No. 49)1065 1090
3. M. Paz, 1994 International Handbook of Earthquake Engineering: "Codes, Programs and Examples" edited by Mario Paz, Chapter 17, Greece by G.C. Manos, Chapman and Hall 0-41298-211-0
4. T. Pauley, M. J. N. Priestley, 1992 Seismic Design of Reinforced Concrete and Masonry Structures J. Wiley & Sons INC USA.
5. Organization of Earthquake Planning and Protection of Greece (OASP), 2001 Guidelines for Level- A earthquake performance checking of buildings of public occupancy», Athens
6. Organization of Earthquake Planning and Protection of Greece (OASP), 2011 Guidelines for Retrofitting in Reinforced Concrete Buildings, Athens.
7. Eurocode-8 2004 2004 Design of structures for earthquake resistance- Part 2: Bridges, DRAFT 3 European Committee for Standardization
8. F.fF. I. B. I.S. C. 2001 Externally Bonded FRP Reinforcement for RC Structures" in Bulletin 14, f.T.G. 9.3, Editor. 165
9. Greek Code for the Design of Reinforced Concrete Structures" 2000in Greek).
10. Provisions of Greek Seismic Code 2000 OASP, Athens, December 1999. Revisions of seismic zonation introduced in 2003
11. ACI, 2008 Guide for the Design and Construction of Externally Bonded FRP Systems for Strengthening Concrete Structures (ACI 440.2R-08) in ACI 440.2R-08 American Concrete Institute: Farmington Hills 45
12. C. Bakis, et al. 2002 Fiber-Reinforced Polymer Composites for Construction- State of the Art Review' Journal of Composites of Construction, ASCE May
13. K. Katakalos, G. C. Manos, C. G. Papakonstantinou, 2012 Comparison between Carbon and Steel Fiber Reinforrced Polymers

with or without Ancorage", 6th International Conference on FRP Composites in Civil Engineering Rome

14. G. C. Manos, V. Kourtides, P. Matsukas, 2007 Investigation of the flexural and shear capacity of simple R/C beam specimens including repair schemes with fiber reinforced plastics FRPRCS-8 World Conference, Greece, July 16-18 2007

15. A. Nanni, 1995 Concrete repair with externally bonded FRP reinforcement Concrete International 17 6 22 26

16. J. G. Teng, J. F. Chen, S. T. Smith, L. Lam, 2002 FRP strengthened RC structures. Chichester: John Wiley and Sons.

17. C. G. Papakonstantinou, K. Katakalos, 2009. Flexural Behavior of Reinforced Concrete Beams strengthened with a hybrid retrofit system Structural Engineering and Mechanics Techno Press 31-5

18. J. F. Chen, J. G. Teng, 2003 Shear Capacity of Fiber-Reinforced Polymer-Strengthened Reinforced Concrete Beams: Fiber Reinforced Polymer Rupture Journal of Structural Engineering 129 5 May 1 ASCE, 5 615 625

19. C. G. Papakonstantinou, K. Katakalos, G. C. Manos, 2012 Reinforced Concrete T-Beams Strengthened in Shear with Steel Fiber Reinforced Polymers 6th International Conference on FRP Composites in Civil Engineering Rome

20. J. F. Chen, J. G. Teng, 2001 Anchorage strength models for FRP and steel plates bonded to concrete Journal of Structural Engineering, ASCE 127 7 784 791

21. A. J. Lamanna, L. C. Bank, D. W. Scott, 2004 Flexural strengthening of R/C beams by mechanically Attaching FRP strips Journal of Composites of Construction, ASCE May/June 203 210

22. J. H. Lee, M. M. Lopez, C. E. Bakis, 2007 Flexural Behavior of reinforced concrete beams strengthened with mechanically fastened FRP strip. FRPRCS-8, World Conference, University of Patras July

23. J. Paterson, D. Mitchell, 2003 Seismic Retrofit of Shear Walls with Headed Bars and Carbon Fiber Wrap Journal of Structural Engineering, ASCE May 606 614

24. G. C. Manos, E. Papanaoum, 2009 Earthquake Behavior of a R/C Building Constructed in 1933 before and after its RepairSTREMAH, Tallin22 24 June

25. J. F. Chen, J. G. Teng, 2003 Shear capacity of FRP strengthened RC beams: FRP debonding Construct Build Mater 17 1 27 41

26. M. Ekenel, A. Rizzo, J. J. Myers, A. ". Nanni, 2006 Flexural Fatigue Behavior of Reinforced Concrete Beams Strengthened with FRP Fabric and Precured Laminate Systems Journal of Composites of Construction, ASCE September/October 433 442

27. A. Khalifa, A. Nanni, 2002 Rehabilitation of rectangular simply supported RC beams with shear deficiencies using CFRP composites Construction and Building Materials 16 3 April 135 146

28. G. C. Manos, K. Katakalos, C. G. Papakonstantinou, 2011 Shear behavior of rectangular beams strengthened with either carbon or steel fiber reinforced polymers Applied Mechanics and Materials (Trans Tech Publications) 82 571 576

29. T. C. Triantafillou, C. P. Antonopoulos, 2000Design of concrete flexural members strengthened in shear with FRP Journal of Composites for Construction 4 4 November ASCE.

30. G. N. J. Kani, 1966 Basic Facts Concerning Shear Failure J. ACI 63 Jun 675 692

31. P. Zararis, 2002 Reinforced Concrete Kyriakides Editions, Thessaloniki (in Greek).

32. T. C. Zsutty, 1968 Beam Shear Strength Prediction by Analysis of Existing DataJournal ACI 65 November 943 951

33. A. Triantafillou, 2003 Software for the design of reinforced concrete elements strengthened with fibre reinforced polymers FRP (in Greek).

34. S. J. Pantazopoulou, J. F. Bonacci, S. Sheikh, M. D. A. Thomas, N. Hearn, 2001 Repair of corrosion-damaged columns with FRP wraps Journal of Composites for Construction 5 1 3 11

35. S. P. Tastani, S. J. Pantazopoulou, et al. 2006 Limitations of FRP Jacketing in Confining Old-Type Reinforced Concrete Members in Axial Compression Journal of Composites for Construction 10 1 February 1,. ASCE.

36. G. C. Manos, V. Kourtides, 2007 Retrofitting of long rectangular R/C Cross-Sections with Partial Confinement employing Carbon Fiber Reinforcing PlasticsFRPRCS-8 Conf. GreeceJuly 16-18128

37. G. Tsonis, 2004 Seismic Assessment and Retrofit of Existing Reinforced Concrete BridgesPh.D. Thesis, Politecnico di Milano.

38. K. Kawashima, 2000 Seismic performance of RC bridge piers in Japan: an evaluation after 1995 the Hyogo-ken nanbu earthquake Prog. Struct. Engng Mater 2 82 91

39. A. V. Pinto, editor 1996 Pseudodynamic and Shaking Table Tests on R.C. Bridges. ECOEST PREC*8 R 8 November

40. G. C. Manos, K. Katakalos, V. Kourtides, C. Mitsarakis, 2007 Upgrading the Flexural Capacity of a Vertical R/C Member Using Carbon Reinforcing Plastics Applied Externally and Anchored at the Foundation FRPRCS-8 World Conference July

41. G. C. Manos, K. Katakalos, V. Kourtides, 2008 Study of the anchorage of Carbon Fiber Plastics (CFRP) utilized to upgrade the flexural capacity of Vertical R/C members 14WCEE Beijing, CHINA

42. G.C. Manos, K. Katakalos, V. Kourtides, 2011 Construction structure with strengthening device and method European Patent Office, Patent WO2011073696 A1) —2011- 06-23

43. G.C. Manos, K. Katakalos, V. Kourtides, 2011 The influence of concrete surface preparation when fiber reinforced polymers with different anchoring devices are being applied for strengthening R/C structural members Applied Mechanics and Materials 82 600 605

Chapter 9

Applying Post-Tensioning Technique to Improve the Performance of FRP Post-Strengthening

Mônica Regina Garcez[1], Leila Cristina Meneghetti[2] and Luiz Carlos Pinto da Silva Filho[3]

[1] *Federal University of Pelotas, Brazil*

[2] *University of São Paulo, Brazil*

[3] *Federal University of Rio Grande do Sul, Brazil*

1. INTRODUCTION

Reinforced concrete structures are, frequently, submitted to interventions aiming to restore or increase their original load capacity. According to Garden & Hollaway [1], the choice between upgrading and rebuilding is based on factors specific to each individual case, but certain issues are considered in every case. These are the length of time during which the structure will be out of service or providing a reduced service, relative costs upgrading and rebuilding in terms of labor, materials and plant, and disruption of other facilities.

Several post-strengthening techniques were developed in the last decades. Most of them are based on the addiction of a structural element to the external face of the element to be post-strengthened.

According to Täljsten [2], the method of post-strengthening existing structures with steel plates bonded to the structure with epoxy adhesive was originated in France, in the nineteen sixties, when L'Hermite (1967) and Bresson (1971) carried out tests on post-strengthened concrete beams. Additionally, Dussek (1974) reported the use of this post-strengthening method in South Africa in the middle 60's. In both cases the post-strengthening was successful and the load bearing capacity was increased. These first investigations in France and South Africa inspired future research in Switzerland (1974), Germany (1980), United Kingdom (1980), Japan (1981) and Belgium (1982). The idea of post-strengthen existing reinforced concrete structures with bonded steel was improved due to the development of synthetic adhesives, based on epoxy resins, suitable to ensure good adhesion and chemical resistance to aggressive agents.

In the last decades, non-corrosive, low-weight and high-resistant materials started to be developed and applied on the construction of new buildings, aiming to produce durable structures. These materials, called Fiber Reinforced Polymers (FRP), started to be investigated in the middle 80´s at EMPA (Swiss Federal Laboratories for Materials Testing and Research), in Switzerland. At that time, the carbon fiber was elected as the most suitable for post-strengthening applications due to its low-weight, high tensile strength, high modulus of elasticity and resistance to corrosion. Since then, many structures were post-strengthened with FRP in Japan, Europe, Canada and United States and nowadays the use of FRP is growing worldwide.

Most of FRP post-strengthening systems used nowadays consist of carbon fibers embedded in epoxy matrices and provide high modulus of elasticity and tensile strength. For bridge repair, carbon fiber is the material best suited in most cases, because the fiber is alkaline-resistant and does not suffer stress corrosion, two very important arguments for such applications. Actually, there are many reasons that make carbon fibers one of the most attractive alternatives for post-strengthening concrete structures. Considering all reinforcing fiber materials used to produce FRP, the carbon fibers have the highest specific modulus and specific strength that provide a great stiffness to the system, being an ideal choice to be applied in structures sensitive to weight and deflection. Compared with steel, carbon fibers can be 5 times lighter and present a tensile strength 8 to 10 times higher.

The main impediment to the massive use of CFRP (Carbon Fiber Reinforced Polymers) regards to the high cost of the carbon fibers. Meier, in 2001 [3], pointed out that the functionality and the mechanical properties of CFRP should be better explored, due to its relatively high cost. Indeed, the use of only 10%-15% of the tensile strength of the CFRP, as it happens in some bonded post-strengthening systems, is not economically viable.

This chapter aims to analyze the efficiency of prestressed CFRP strips used to post-strengthen reinforced concrete beams, by means of cyclic and static loading tests, as an alternative to better use the tensile strength of these materials.

2. REINFORCED CONCRETE ELEMENTS POST-STRENGTHENED WITH PRESTRESSED FRP STRIPS

The aim in prestressing concrete beams may be, according to Garden and Mays [4], either to increase the serviceability capacity of the structural system of which the beams form a part or to extend its ultimate limit state.

According to El-Hacha [5], FRP are well suited to prestressing applications because of their high strength-to-weight ratio that provides high prestressing forces, without increase on the self-weight of the post-strengthened structure. The prestressing technique may improve the serviceability of a structural element and delay the onset of cracking. When prestressed FRP are used, just a small part of the ultimate strain capacity of the material is used to prestress the FRP, the remaining strain capacity is available to support external loads and also to ensure safety against failure modes associated to peeling-off at the border of flexural cracks and at the ends of the post-strengthening.

Several FRP prestressing systems are currently available consisting of rods, strands, tendons or cables of FRP. However, in some cases, it may be advantageous to bond FRP sheets or strips onto the structural element surface in a prestressed state. According to fib Bulletin 14 [6], prestressing the FRP prior to bonding has the following advantages:

- Provides stiffer behavior as at early stages most of the concrete is in compression and therefore contributing to the moment of resistance. The neutral axis remains at a lower level in the prestressed case if compared to the unstressed one, resulting in greater structural efficiency.
- Crack formation in the shear span is delayed and the cracks, when they appear, are more finely distributed and narrower. Thus, serviceability and durability are improved, due to reduced cracking.
- The same level of strengthening is achieved with smaller areas of stressed FRP, compared to unstressed ones.
- Prestressing significantly increases the applied load at which the internal steel reinforcement begins to yield if compared to an unstressed structural member.

On the other hand, prestressing FRP systems are more expensive than the non-prestressing ones, due to the greater number of operations and the equipment that is required to prestress the FRP.

2.1. Losses of Prestressing Force

Prestressed FRP bonded to concrete structures are sujected to prestress losses, as it happens in any prestressing system. Such prestress losses may be instantaneous, due to immediate elastic deformation of concrete, or time dependent, due to creep and shrinkage of concrete and relaxation of the FRP.

Immediate elastic deformation of the concrete may reach 2% to 3%, according to fib Bulletin 14 [6], and happens when the prestress force is transferred into the concrete beam. If prestress is applied by reacting against the structural member there will be no loss. It happens because if the prestressing device if fixed on the structural element that will be post-strengthened, a compensation occurs: as the FRP is being stressed, the concrete is being compressed. However, FRP elements that have already been prestressed will experience a loss of prestress due to the shortening of the beam upon the prestressing of subsequent FRP elements. In such cases it is necessary to determine the average loss of prestress per FRP element.

Time dependent losses, due to creep and shrinkage of concrete, according to the fib Bulletin 14 [6], reach about 10% to 20% and are similar to the ones of conventional prestressing.

Prestressing losses due to relaxation of FRP depends, according to ACI 440.4R-04 [7], on the characteristics of the FRP composite. The document also informs that losses due to relaxation of fibers may be neglected when CFRP are used, since the relaxation of carbon fibers is very low. Losses of 0,6% to 1,2% must be considered due to the relaxation of the polymer and losses of 1% to 2% must be considered due to the straightening of fibers.

Results of a research program developed by Triantafillou et al. [8] indicate that, when prefabricated CFRP are used, prestress losses of 10% must be considered, due to the instantaneous and time dependent losses at the concrete and adhesive and also due to the relaxation of the CFRP.

Garden and Mays [4] consider that prestressed FRP also suffer prestress losses due to the shear transferred through the adhesive and into the concrete by the FRP tension. This shear action is sufficient to fracture the concrete even at low prestress levels so it is necessary to install anchorages at the ends of the FRP element to resist this action.

2.2. Maximum Prestressing Force

Figure 1(a), by Triantafillou et al. [8], shows the premature failure of a concrete beam post-strengthened with a CFRP strip, without any anchorage system, immediately after the complete release of the prestressing force. Horizontal shear cracks propagated from both ends of the CFRP strip through the concrete layer and stopped at a certain length. Figure 1 (b) shows that this failure mode may be prevented if anchorage systems are used at the ends of the strips. The authors suggest that the maximum prestressing force that avoids the need of anchorage systems provide very low prestressing levels, 15% to 20%, depending on the cross section of the CFRP strip.

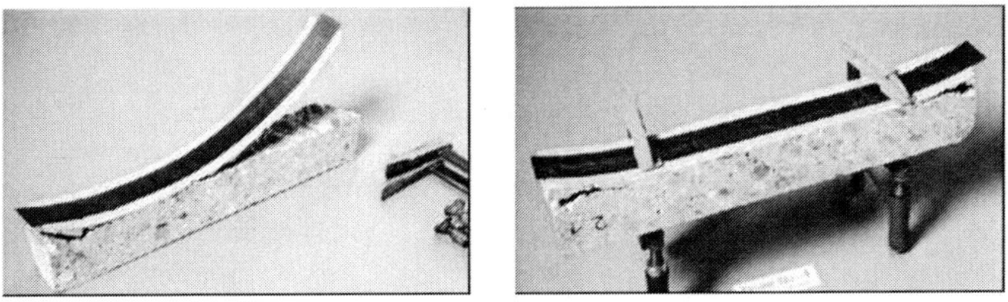

(a) (b)

Figure 1. (a) Premature failure of a prestressed CFRP strip without anchorage; (b) Action of an anchorage system (Triantafillou et al. [**8**]).

Thus, the addition of anchors at the end of the prestressed FRP sheets or strips reduces the shear deformation that occurs within the resin or adhesive layer upon releasing the prestressing force and reducing the shear stresses transferred to the base of the concrete section. Thereby, anchorage systems minimize the possibility of premature failures (El-Hacha [5]).

According to El-Hacha et al. [9], prestressing levels of at least 25% of the FRP tensile strength may be necessary to achieve a significant improvement in terms of the structural stiffness and load carrying capacity.

Meier [10] suggests that a prestress level as high as 50% of the CFRP strength might be necessary to increase the ultimate strength by delaying the premature failure. Experimental results presented by Deuring [11] showed that increasing the level of prestress in the CFRP from 50% to 75% reduced the strength of the beam because the highly prestressed laminates had little strain capacity remaining and the CFRP presented premature failure.

It is important to have in mind that, when post-strengthening is prestressed the modulus of elasticity of the FRP is of great significance, since the FRP element needs to be stiffer to hold up a significant loading that, before the post-strengtnening, was made only by the steel reinforcement (El-Hacha, [5]).

2.3. Prestressing Techniques

Various approaches to prestress FRP have been proposed by researches and used experimentally. These methods are based on directly or indirectly prestress the FRP prior to bonding and are described bellow.

2.3.1. Cambered beam prestressing technique

In this method, developed by Ehsani & Saadatmanesh [11], no tension is directly applied to the fibers, but the FRP sheets are indirectly prestressed by cambering the beam to be post-strengthened before bonding them to the bottom face of the concrete beam.

The beam is first deflected upward by means of hydraulic jacks, as one can see in Figure 2 (a). The beam is then held in the deflected position until the adhesive is completely cured. After the cure of the adhesive, the FRP is completely bonded to the lower face of the beam and the jacks may be removed, as showed in Figure 2 (b). Once the jacks are removed, the beam will deflect downward and tensile stresses will be induced in the lower face of the beam.

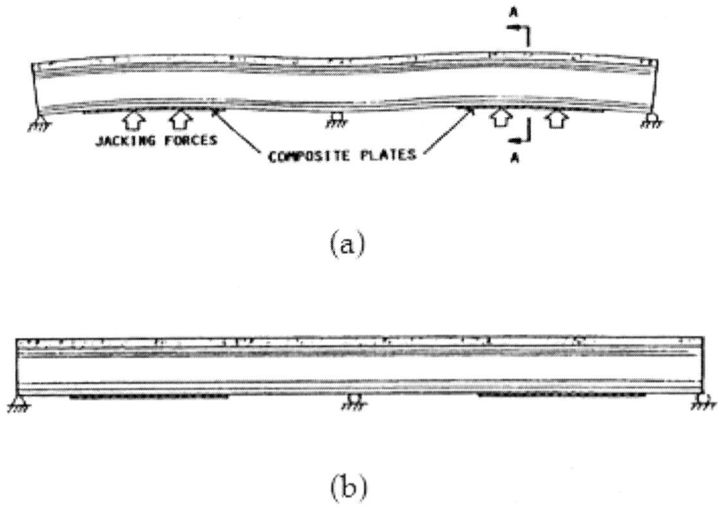

Figure 2. Sequence of prestressing procedures: (a) Camber by jacking; (b) Remove jacks when epoxy is cured (Ehsani & Saadatmanesh [**11**]).

The level of prestress will depend on the length of the beam and the degree of camber induced in the beam. According to Ehsani & Saadatmanesh [11], the level of prestress in this method is not high and since the highest prestress will be present at the midspan of the beam, anchorage systems are not required.

According to El-Hacha [5], in field applications, the effort required to camber the bridge near midspan is extensive relative to the low prestressing force induced in the FRP.

2.3.2. Frp prestressed against the strengthened element

In this prestressing system, developed at Queen's University and Royal Military College of Canada and presented by Wight et al. [12], the sheets are tensioned by reacting directly on the beam. The prestressed sheets are bonded to the lower face of the beam and the ends of the sheets are attached to the beam by a mechanical anchorage system. Multiple layers of prestressed sheets may be applied to the beams in successive layers, when required, due to the limits on the tensile capacity of individual sheets or due to the need to limit the load at an anchorage location.

In this process, the beams are inverted to receive the prestressing system. The mechanical prestressing and anchorage system used for the reinforced concrete beams are shown in Figure 3. According to Wight et al. [12] the mechanical anchorage system consists of steel roller anchors bonded to the sheets and steel anchor assemblies fixed to the beam. The roller anchors that grip the sheet consists of two stainless-steel rollers bonded to each end of the sheet. Prior to prestressing operations, the sheet is wrapped and bonded round the roller. To prestress the sheets, the roller at one end of the FRP sheet is fixed to the beam and the roller at the other end is movable. During prestressing, the movable roller is attached by steel prestressing strands to a hydraulic jack that reacted against the beam. The prestress is applied to the sheet, and the sliding roller is then attached, in its extended position, to a second permanent anchorage assembly. Subsequent layers may be added to the beam, using the same technique, until the desired thickness of FRP is achieved. The authors suggest that a weight perpendicular to the beam surface may be used to bring the sheet into contact with the beam surface.

El-Hacha at al. [9] used this technique to post-strengthen damaged concrete beams under severe environmental conditions. Results presented by the authors suggest that keeping the mechanical anchorage system in place prevented failures associated with high shear stresses at the ends of the sheets. Anchorages also prevented tensile fracture in the concrete cover thickness upon transfer of sheet prestress into the concrete. The prestressing system used to post-strengthen the beams improved the serviceability, controlling the formation of new cracks, delaying the formation of new cracks and limiting deflections in the beams tested. Furthermore, the prestressed CFRP sheets contributed to the load carrying capacity of the beams and significantly redistributed the stress from the internal steel reinforcement to the CFRP sheet.

2.3.3. Technique of prestressing frp prior to bonding

In this method, studied by Triantafillou & Deskovic [13, 14], Deuring [15], Quantril & Hollaway [16] and Garden & Mays [4], the FRP sheet is first pretensioned and applied on the tensile face of the beam as one can see in Figure 4 (a) and (b). Aluminium tabs are epoxy bonded on both faces of the FRP to provide stress distribution in the end regions and then each end of the FRP is sandwiched between two predrilled steel plates bolted on at each end. The system is loaded into a prestressing frame. A vacuum bag technique may be necessary to support the external FRP during bonding.

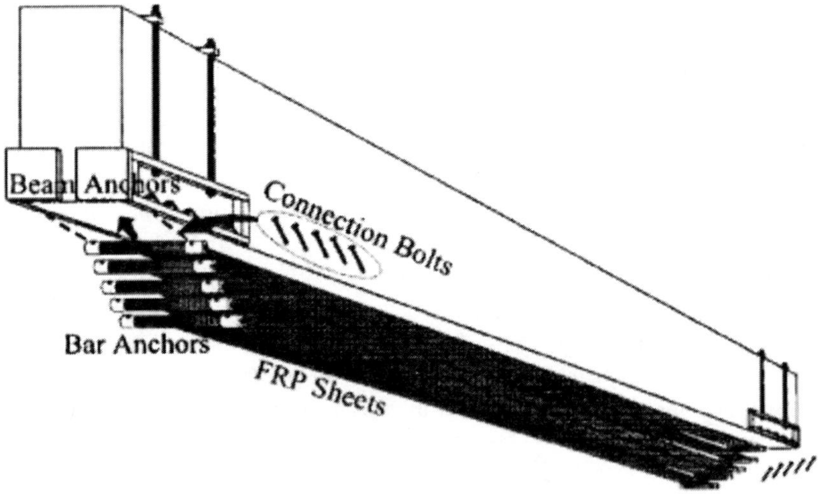

Figure 3. Prestressing system (Wight et al. [**12**]).

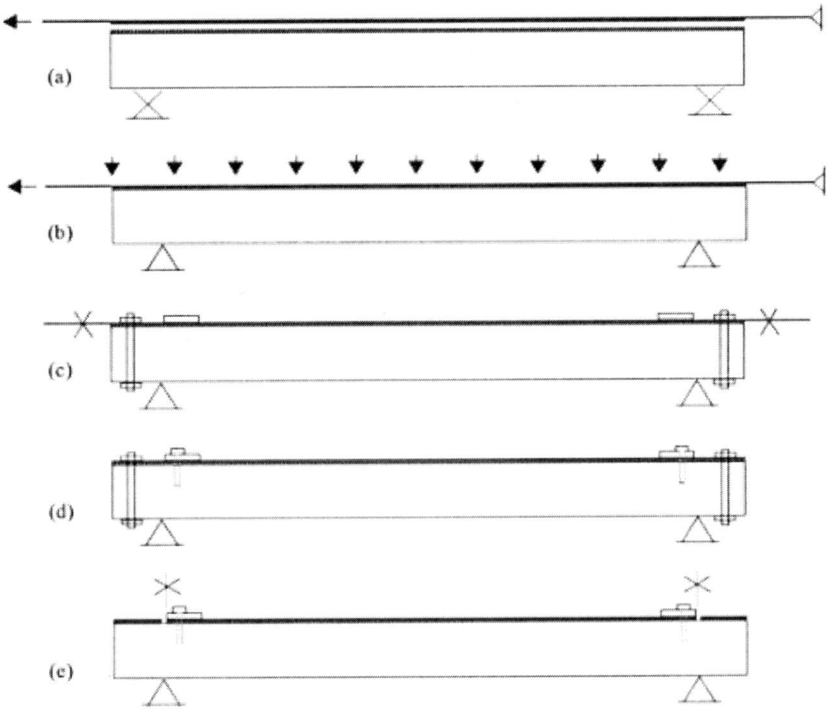

Figure 4. Sequence of prestressing procedures: (a) Pretensioning of the strip; (b) Application of the pretensioned strip on the tensile face of the beam; (c) Cutting of the sheet to transform it into a prestressing element; (d) Placement of the steel bolts; (e) End damps are removed (Garden & Mays [**4**]).

After the adhesive is fully cured, steel clamps are installed at each end of the beam to ensure adequate anchorage and the two ends of the FRP are cut. Then the sheet is transformed into a prestressing element as showed in Figure 4 (c).

In the next step, showed in Figure 4(d) holes are drilled through the FRP and the adhesive into the concrete beam to receive steel bolts that are bonded into the holes and allowed to cure.

Figure 4 (e) shows the last stage of the procedure, when the end clamps are removed and the FRP is cut through at each end. Then, the bolted steel endplates will play the role of the anchorage system.

Quantril & Hollaway [16] noted that cracking under the action of a given external load was found to be much less extensive and less developed than for an identical non-prestressed specimen. Besides that, prestressing produced significant increases in the load which causes yield of the internal steel over a non-prestressed specimen. Gains may be still larger if FRP action can be maintained past steel yielding to loads approaching failure. Applying the prestress prior to bonding also affects the mode of failure of the specimen, reducing the amount of shear cracking which could initiate failure in the shear spans. The greater the level of prestress, the better is the confinement effect on the development of shear cracking what increases the failure load for cases governed by shear failure. Results also suggest the levels of ductility and stiffness may be increased as well as the maximum strains in the FRP at a given load level. Despite all the advantages presented by the authors, field application of this technique would probably require methods and procedures adaptations due to working restrictions such as the overhead position and limited access of most structural elements.

According to Garden & Mays [4] the level of prestress that can be applied is limited by the tensile strength of the FRP and should not precede either yielding of the internal steel or compressive failure of the concrete to ensure adequate ductility. Results showed by the authors suggest that the level of prestress may also be limited by the strength of the plate and anchorages, by the horizontal shear strength of the adhesive-FRP interface and by the bottom layers of the concrete.

One potential benefit of this technique is the reduction of FRP material associated costs since the same strength levels can be reached with reduced area fraction (Triantafillou & Deskovic [14]). According to Triantafillou et al. [8] the method can also lend itself to prefabrication because of its simplicity and the important properties offered by FRP materials.

2.3.4. Prestressing method developed by stoecklin & meier [17]

Stoecklin & Meier [17] developped, at EMPA (Swiss Federal Laboratories for Material Testing and Research) a method to apply prestressed FRP strips to concrete structures. In this method, the FRP strip is first prestressed then bonded at the beam that will receive the post-strengthening. Since it is very complicated to grab and prestress the FRP strip, due to its anisotropic behavior, a prestressing device was designed, as one can see in Figure 5. The prestressing device consists of two wheels which are connected to a beam of the required length, as shown in Figure.

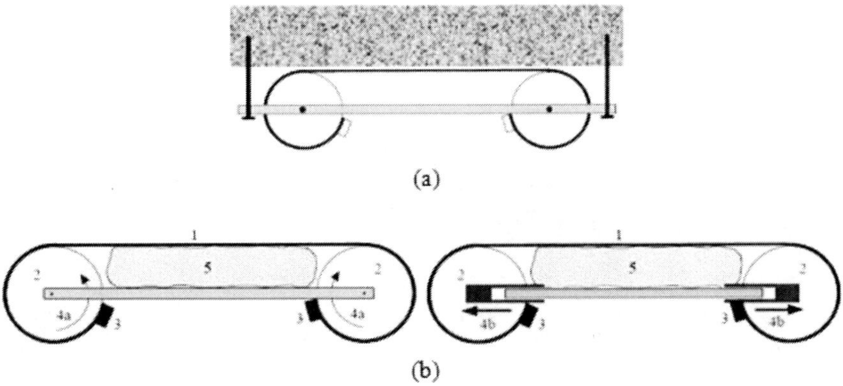

Figure 5. Prestressing device developed by Stoecklin & Meier [**17**]: (a) Placement of the prestressing device under the beam; (b) Two ways of prestressing a CFRP strip.

The FRP strip (1) is wrapped around the wheels (2) and clamped at its ends (3) as shown in Figure 5 (b). The strip can be prestressed by rotating one or both wheels (5a) or displace the wheels (5b). As one can see in Figure 5(a), the prestressing device with the prestressed FRP strip is temporarily mounted to the structure and can be pressed against the structure with a constant pressure by means of an air-cushion (5) between the FRP strip and the beam. (Stoecklin & Meier [17]).

In a new version of the prestressing device developed by Stoecklin & Meier [17], two separate prestressing units at each end of the strip are directly mounted to the structure, what means that the FRP strip is prestressed against the structure, as shown in Figure 6.

To overcome anchorage problems at the ends of the FRP strips, the prestressing force can be reduced gradually from the mid-span to both ends of the FRP strips.

As described by Meier et al. [18], gradual anchoring is achieved by first bonding a fully pretensioned section in the middle of the FRP strip at mid-span. A system of electric heating may be used to speed up curing of the adhesive in the bonded section within the pot life of the adhesive. After curing the central part of the FRP strip at mid-span, the prestressing force is slightly reduced and another section is bonded at each side of the strip also using the electric heating system to speed up curing the adhesive.

Figure 6. New version of the prestressing device developed by Stoecklin & Meier (Meier [**10**]).

This process is repeated in several stages until the entire length of the strip is bonded and the prestressed level at the ends of the strips has been reduced to a low level, as one can see in Figure 7. In this way, anchorages are not required at the end of the prestressed strip.

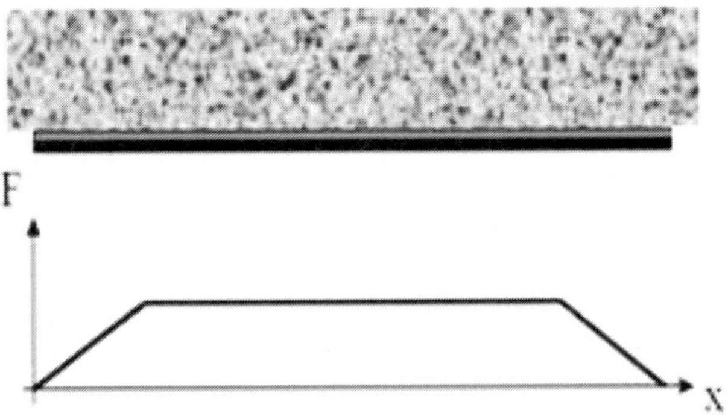

Figure 7. Gradual anchorage of prestressed CFRP strips (Stoecklin & Meier [**17**]).

In the prestressing method developed by Stoecklin & Meier [17] the strip is prestressed before the application at the beam. In such case, prestressing is applied by reacting against the structural member, since the prestressing device is mounted to the structure.

When the first FRP strip is prestressed, using the device developed by Stoecklin and Meier [17], imediate losses of prestress due to elastic deformation of the concrete, that happen when the prestress force is released, can be neglected, since prestressing is applied by reacting against the structural member. However, strips that have already been prestressed will experience a loss of prestress due to the shortening of the beam upon the prestressing of subsequent FRP strips.

2.4. Failure Modes Of Reinforced Concrete Beams Post-Strengthened With Prestressed Frp Submitted To Static Loading

According to Hollaway [19], the anisotropic behavior of the composite materials leads to a complex rupture mechanism that may be characterized by extensive damages on the composite material when submitted to static and cyclic loading. The level of damage, however, depends on the properties of the composite material and on the applied loading.

Failure modes of reinforced concrete structures post-strengthened with FRP include crushing of concrete, yielding of steel reinforcement or tensile failure at the FRP.

Teng et al. [20] report that failure modes of reinforced concrete beams post-strengthened with FRP can be broadly classified into two types: those associated with high interfacial stresses near the ends of the bonded FRP and those induced by a flexural or flexural-shear crack away from the ends, which is also referred to as intermediate crack-induced debonding. Thus premature failures, in general, are associated to:

- High interfacial stresses near to the ends of the bonded FRP, also called peeling-off.
- Flexural or flexural-shear crack away from the ends, as shown in Figure 8.

Concrete structures post-strengthened with prestressed FRP also show premature failures as described by Teng et al. [20]. However, in prestressed systems, the high strength of the FRP used to post-strengthen structures is much better used, and, depending on the configuration of the post.-strengthening, tensile failure of the FRP may be achieved.

Garden and Hollaway [1] presented, in 1998, a specific study regarding the failure modes of reinforced concrete beams post-strengthened with prestressed FRP, with prestressing levels ranging from 25% to 50% of the FRP strength. Results showed that a high prestress level was required to enable the ultimate capacity of strip to be reached, before shear displacement reached its critical value.

Figure 8. Rupture of FRP close to a flexural crack tip.

According to Garden and Mays [4], the level of prestress that can be applied will be limited by the tensile strength of the FRP. Tensile failure of the FRP should not precede either yielding of steel reinforcement or crushing of concrete, to ensure adequate ductility. Results of the experimental program showed that the level of prestress may also have to be limited by the strength of the anchorage devices, by the horizontal shear strength of the adhesive/FRP interface and by the bottom layers of concrete.

2.5. Failure Of Post-Strengthened Beams Submitted To Cyclic Loading

Fatigue may be defined as a permanent and progressive damage process that induces gradual and cumulative crack growth and might, ultimately, result in the complete fracture of the elements subjected to cyclic loads, if the stress variation and the number of load cycles are large enough. This term was established by the first researchers of the theme due to its nature: a progressive damage process caused by cyclic loads, difficult to observe, that changes the ultimate capacity of the material (Meneghetti et al. [21]).

The usual fatigue failure mechanism for post-strengthened RC beams, when subjected to cyclic loads, is marked by the rupture of one of the steel rebars, followed by a stress redistribution that overloads the remaining bars.

Meier U. [22] highlights that the steel rebars fail before the FRP post-strengthening, as can be seen inFigure 9 that shows the steel rebars of a post-strengthened concrete beam after a fatigue failure. However, cyclic loads can also damage the adhesive and affect the interface concrete-adhesive and adhesive-FRP, leading to premature failures.

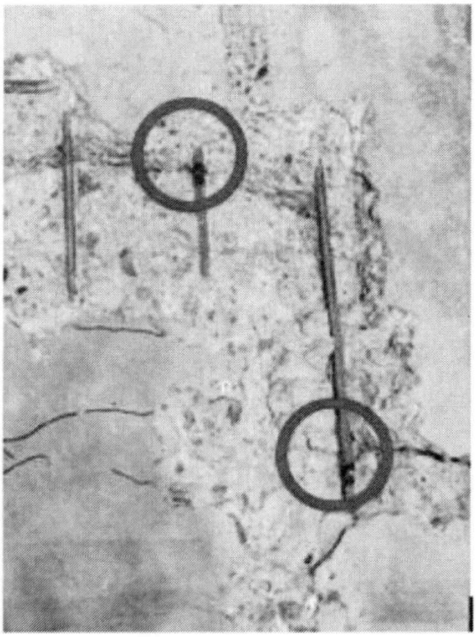

Figure 9. Failure of concrete beam after fatigue loading test.

According to Ferrier et al. [23], the performance and the durability of a concrete structure post-strengthened with FRP, when subjected to cyclic loads, depends not only on the FRP fatigue behavior but also depends on the interface concrete-adhesive and adhesive-FRP. Authors point out the importance of understanding the behavior of these materials under cyclic loading, since a typical reinforced concrete highway bridge deck with a design life of 40 years may experience a minimum of $58x10^8$ loading cycles of varying intensities.

3. EXPERIMENTAL ANALYSIS OF REINFORCED CONCRETE BEAMS POST-STRENGTHENED WITH PRESTRESSED FRP

3.1. Description Of Specimens

Aiming to analyze the behavior of concrete beams post-strengthened with prestressed CFRP strips under static loading three beams were tested: VT, VFC_NP_01 VFC_PE_01 (Table 1). Regarding the cyclic loading tests two beams were tested: VFC_PC_01 and VFC_PC_02 (Table 2). Stress levels applied at beam VFC_PC_01 were 50% and 80% of the yielding stress observed at beam VFC_PE_01, tested under static loading. Stress levels applied at beam VFC_PC_02 were more reasonable, 50% and 60% of the yielding stress observed at beam VFC_PE_01.

Table 1. Description of experimental program – static loading.

Beam	Post-strengthening	Prestressing level
VT	-	-
VFC_NP_01	Two CFRP non-prestressed strips	-
VFC_PE_01	Two CFRP prestressed strips	35% of ε_{fu}

Table 2. Description of experimental program – cyclic loading.

Beam	Post-strengthening	Test	Prestressing level applied on the strips	Stress range of fatigue loading
VFC_PC_01	Two CFRP prestressed strips	Bending cyclic loading	35% of ε_{fu}	50% to 80%
VFC_PC_02				50% to 60%

3.2. Reinforced Concrete Beams

The reinforced concrete beams were rectangular, 6500mm long, 1000mm wide, and 220mm deep. All beams were reinforced with seven bottom 15mm steel bars (ρ = 0.0041). The shear reinforcement consisted of 8mm steel stirrups spaced each 90mm (11.17cm²/m). Geometry and reinforcement details for the beam are shown in Figure 10.

Aggregates used to produce the concrete were the ones available in the Switzerland region and the cement was the Portland CEM I 42.5 (95% of clinquer, and 5% of other components), equivalent to the Brazilian CPI. The average compressive stress (cube strength) of the concrete, after 28 days, was 44MPa.

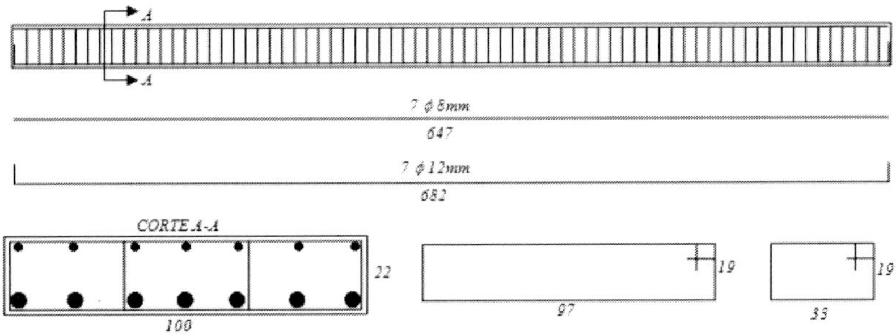

Figure 10. Geometry and reinforcement details of tested beams.

Table 3. Post-strengthening system characteristics.

	Carbodur S 512 Strip	
Thickness (mm)	1.2	
Width (mm)	50	
Tensile Strength (MPa)	2,800	
Ultimate Strain (°/₀₀)	17	
Young's Modulus (MPa)	165,000	
Temperature Resistance (°C)	150	
Fiber Volumetric Content (%)	68	
Density (g/cm³)	1.60	
	Sikadur®-30 Resin	Sikadur® 30LP Resin
Components	3Part A:1Part B	2Part A:1Part B
Pot life at 25ºC (min)	-	60
Pot life at 35ºC (min)	40	-
Pot life at 55ºC (min)	-	30
Tg (ºC)	62	107
Young's Modulus (MPa)	12,800	10,000

The 8mm steel bars had average yield stress, yield strain, ultimate stress and modulus of elasticity of 554MPa, 2.51º/oo, 662MPa and 220 Gpa, as indicated by tensile tests. The 12mm steel bars had average values of: 436MPa, 1.98º/oo, 688MPa e 215 GPa.

3.3. Post-Strengthening System

Sika® Carbodur (Carbodur S 512 and Sikadur®-30) was the CFRP system used to post-strengthen the beams. However, Sikadur®-30LP adhesive was used to bond the prestressed strips to the concrete, due to its extended pot life. Table 3 shows the characteristics of the strips and adhesives, provided by manufacturer.

3.4. Post-Strengthening Procedure

The application of Sika® Carbodur system demands a surface preparation for the concrete and the strip. The concrete surface must be clean and free from grasses and oil, dry and have no loose particles. Considering the application of prestressed strips, after concrete and strip surface preparation, the strip is clamped, prestressed and covered with Sikadur®-30LP adhesive. Then, thermocouples are settled at the strip aiming to control the temperature applied to accelerate the cure of the adhesive. Figure 11 shows the application of epoxy adhesive on the strip and the procedure to clamp the strip on the prestressing device.

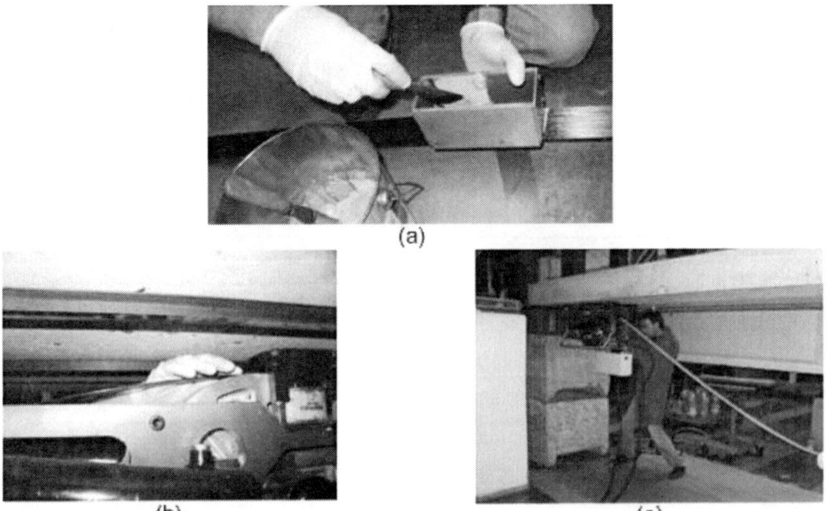

Figure 11. (a) Application of epoxy adhesive; (b) e (c) Procedure to clamp the strip on the prestressing device.

A gradual anchorage system was applied: after the cure of the adhesive at the middle part of the beam, the prestressing force was marginally reduced and the following areas were bonded. The force was reduced further and the adjacent areas were bonded. This procedure was repeated until there was no remaining prestressing force at the ends of the strip. With the gradual reduction, the level of prestress applied at the ends of the strip is very low or close to zero, eliminating the need of additional anchorage systems.

The maximum prestressing force applied to prestress the strip was 60kN, at mid-span. Then, prestressing force was gradually reduced, to 48kN, 36kN, 24kN, 12kN and, finally, zero, at the ends.

3.5. Test Procedure

The experimental program was developed at EMPA – Switzerland. Loading was applied according to a six point bending test scheme: simple supported beam and four vertical loads, spaced by 12000mm, symmetrically applied along the 6000mm span. Load was applied by two 100kN hydraulic jacks. During the tests, values of deflection at mid-span and specific strain at steel, concrete and FRP were continuously recorded by a computer controlled data acquisition system.

Table 4. Ultimate Loads and failure modes of post-strengthened beams tested under static loading.

Beam	Post-strengthening	Ultimate load	Failure Mode
VT	-	100.14kN	Yielding of steel followed by concrete crushing
VFC_NP_01	Two 1.2mm x 50mm non-prestressed strips	127.25kN	Premature failure (peeling-off)
VFC_PE_01	Two 1.2mm x 50mm prestressed strips	162.41kN	Premature failure (peeling-off)

Figure 12. Beam VFC_NP_01: (a) During test; (b) After premature failure of strips.

3.6. Behavior Of Post-Strengthened Beams Tested Under Static Loading

3.6.1. Loads and failure modes

Table 4 shows that the flexural capacity of beam VFC_NP_01, post-strengthened with two non-prestressed CFRP strips, increased 27% when compared to the control beam. On the other hand, post-strengthening of beam VFC_PE_01, two prestressed CFRP strips, increased 62.41% the load bearing capacity of the beam.

Results of beams VFC_NP_01 and VFC_PE_01 can be explained by the principles of prestressing. When the prestressing force applied on the CFRP strips is released, compressive stresses are induced on the concrete. Such compressive stresses delay the concrete cracking and the yielding of the steel reinforcement. Thus, the load bearing capacity of the post-strengthened element is increased.

Premature failures (peeling-off) of beams VFC_NP_01 and VFC_PE_01 occurred due to the high interfacial stresses near to the ends of strips. Peeling-off failures are catastrophic and happen without any previous advice. Figure 12 shows two CFRP strips of beam VFC_NP_01 after peeling-off. Both strips are completely detached from the beam; however, strips do not present any damage, once the failure occurred at the concrete/adhesive interface.

3.6.2. Displacements at mid-span

Figure 13 shows that all post-strengthened beams present similar behavior regarding stiffness until concrete cracking. Results indicate that in such cases the action of the post-strengthening begins just when the structural element is already cracked.

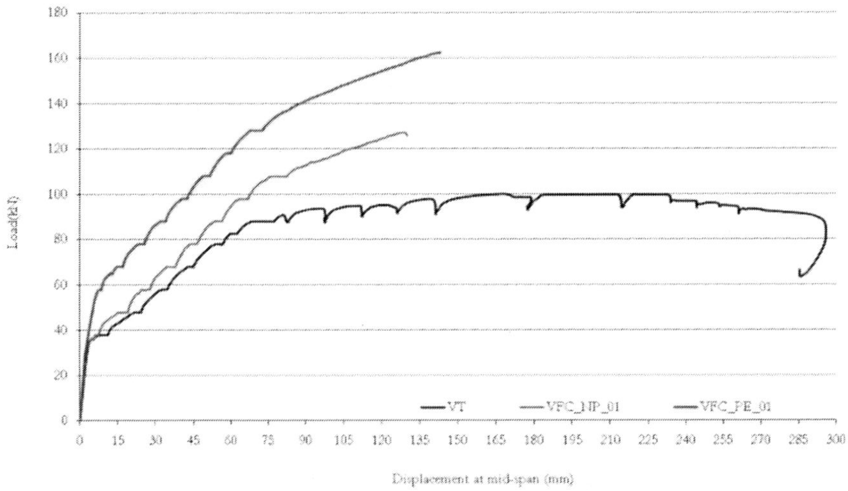

Figure 13. Load vs Displacement response of the beams tested under static loading.

Beam VFC_NP_01 post-strengthened with non-prestressed strips and the control beam, showed no significant difference regarding stiffness. However, beam VFC_PE_01 showed a stiffer behavior, when compared to beans control and VFC_NP_01, due to the increase of the cracking load and the later yielding of the reinforcement steel.

Control beam and beam VFC_NP_01, post-strengthened with non-prestressed CFRP strips present similar values of loading and displacement at mid-span at concrete cracking, due to the fact that the post-strengthening begins to act just after concrete cracking.

Cracking load of beam VFC_PE_01 was 57% higher than the ones of control beam and beamVFC_NP_01. Differences at the first stage of the loading versus displacement response at mid-span happen because prestressing leads to the development of compressive stresses at the bottom face of the beam. When non-prestressed strips are applied, the bottom face of the beam is already tensioned at the beginning of the loading process. Thus, when post-strengthening is prestressed, the loading needed to crack the concrete is significantly higher.

Differences regarding yielding of the reinforcement steel were significant. Increasing of 22% for beam VFC_NP_01 and 45% for beam VFC_PE_01, related to control beam, was observed.

3.6.3. Anchorage system

The gradual anchorage system worked properly and allowed the use of 83% of the tensile strength of the strips. Advantages of the gradual anchorage include the fact that the same device is used to prestress the strip and to gradually reduce the prestress load from the mid-span to the ends of the strip. The result is a prestressed CFRP strip without any external anchorage systems.

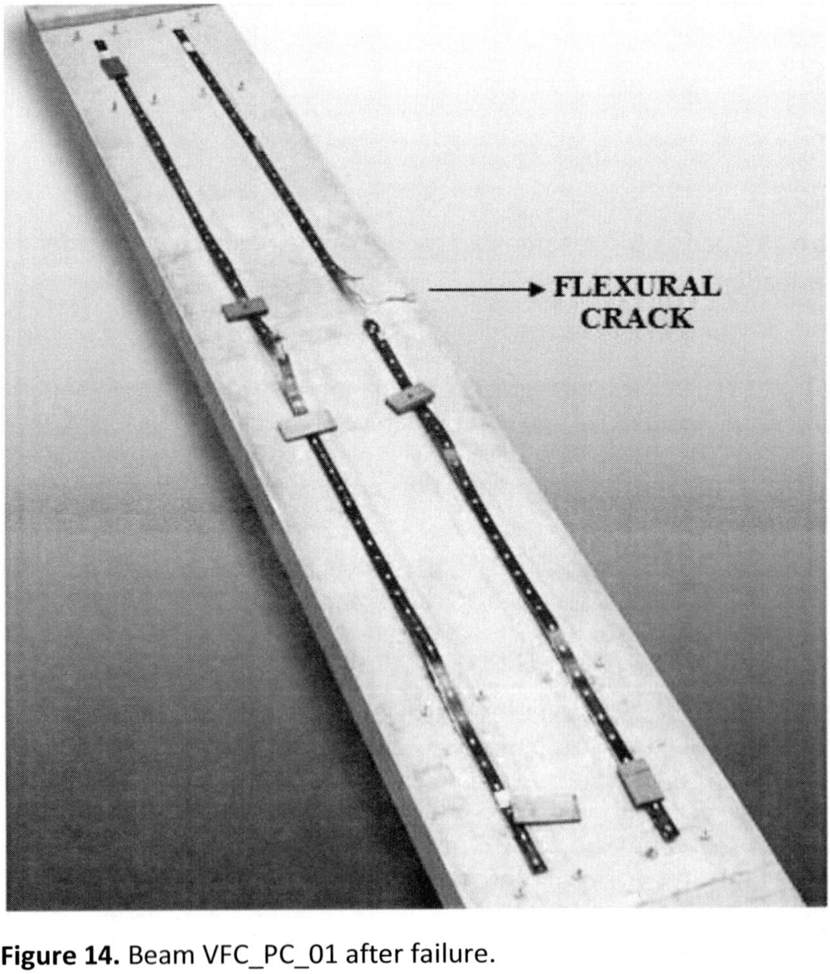

Figure 14. Beam VFC_PC_01 after failure.

3.7. Behavior Of Post-Strengthened Beams Tested Under Cyclic Loading

3.7.1. Beam vfc_pc_01

Beam VFC_PC_01 was submitted to stress levels of 50% and 80% of the yielding stress observed at beam VFC_PE_01, tested under static loading. Maximum and minimum applied loads were, respectively, 80kN and 40kN, which, added to the self weight, 28 kN, resulted applied loads of 108kN and 68kN (66% and 42% of the ultimate capacity of beam VFC_PE_01). Aiming to produce the first cracks, beam VFC_PC_01 was first pre-loaded up to108kN. Then, cyclic loading was applied at a frequency of 4Hz.

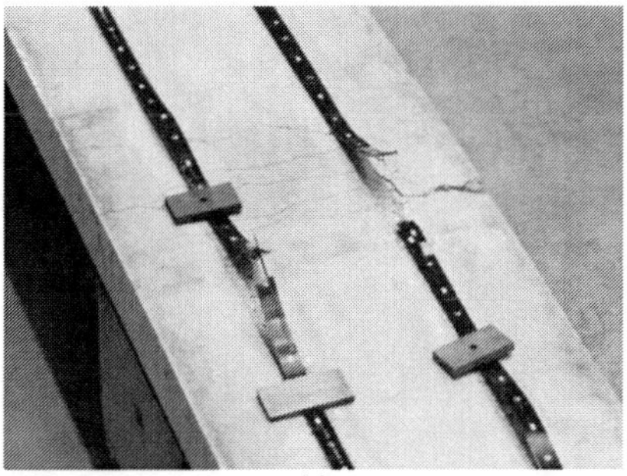

(a)

(b)

Figure 15. Failure of post-strengthening strips: (a) Next to a flexural crack; (b) Distant from flexural cracks.

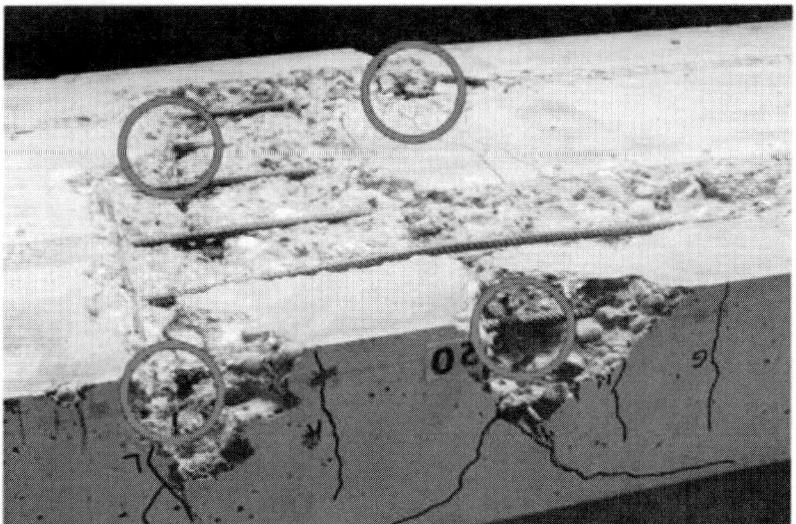

Figure 16. Fatigue failure of steel rebars.

When 282,000 cycles were reached, a crack of about 2.2mm was observed, approximately at mid-span, reaching about 90% of the cross-section. After 331,300 cycles the machine automatically stopped, when the deflection limit was reached. It was not observed any sign of apparent failure at the strips. Larger displacement limits were settled and the test was restarted. However, the post-strengthening failed before the maximum load of 108kN was reached. When the first strip debonded, the sudden release of the pre-tensioning force caused a compressive failure at the CFRP, as a secondary failure (Figure 14). The secondary failure occurred in a region that was damaged due to the presence of the flexural crack showed in Figure 14, which reached about 90% of the cross-section.

Figure 15 shows one of the CFRP strips after the failure of the post-strengthened beam. Flexural crack showed in Figure 14 induced the identification of a steel rebar broken due to fatigue. After the test, the concrete was removed from the bottom of the beam and all steel rebars were inspected. Figure 16confirms the existence of more steel rebars broken also due to fatigue.

Figures 17 and 18 show strains in the concrete and in the CFRP strips, obtained by deformeters, which gauge points were placed along the bottom of the beam. Measurements were made during pre-loading, after 30,000 cycles, and after 100,000 cycles, with the beam subjected to the maximum load (108kN). It is shown that the measurements made in the CFRP strips allowed the construction of well defined curves. However, due to the crack growing between the gauge points, this behavior could not be observed in the measurements made in the concrete. Nevertheless all obtained responses followed an expected pattern.

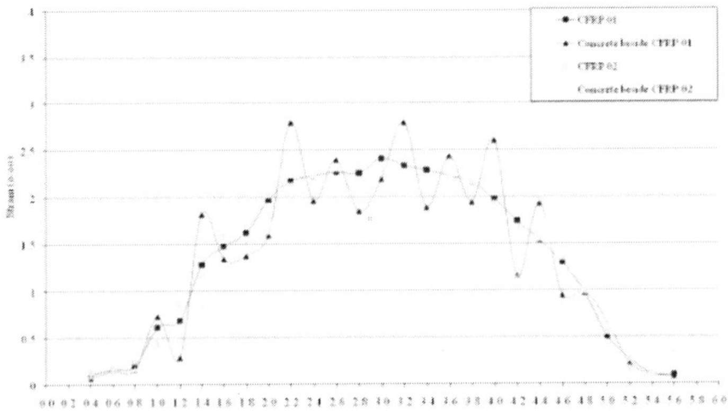

Figure 17. CFRP and concrete strains of beam VFC_PC_01, submitted to 108kN, during pre-loading.

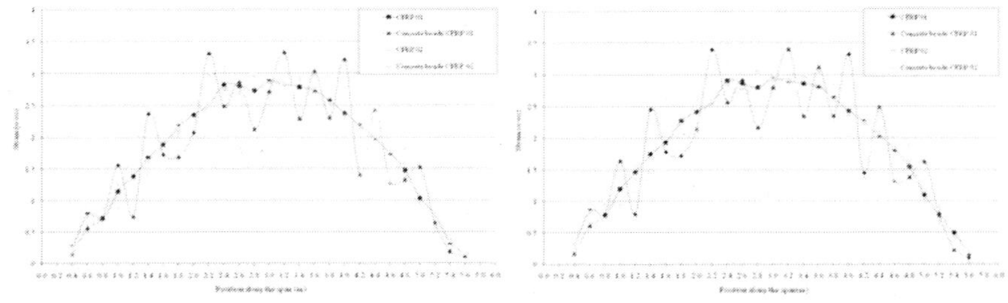

(a) (b)

Figure 18. CFRP and concrete strains of beam VFC_PC_01, submitted to 108kN after (a) 30,000 cycles and (b) 100,000 cycles.

Strains in the CFRP strips, at mid-span, during pre-loading, varied from 2.00º/oo up to 2.50º/oo. At the end of 30,000 cycles, strains increased to levels that varied from 2.50º/oo up to 3.00º/oo. From 30,000 cycles up to 100,000 cycles, it was not observed any significant variation in the strains. Strains measured, added to the strain applied to prestress each strip (5.95º/oo), give for each strip a total strain of of 8.45º/oo and 8.95º/oo. It is also noted that, at mid-span, where most of the cracks could be found, strains measured in the concrete and in the FRP are quite different. It happens because the FRP strip acts as a belt, blocking the concrete crack opening. Therefore, it can be observed that several points along to the beam are subjected to different strains. Such points can, eventually, be related to the occurrence of premature failures. Strains at a distance of 1,2 m from both beam ends, out of the loading region, are not greater than 1.50°/oo, however, these values increased about 100% from preloading up to 100,000 cycles.

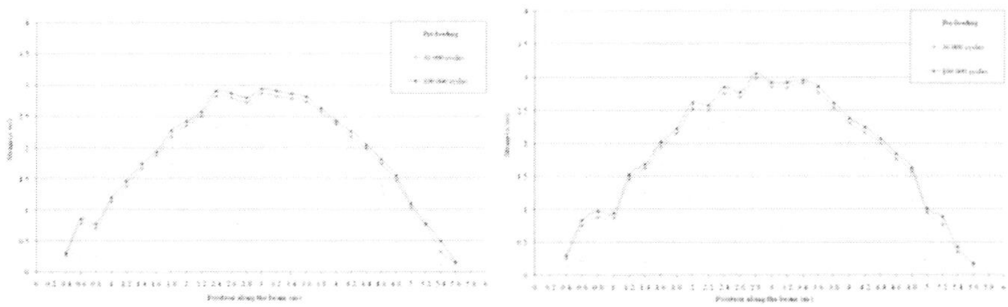

(a) (b)

Figure 19. CFRP strains of strip 01 (a) and 02 (b) of beam VFC_PC_01 during cyclic loading.

Figure 20. Cracks at mid-span of beam VFC_PC_01 after 100,000 cycles.

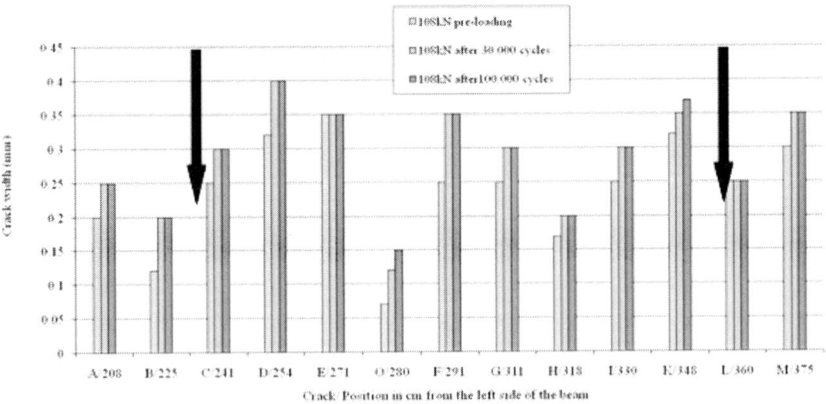

Figure 21. Cracks at mid-span of beam VFC_PC_01 after 100,000 cycles.

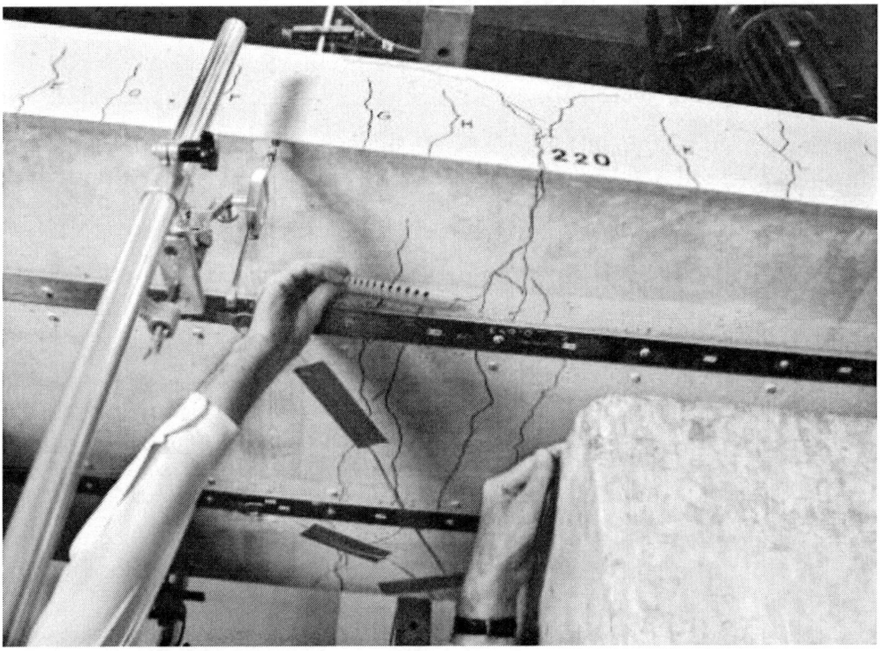

Figure 22. 2,2mm width crack at mid-span of beam VFC_PC_01 after post-strengthening failure.

Figure 19 shows the strains in the CFRP strips measured during pre-loading, after 30,000 cycles and after 100,000 cycles. Significant variations were not observed in the range from 30,000 to 100,000 cycles. However, strains increased about 0.50º/oo from the pre-loading up to 10,000 cycles. Figure 19also shows that up to 100,000 cycles the behavior of both strips is similar. However, some variations can be observed at mid-span, due to the high cracking.

The strategy adopted to monitor the crack growing at mid-span of beam VFC_PC_01 can be observed at Figure 20. Results of crack openings and the respective position from the left end of the beam are shown in Figure 21.

A high concentration of cracks can be observed at the mid-span of beam VFC_PC_01, between the four loading points. Figure 21 shows the crack openings at mid-span between the two central loading points (signaled on the figure by two vertical arrows). It is noteworthy that, before 100,000 cycles, crack openings did not reached 0.05mm.

The highest crack opening after 100,000 cycles, named D, was 0.4mm, located 254cm from the left side of the beam. However, apparently the post-strengthening failed due to a 2.2mm crack opening, named I, after 331,300 cycles. It is important to notice that, after 100,000 cycles, this crack opening was about 0.3mm (Figure 22).

The decision of testing beam VFC_PC_01 under a high stress variation led to the fatigue failure before 5,000,000 cycles, that was considered the pattern of infinite fatigue life. Stress levels applied to the beam VFC_PC_02, however, are more consistent with the ones usually found in real structures. Results of beamVFC_PC_02 will allow a more detailed analysis of the CFRP prestressing technique used, as well as of the gradual anchorage system.

3.7.2. Beam vfc_pc_02

Beam VFC_PC_02 was submitted to stress levels of 50% and 60% of the yielding stress observed at beam VFC_PE_01, tested under static loading. Maximum and minimum applied loads were, respectively, 50kN and 40kN, which, added to the self weight of 28 kN, resulted in applied loads of 78kN and 68kN (48% and 42% of the ultimate capacity of VFC_PE_01). Aiming to produce the first cracks, beam VFC_PC_02 was pre-loaded up to78kN. Then, cyclic loading was applied at a frequency of 4Hz.

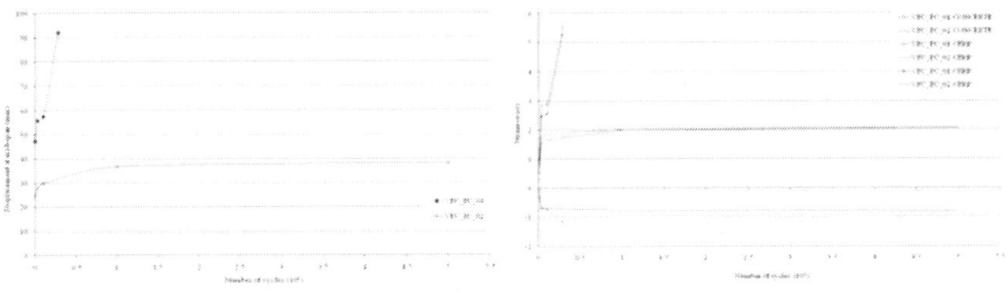

(a) (b)

Figure 23. Beams VFC_PC_01 and VFC_PC_02: (a) Displacement at mid-span vs number of cycles (b) Concrete and CFRP strains vs number of cycles.

Figure 23 (a) shows that vertical displacements at mid-span, for beam VFC_PC_02, measured with the beam subjected to the maximum load (78kN) varied 12.70mm from preloading up to 1,000,000 cycles. From this point, up to the end of the test, vertical displacements increased just 1.49mm. Data of beam VFC_PC_01 showed a large increase in the vertical displacements at mid-span after 100,000 cycles, probably due to the fatigue failure of the steel rebars. Strains in the concrete and in the CFRP strips (Figure 23 (b)) behave similarly to the displacements at mid-span, where most of the variations occurred before 1,000,000 cycles, and, after that, showed stability up to

5,000,000 cycles. Beam VFV_PC_01 also showed a similar behavior between strains and vertical displacements at mid-span, however, with a significant increasing after 100,000 cycles, probably due to the fatigue failure of the steel rebars.

Figures 24 to 26 show the strains in concrete and in CFRP strips, obtained by deformeters, which gauge points were placed along the bottom of the beam. Measurements were made during pre-loading, after 30,000 cycles, after 100,000 cycles, after 1,000,000 cycles and after 5,000,000 cycles, with the beam subjected to the maximum load (78kN).

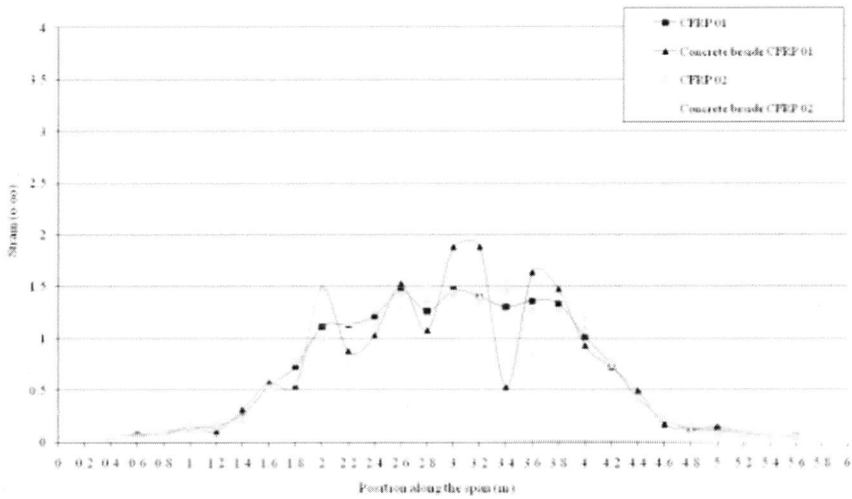

Figure 24. CFRP and concrete strains of beam VFC_PC_02, submitted to 78kN, during pre-loading.

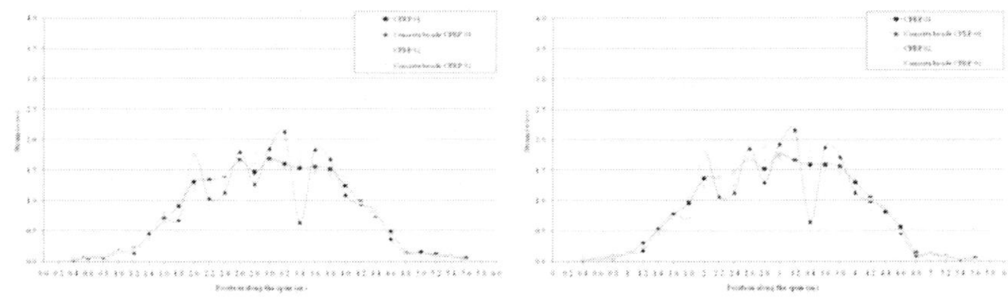

(a) (b)

Figure 25. CFRP and concrete strains of beam VFC_PC_02, submitted to 78kN after (a) 30,000 cycles and (b) 100,000 cycles.

It can be noticed that the strains obtained by deformeters, placed along the bottom of the beam VFC_PC_02, varied from the pre-loading up to 1,000,000, tending to stabilize after 5,000,000, cycles. Portions located between the loading points (1,2 m to 4,8 m from the beam end) clearly show the presence of cracks in the concrete. From the pre-loading up to 30,000 cycles, it was not observed any significant variation in the strains along the gradual anchorage zone (1.2m from the both beam ends). Strains increased after 100,000 cycles, and, after 5,000,000 cycles, the level of strains of the beginning of the test could be observed just along the first 0.6m from both beam ends. Figures 24 to 26 show that strains in the anchorage zones of beam VFC_PC_01 were higher than the ones of beam VFC_PC_02.

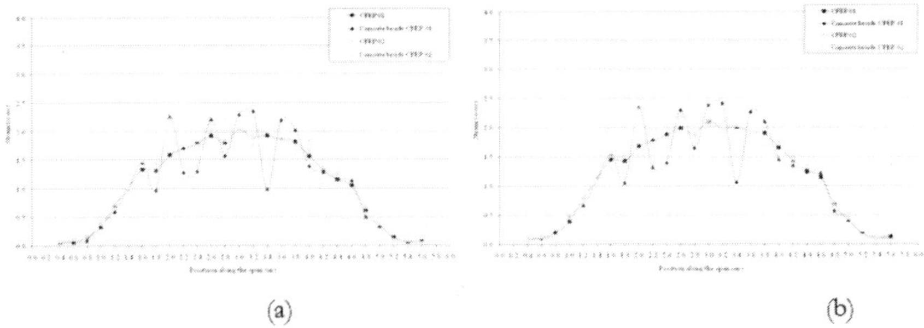

Figure 26. CFRP and concrete strains of beam VFC_PC_02, submitted to 78kN after (a) 1.000.000 cycles and (b) 5.000.000 cycles.

Data of Figures 24 to 26 show that strains in the CFRP strips, between the two central loading points, placed at 2.4m and 3.6m from the beam ends, measured at pre-loading, varied from 1.00 º/oo up to 1.50º/oo. After 30,000 cycles strains increased up to 2.00 º/oo and no significant variation were observed from 30,000 cycles up to 100,000 cycles. Measurements after 1,000,000 and 5,000,000 cycles registered a maximum strain of 2.11º/oo. Such strain, added to the strain applied to prestress each strip (5.95º/oo), give for each strip a total strain of of 8.06º/oo. Strains in the strips of beam VFC_PC_02 were smaller to the ones of beam VFC_PC_02, since, for the second beam, the maximum load and the difference between the maximum and the minimum load were smaller. Results of beam VFC_PC_02 indicate that, up to 5,000,00 cycles, it was not observed any damage on the post-strengthening system, due to the application of the cyclic loading.

Figure 27 (a) shows the strains in the CFRP strips, measured from the pre-loading up to 5,000,000 cycles. The most significant variations occurred up to 1,000,000 cycles. Strains in the CFRP strips varied about 0.85º/oo from the pre-loading up to 5,000,000 cycles. The greatest differences regarding strains were found at 1.8m, 2.8m and 4.6m from the left side of the beam (Figure 27 (a)), and at 2.6m from the left side of the beam (Figure 27 (b)).

Results indicate the existence of a kind of progressive strain at the anchorage regions, which can, ocassionally, generate adherence problems regarding long-term fatigue. Such effect should be better investigated, however, the long time demanded to realize fatigue tests, sometimes, inhibits this initiative.

Crack growing at mid-span of beam VFC_PC_01 can be observed at Figure 28, which shows the results of all crack opening measurements made, from pre-loading up to 5,000,000 cycles. Figure 28shows the results of crack openings and the respective position from the left side of the beam, at mid-span, between the two central loading points (signaled on the figure by two vertical arrows).

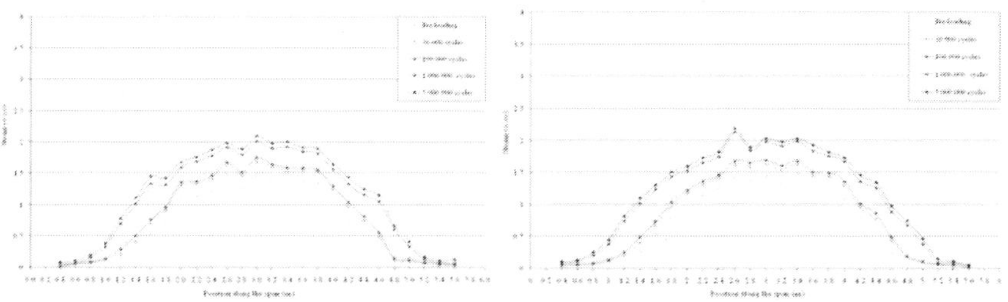

(a) (b)

Figure 27. CFRP strains of strip 01 (a) and 02 (b) of beam VFC_PC_02 during cyclic loading.

First cracks at mid-span appeared during pre-loading, reaching less than 0.15mm. From this point up to 100,000 cycle, crack openings increased, but did not exceed 0.20mm. After 1,000,000 cycles the maximum crack opening was 0.22mm, and after 5,000,000, this value was not exceeded. Cracking at the gradual anchorage regions appeared just after 100,000 cycles, however, the maximum cracking opening observed was 0.05mm. From 100,000 up to 5,000,000 cycles, the maximum crack opening measured at these regions was 0.10mm. Results of crack openings obtained from beams VFC_PC_01 and VFC_PC_02 cannot be compared directly, due to the difference regarding the maximum and minimum loads applied to generate the cyclic loading. As the maximum load applied on the beam VFC_PC_01 (108kN) was higher than the one applied on the beam VFC_PC_02 (78kN), beam VFC_PC_01 showed higher values of crack openings once pre-loading. Values of crack openings obtained after 5,000,000 cycles, for beam VFC_PC_02, were reached by beam VFC_PC_01 after just 282,000 cycles.

4. CONCLUSIONS

4.1. Post-Strengthened Beams Tested Under Static Loading

Results obtained with the development of the research program allowed the investigation of changes on the behavior of post-strengthened elements due to prestressing. The increasing on the load bearing capacity of the beam post-strengthened with prestressed strips, higher than the one of the beam post-strengthened with non-prestressed strips, highlights the efficiency of the prestressing technique. All post-strengthened beams showed vertical displacements at mid-span lower then the ones of the control beam. However, the stiffer behavior showed by all post-strengthened beams was evidenced only after concrete cracking. Due to the increasing of the concrete cracking load and the later yielding of the reinforcement steel, the beam post-strengthened with prestressed CFRP strips showed a stiffer behavior when compared to the one post-strengthened with non-prestressed CFRP strips. Gradual anchorage worked properly, dismissed the use of any external anchorage system and allowed the use of 83% of the tensile strength of the strip.

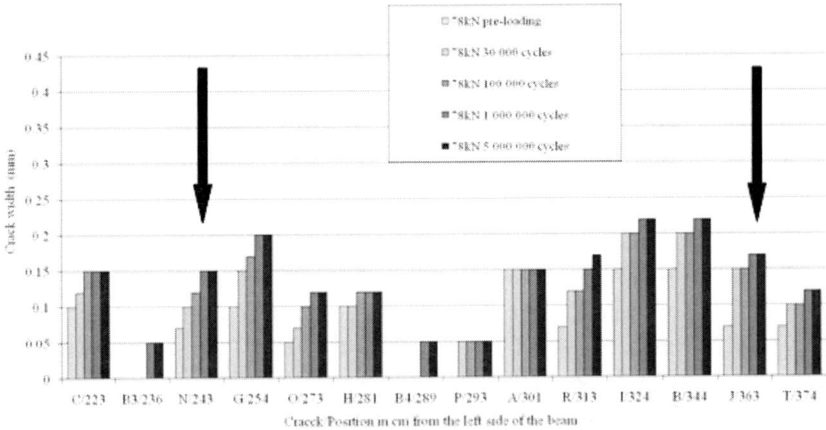

Figure 28. Cracking at mid-span of beam VFC_PC_2 during test.

4.2. Post-Strengthened Beams Tested Under Cyclic Loading

Results of the reinforced concrete beams tested under cyclic loading show that when these structures are post-strengthened with prestressed CFRP strips, damages that occur due to fatigue are mainly related to the level of stress at the steel rebars. Experimental results showed that the damage, which led to the rupture of the steel rebars, is related to the level of stress during loading, and, that it is not related to the type of post-strengthening. Tests showed that an increasing of 20% in the maximum stress of the steel rebars significantly reduced the fatigue life time of the post-strengthened element, decreasing about 15 times the number of cycles up to failure. These results emphasize the importance of proceeding the monitoring of structures that are usually submitted to cyclic loading, such as highway and railway bridges. In some cases, when these structures were designed to support traffic loads smaller to the ones that they are submitted nowadays, the use of post-strengthening may increase their lifetime, since the use of post-strengthening may lead to a reduction in the stress level of the steel rebars.

5. ACKNOWLEDGEMENTS

Authors would like to acknowledge CNPq (Portuguese acronym of the Brazilian Ministry of Science's National Research Council) and CAPES (Portuguese acronym of the Higher Education Human Resources Development Agency of the Brazilian Ministry of Education) for providing the financial support needed to develop this project. Authors would also like to express their appreciation for the technical support given by the research team EMPA (Swiss Federal Laboratories for Materials Testing and Research), in Switzerland. Finally, authors gratefully acknowledge Prof. Urs Meier for his support and assistance during the PhD. studies of the first author of this paper, developed at EMPA – Switzerland.

REFERENCES

1. H. N. Garden, L. C. Hollaway, 1998 An experimental study of the failure modes of reinforced concrete beams strengthened with prestressed carbon composite plates Composites Part B 29 B 411 424

2. B. Täljsten, 1994 Strengthening of existing concrete structures with epoxy bonded plates of steel or fibre reinforced plastics PhD thesis. Lulea University

3. U. Meier, 2001 Polyfunctional use of advanced composite materials with concrete. N. Banthia, N, Sakai K, Gjørv OE. (eds.) CONSEC01: proceedings of the Third International Conference on Concrete Under Severe Conditions, CONSEC01, Vancouver, Canada. Vancouver: Vancouver B. C

4. H. N. Garden, G. C. Mays, 1999 Strengthening of Reinforced Concrete Structures Using Externally bonded FRP Composites in Structural and Civil Structures Cambridge: CRC Press

5. El-Hacha RMA. 2000 Prestressed CFRP for strengthening concrete beams at room and low temperatures PhD thesis, Queen's University

6. Fédération Internationale du Betón. 2001 Externally bonded FRP reinforcement for RC structures- Bulletin 14 Lausane: FIB;. Technical Report

7. American Concrete Institute. 2004 Prestressing Concrete Structures with FRP Tendons. Farmington Hills: ACI, 35 ACI 440.4R-04.

8. T. C. Triantafillou, N. Deskovic, M. Deuring, 1992 Strengthening of concrete structures with prestressed fiber reinforced plastic sheets. ACI structural Journal: 89 3 235 244

9. R. El -Hacha, G. Wight, M. F. Green, 2004 Prestressed carbon fiber reinforced polymer sheets for strengthening concrete beams at room and low temperatures. Journal of Composites for Construction: 3 13

10. U. Meier, 2005 Design and rehabilitation of concrete structures using advanced composite materials. Klein GMB. (ed.) PRECONPAT05: proceedings of the Pre-Congresso Latino-Americano de Patologia da Construção, PRECONPAT05, Porto Alegre, Brazil. Porto Alegre: UFRGS

11. M. R. Ehsani, H. Saadatmanesh, 1989 Behaviour of externally prestressed concrete girders. Iffland JSB. (ed.) STRUCTURES CONGRESS 89: proceedings of the Seventh Annual Structures Congress, STRUCTURES CONGRESS 89, San Francisco, United States. New York: ASCE

12. R. G. Wight, M. F. Green, MA Erki, 2001 Prestressed FRP sheets for post-strengthening of reinforced concrete beams. Journal of Composites for Construction: 214 220

13. T. C. Triantafillou, N. Deskovic, 1991 Innovative prestressing with FRP sheets: Mechanics of short-term behavior. Journal of Engineering Mechanics: 117 7 1652

14. T. C. Triantafillou, N. Deskovic, 1992 Presressed FRP sheets as external reinforcement of wood members. Journal of Structural Engineering: 118 5 1270

15. M. Deuring, 1993 Verstärken von Stahlbeton mit gestpannten Faserverbund-Werkstoffen. Dübendorf: EMPA 276 Report 224

16. R. J. Quantryll, L. C. Hollaway, 1998 The flexural rehabilitation of reinforced concrete beams by the use of prestressed advanced concrete plates. Composites Science and Technology 58 1259

17. I. Stöcklin, U. Meier, 2003 Strengthening of concrete structures with prestressed and gradually anchored CFRP. Tan KH. (ed.) FRPRCS-6:

proceedings of the Sixth International Symposium on FRP Reinforcement for Concrete Structures, FRPRCS-6, Singapore. Singapore:World Scientific

18. U. Meier, I. Stöcklin, G. P. Terrasi, 2001 Making better use of the strength of advanced materials in structural engineering. Teng JG. (ed.) CICE01: proceedings od the FRP Composites in Civil Engineering, CICE01, Hong Kong, China. Hong Kong: Elsevier

19. L. Hollaway, 1993 Polymer Composites for Civil and Structural Engineering. Hong Kong: Blackie Academic and Professional

20. J. G. Teng, S. T. Smith, J. Yao, J. F. Chen, 2003 Intermediate crack-induced in RC beams and slabs. Construction and Building Materials: 17 447

21. L. C. Meneghetti, M. R. Garcez, Filho. L. C. P. Silva, F. P. S. L. Gastal, 2011 Fatigue life regression model of reinforced concrete beams strengthened with FRP. Magazine of Concrete Research: 63 539

22. U. Meier, 1995 Strengthening of structures using carbon fibre/epoxy composites. Construction and Building Materials: 9 6 341 351

23. E. Ferrier, D. Bigaud, P. Hamelin, L. Bizindavyi, K. W. Neale, 2005 Fatigue of CFRPs externally bonded to concrete. Materials and Structures: 38 1 39 46

Index